NMP

2

BLUE TRACK

Mathematics for Secondary Schools

NMP
2
BLUE TRACK

Mathematics for Secondary Schools

This book was written by

Norman Blackett
Eon Harper
Dietmar Küchemann
Michael Mahoney
Sally Marshall
Edward Martin
Heather McLeay
Peter Reed
Sheila Russell
Peter Taylor
David Womack

NMP Director
Eon Harper

NMP Research
Edward Martin

Longman

Preface

NMP was founded in 1981 to consider the emerging needs of pupils studying mathematics in secondary schools. Its primary consideration has been the research and development of materials that reflect the recommendations of the Committee of Inquiry into the Teaching of Mathematics in Schools (the Cockcroft Report) and the requirements of the GCSE National Criteria in Mathematics. Careful account has also been taken of the findings of the APU Secondary Surveys in Mathematics, of the outcomes of the CSMS research programme, of the HMI discussion document *Mathematics from 5 to 16*, and of the need to match curriculum and assessment in such areas as practical work, problem solving, investigation, pupil discussion, oral work, written and calculator work, and extended assignments. All of these are important features of the materials, which have undergone extensive trials in schools.

The texts all assume the use of a calculator as a material aid to calculation, and its use should be restricted only when indicated. A number of software programs have been developed to enhance certain aspects of the texts, but are not essential to their successful use.

This book has been written for Year 2 of secondary schools. Each chapter in the book has three different types of material: ● core, ■ support (indicated with a blue tab in the margin) and ▲ extension (indicated with a grey tab). All students should complete the core material before going on to the next chapter. Some will find it helpful to complete the support material too, and others will gain much value from the extension material. Students should not be alarmed that they seem to be 'missing out' some of the material in the book.

Detailed explanations of all the features of the book are given in the Year 2 Blue Track Teachers' Handbook, together with an indication of the software related to various aspects of the text.

LONGMAN GROUP UK LIMITED,
Longman House, Burnt Mill, Harlow,
Essex CM20 2JE, England
and Associated Companies throughout the world.

© Longman Group UK Limited 1987

First published 1987
ISBN 0 582 20723 1

Set in 12/14 pt Univers light

Printed in Great Britain by
Butler & Tanner Ltd, Frome and London

Contents

Acknowledgements

The materials were researched and evaluated at the University of Bath. NMP thanks the mathematics departments of the following schools for their assistance.

Castell Alun Comprehensive School, Wrexham, Clwyd
The Corsham School, Corsham, Wiltshire
Edgecliff Comprehensive School, Kinver, Staffordshire
Frome College, Frome, Somerset
The George Ward School, Melksham, Wiltshire
Grange Middle School, Keighley, West Yorkshire
Highdown School, Reading, Berkshire
Hope High School, Salford, Manchester
Kingsway County High School, Chester
Leiston High School, Leiston, Suffolk
Manvers Pierrepont High School, Nottingham
St John's School, Marlborough, Wiltshire
St Laurence Comprehensive School, Bradford-on-Avon, Wiltshire
Smestow School, Wolverhampton
Wyke Manor School, Bradford, West Yorkshire

The publishers are grateful to the following for their advice and assistance.

Derek Foxman	Head of Mathematics Department, National Foundation of Educational Research
Arnold Howell	Senior Author, *Mathematics for Schools*
Professor Celia Hoyles	Institute of Education, University of London
John Mason	Faculty of Mathematics, Open University
Professor Ray Ogden	George Sinclair Professor of Mathematics, University of Glasgow
Richard Strong	County Inspector of Mathematics, Somerset

We are also grateful to the following for permission to reproduce photographs and other copyright material.

British Airways, page 151 above; British Petroleum, page 224; Camera Press, pages 9 above (photo Frank Lane Agency/W. W. Roberts), 121 (photo Ken Lambert), 149 (photo DPA/Heirler), 151 below (photo DPA/Baum), 226 (photo *The Times*), 236 above left (photo Klaus D. Francke) and 290 below left (photo Zentralfoto); J. Allan Cash, pages 197, 236 above centre, above right, below centre and below right and 290 above right; Gremlin Graphics Software, page 17 above; *The Guardian*, page 156 (photo Martin Argles); Highlands and Islands Development Board, page 236 below left; Frank Lane Agency, pages 79 (photo Kathrin Dietl) and 290 above left (photo Gandner-Dower); Methuen Children's Books, *The Pooh Cookbook*, page 15; George Philip and Son, *Modern School Atlas*, 1983, graphs page 197 below right; Science Museum, page 192; Sealink British Ferries, page 14; SNCF, page 290 below right (photo Patrick Olivain); Sally Anne Thompson Animal Photography, page 9 below; *The Times*, share prices page 240; Tipp-Ex Vertrieb GmbH, page 203; Simon Warner, page 222.

1 Enlarging and reducing

A Horace's maths lessons are always in room 2.
He is often late.
He says it's because there is no number on the door.

Horace's teacher says, 'Design some **2**s like this one.
The best design will be stuck on the door.'

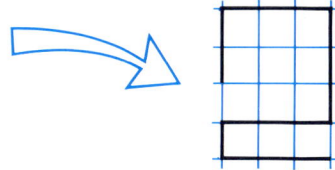

Here are some of the class's designs.

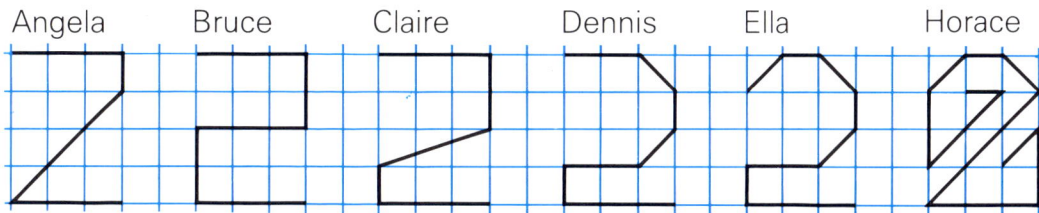

Angela   Bruce   Claire   Dennis   Ella   Horace

1   Draw your own design on squared paper.
Make it 3 units long and 4 units high.

2   The teacher asks the class to
enlarge their designs.
'Double the length of every line,'
she says.

Angela and Bruce draw these.

Enlarge your own design.
Double the length of every line.

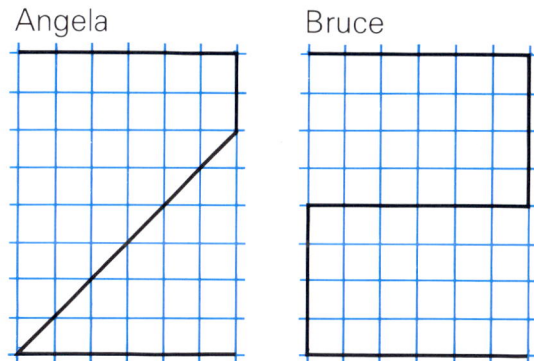

Angela   Bruce

3   Claire, Dennis, Ella and Horace draw these.

a) Two of them have not doubled each length.
Who are they?

b) Draw their designs, enlarged correctly.

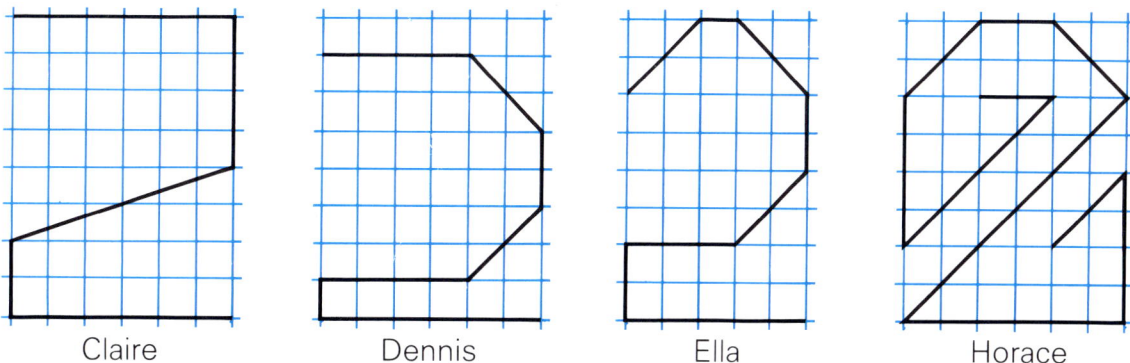

Claire   Dennis   Ella   Horace

4 The teacher decides that this is the best design.

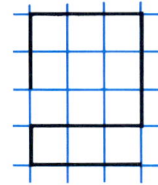

She asks the class to draw an **enlargement** of it.
'Double the length of every line,' she says.
'The neatest correct drawing will be stuck on the door.'

a) Here are some of the class's drawings.

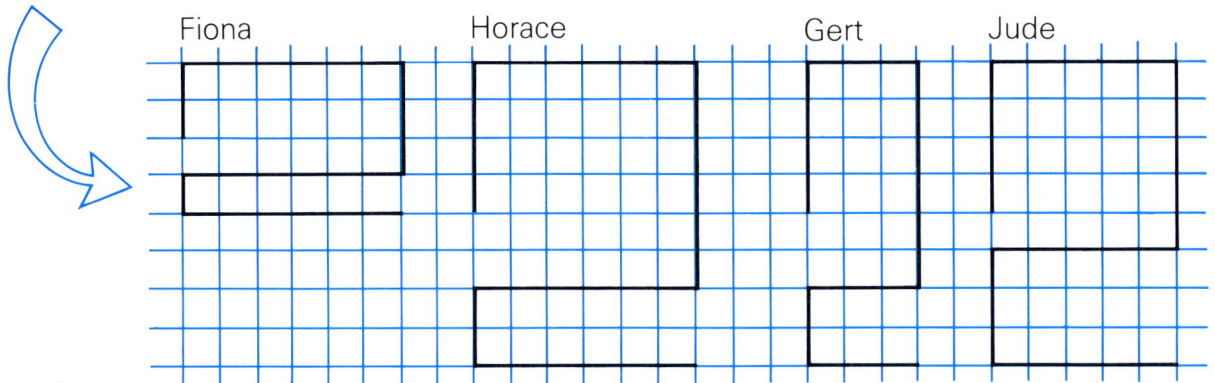

Fiona   Horace   Gert   Jude

Only one of the drawings is a correct enlargement.
Which one is it?

b) Fiona has doubled the width, but not the height.
  (i) What mistake has Gert made?
  (ii) What mistake has Jude made?

5 The teacher sticks Horace's **2** on the door.
The next day Horace is late again.

'The number on the door is too small,' he says.
'We'll make it bigger,' says the teacher.
'Make another enlargement,' she tells the class.
'This time, multiply the length of every line by 3.'

Here are two of
the pupils' drawings.

Lou

a) Only one drawing
    is correct.
    Which is it?

Karen

b) What mistake has the
    other pupil made?

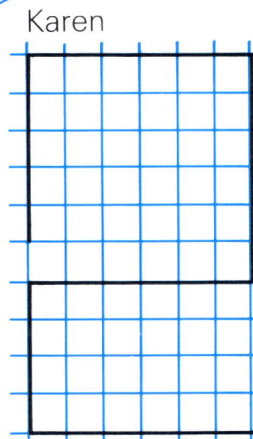

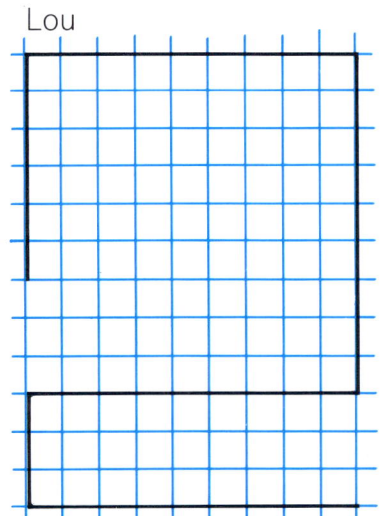

6 The teacher sticks Lou's drawing on the door.
The next day Horace is late again.
'The lines are too thin,' he says.
The teacher draws a 'solid' **2**.
'Enlarge this by any amount you like,' she says.

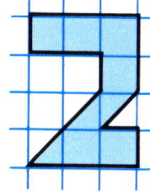

Here are some of the class's drawings.

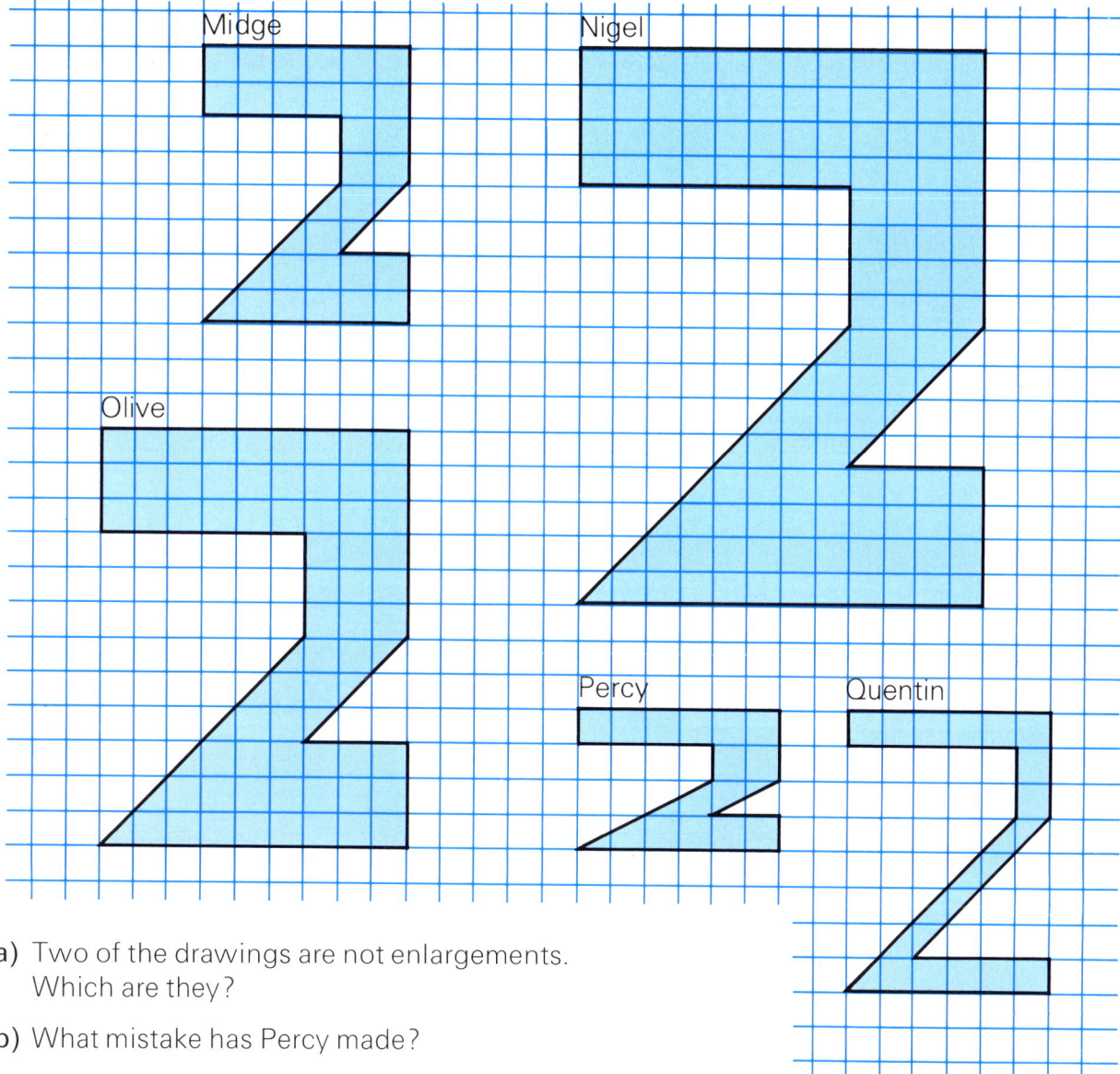

Midge

Nigel

Olive

Percy   Quentin

a) Two of the drawings are not enlargements.
Which are they?

b) What mistake has Percy made?

Take note

Midge has doubled the length of every line.
We say she has used a **scale factor** of × 2.

c) What scale factor has Nigel used?

d) What scale factor has Olive used?

● 3

7 Copy this solid **2** onto squared paper.
Enlarge your **2** by a scale factor of ×5.

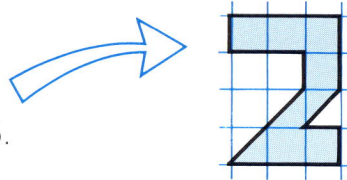

8 Horace has drawn this **H** on 1 cm squared paper.
He enlarges it by a scale factor of ×5.

a) How high is his enlarged **H**?

b) How wide is it?

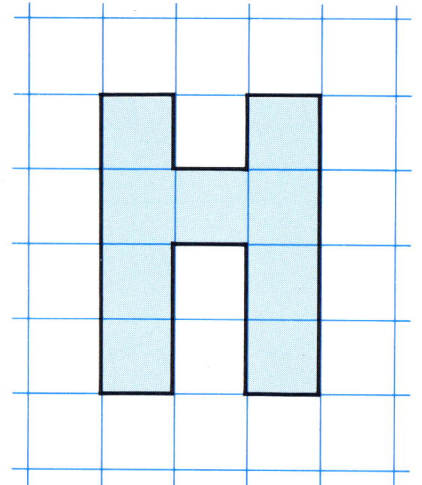

9 Horace decides to draw another enlargement of this **H**.
He draws it 12 cm wide.

a) How high should it be?

b) What scale factor is he using?

10 Horace starts to draw a third enlargement of his **H**.

a) How high will the enlargement be?

b) How wide will it be?

c) What scale factor is he using?

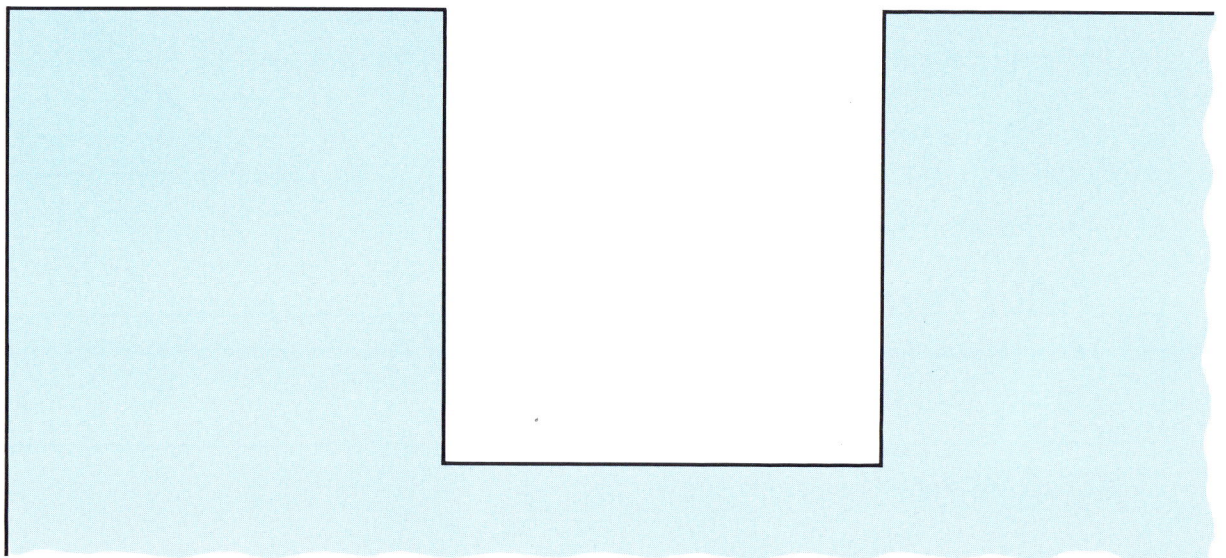

Reducing

B Horace wants to be a postman when he grows up.
He decides to design a postage stamp.

The stamp commemorates Pythagoras.
Pythagoras is Horace's favourite mathematician.

1 Horace's design is 9 cm wide.

   a) Measure its height.

   b) The Post Office decides to print it this size.

   How many times shorter and narrower is this than Horace's design?

30ᵖ

$$a^2 + b^2 = c^2$$

c   b
a

PYTHAGORAS

2 The Post Office has **reduced** Horace's design.
It used a scale factor of ÷ 3.
Copy and complete this table.

Scale factor ÷ 3	Length on Horace's design	Length on actual stamp
Height of Queen's head	24 mm	
Width of zero	9 mm	
Length of word 'Pythagoras'	6 cm	
Side **a** of triangle	24 mm	
Side **b** of triangle		
Side **c** of triangle		

3   This is another of Horace's stamp designs.

These are **reductions** of it.
Write down the scale factor for each one.

a)

b)

c)

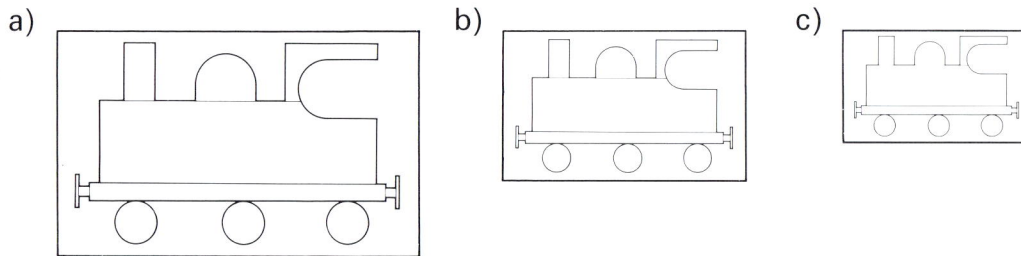

4

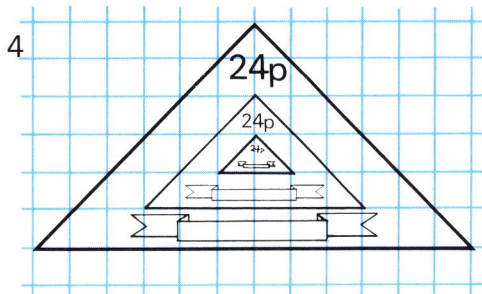

This is the design which Horace likes best.
Each smaller stamp is a reduction of the
large stamp.
What scale factor is used for each?

Think it through

5   Horace enlarges this **H** using a
    scale factor of × 10.
    Midge reduces Horace's enlargement.
    She uses a scale factor of ÷ 4.

a) How high is Midge's **H** ?

b) How wide is it?

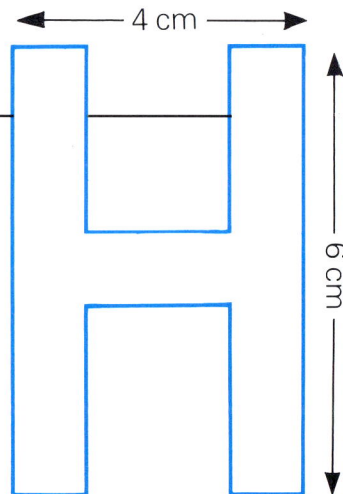

4 cm

6 cm

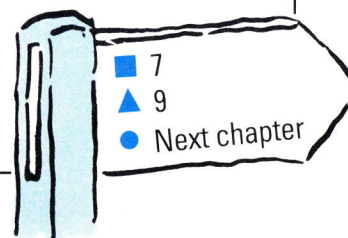

■ 7
▲ 9
● Next chapter

Portraits

A
B
Horace misses his next maths lesson.
He goes to room Z by mistake.
Room Z is the art room.

The class are drawing self-portraits.
Horace draws this.

'Ah!' says the teacher. 'It's Horace.'

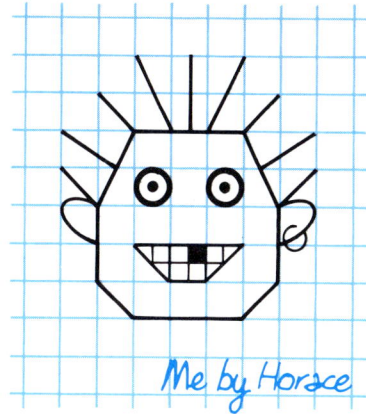

Me by Horace

1   The teacher asks Horace to enlarge
his drawing.
Horace starts by drawing this.

The outline is correct.
However, the mouth is in the wrong position.

Copy the drawing with the mouth
in the correct position.
Use squared paper.

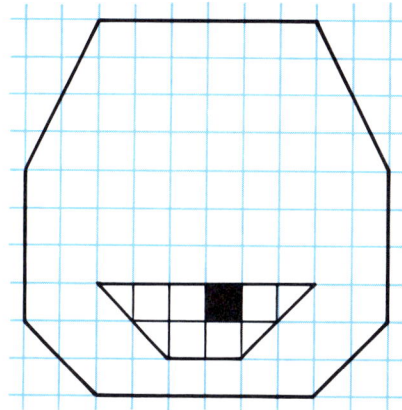

2   Horace now draws
this.
The eyes have been enlarged correctly,
but not the ears.
Add the eyes and ears correctly
to your drawing.

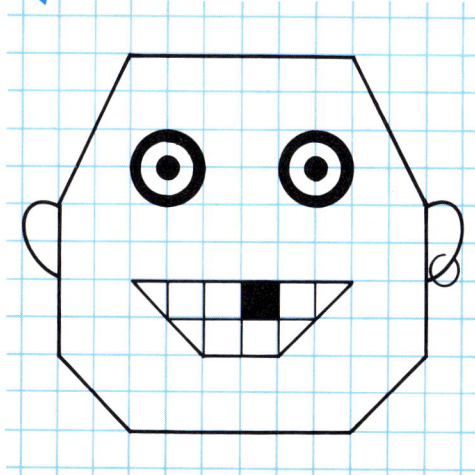

3   Horace adds the hair,
but one strand is wrong.

Add the hair correctly
to your drawing.

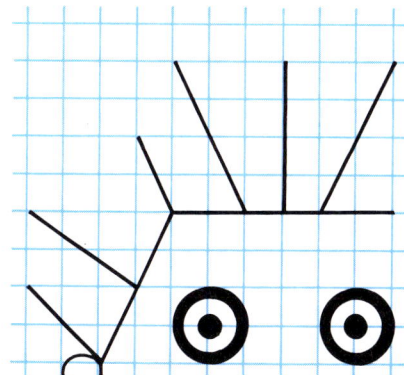

4   What scale factor is Horace using to enlarge his self-portrait?

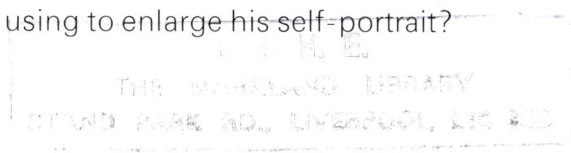

5 Here are some more portraits by Horace.
Copy and complete the enlargements on squared paper.
Write down the scale factor for each enlargement.

a)

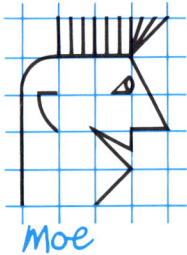

b)

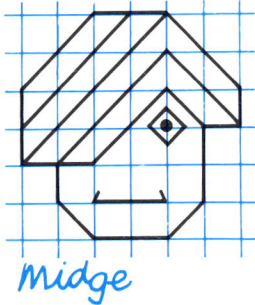

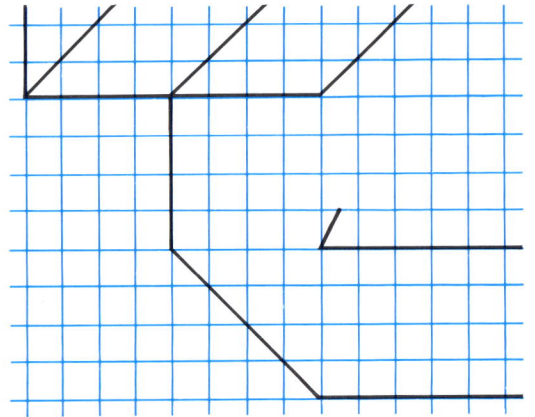

6 This is a portrait of Maisie by Horace.
Copy and complete the reduction on squared paper.
Write down the scale factor.

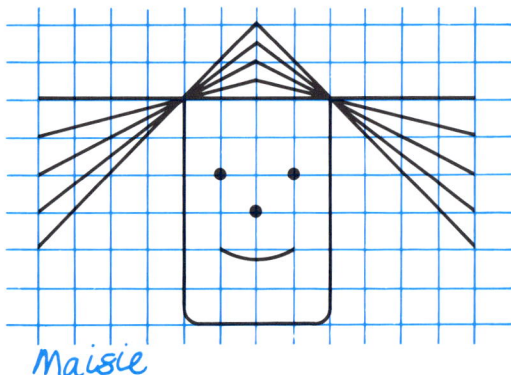

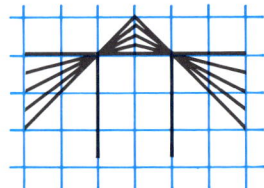

▲ 9
● Next chapter

Combining scale factors

C 1 The scale factor from **A** to **B** is ×2.
The scale factor from **B** to **C** is ×3.

A

B

x 2

x 3

C

a) Which of these is the scale factor
from **A** to **C**?

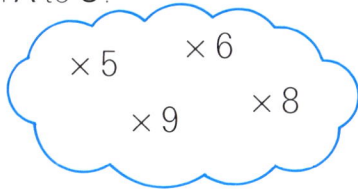

×5   ×6
×9   ×8

b) Check your answer to (a) by measuring.

c) What is the scale factor from
   (i)   **B** to **A**?
   (ii) **C** to **B**?
   (iii) **C** to **A**?

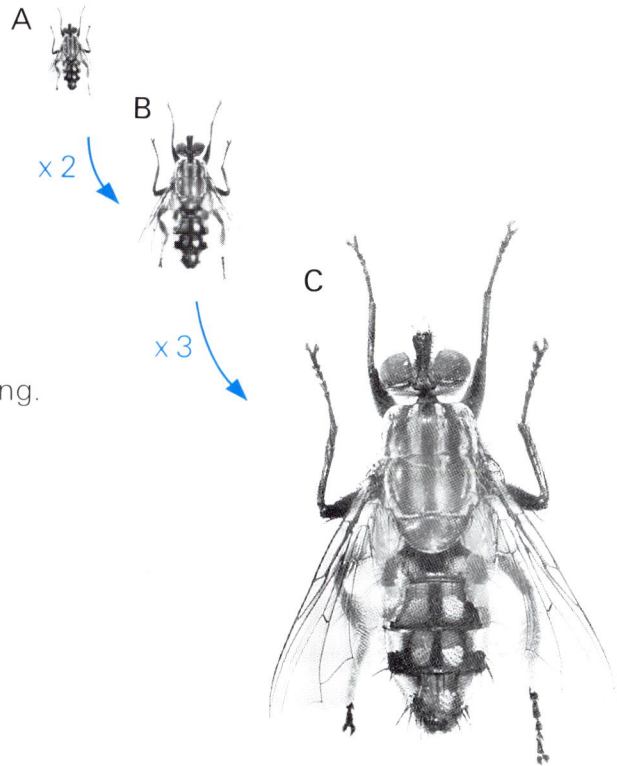

2 The scale factor from **A** to **B** is ×4.
The scale factor from **B** to **C** is ÷2.

a) What is the scale factor from
   A to **C**?

b) What is the scale factor from
   (i)   **B** to **A**?
   (ii) **C** to **B**?
   (iii) **C** to **A**?

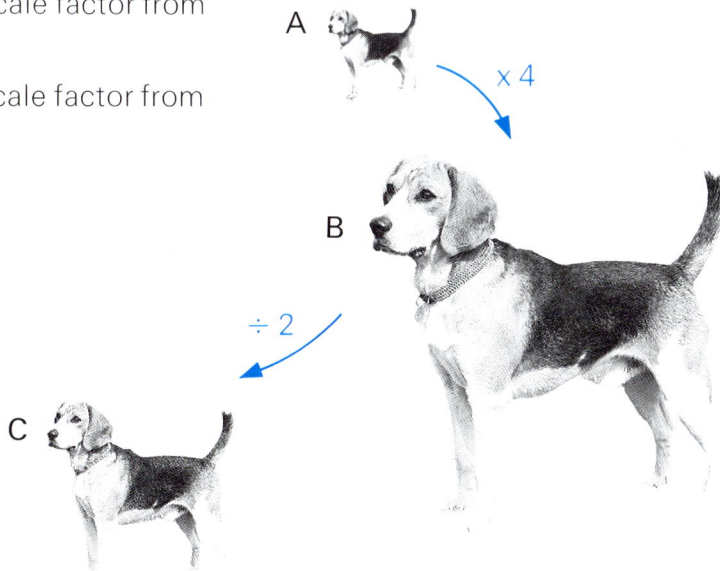

A

x 4

B

÷ 2

C

Think it through

3 A photographer enlarges a portrait using a scale factor of ×12.
Then she reduces the enlargement using a scale factor of ÷4.
What scale factor connects the original portrait and the final print?

Enlarging areas

D _**EXPLORATION**_

1 a) Draw an enlargement of this rectangle.
Use a scale factor of × 2.

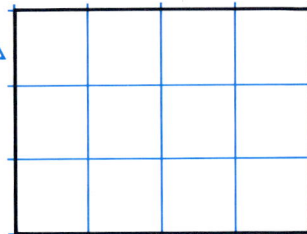

b) Write down the area of
(i) the original rectangle.
(ii) the enlarged rectangle.

c) Think about other enlargements of the rectangle.
Copy and complete this table.

Scale factor of enlargement	Area of enlarged rectangle
× 2	48 cm²
× 3	
× 4	
× 5	
× 10	

d)

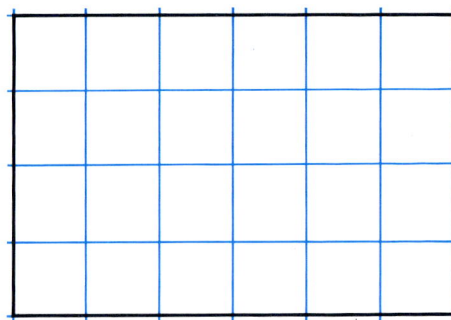

The rectangle has been enlarged
by a scale factor of × 2.
Which of these
tells you how much the
area has been enlarged?

We call it the area enlargement factor.

A × 2   B × 4   C × 8   D × 16

e) Investigate for other enlargements of the rectangle.
Compare the enlargement factor for lengths with the
area enlargement factor.
Look for a general rule which connects them.
Write down what you discover.

● Next chapter

2 Solving problems

A *With a friend*

1 a) Meg is trying to do this problem.
       Decide between you why she won't succeed.
       Each of you write down what you decide.

> Lenny bought a magazine.
> The shopkeeper gave him
> 5p change.
> How much was the magazine?

   b) Meg will be able to work out
       how much the magazine costs . . .
       if she is told how much Lenny
       gave the shopkeeper.
       Here are some more impossible problems.
       Decide between you what is missing.
       Each of you write down what you decide.

A

What does each beefburger weigh?

B Midge buys some ice creams.
   They cost £1.60 altogether.
   How much are they each?

D

C This is Winston's fish tank.
   What is its capacity?

20 cm

3 m   4 m

The painter's plank is made
from two shorter planks.
How long is it?

2 Here is the missing information for the problems in question 1.
Use it to solve each problem.

The fish tank is 40 cm long and 15 cm wide.

Lenny gave the shopkeeper 50p.

There are ten beefburgers in the bag.

Midge bought four ice creams.

The planks overlap by 1 metre.

3 a) Write down a problem of your own which has
something to do with this picture.
Your problem must be impossible
to solve.

You must leave out some important information.

b) Write down what information you left out of your problem.

4 This is a plan for a garden pond.

The pond could be marked out if one more piece of information was given.

Write down what information could be given.

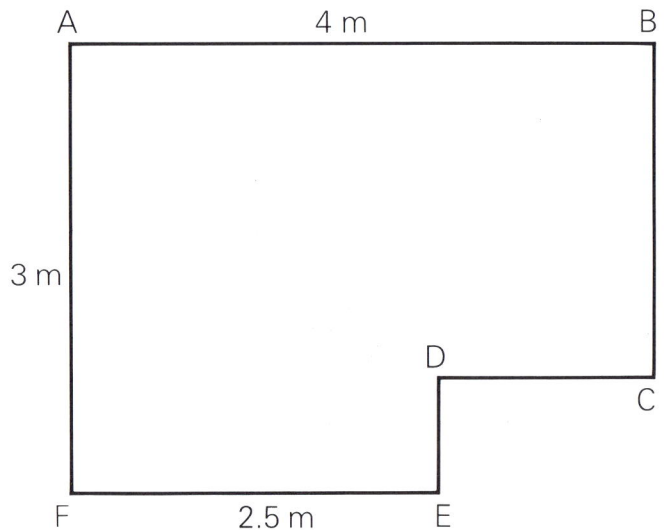

Sifting information

B 1 Sometimes we are told more than we need to know!
Then we have to pick out what is important.
You will have to pick out the important information for these questions.

a) This is an order form for school photographs.

514

THERE IS NO OBLIGATION TO PURCHASE THESE PHOTOGRAPHS

The prices of the photographs, complete with mounts, and inclusive of V.A.T. are:

Large print £2.00
Small prints 75p
FOUR small prints £2.00
ALL FIVE prints for £3.20

Use the enclosed envelope to return the purchase money to the School as soon as possible.

Only the exact money must be sent, as change cannot be given.

Do not include in the envelope an order or the money for re-prints.

Please return any unwanted photographs to the School as soon as possible.

IF ORDERED WITHIN 14 DAYS

Additional mounted copies of your photographs may be had by sending this order direct to the address overleaf.

PLEASE DO NOT send this order to the School, NOR MUST YOU include the money for it with that sent to the School for purchase of the original photographs.

Complete the following details and enclose for identification, one of your photographs (which will be returned with your order), together with the appropriate money. Delivery will be approximately FIVE weeks.

SCHOOL NAME and FULL ADDRESS of the School (in BLOCK LETTERS).

_____

_____

_____Approx. date taken _____

Prints may be ordered in the following sizes and styles: indicate the number required in the appropriate boxes. We regret we are unable to guarantee exact colour-matching of the copies with the originals.

Each 10×8 print	£4.50 ☐
Each 10×8 panel of 4 (5×4 prints)	£4.50 ☐
Each 8×6 print	£4.00 ☐
Each 8×6 panel of 4 (4×3 prints)	£4.00 ☐
Each 7×5 print	£3.50 ☐
Each 7×5 panel of 4 ($3\frac{1}{2} \times 2\frac{1}{2}$ prints)	£3.50 ☐

PHOTOGRAPH FRAMES in Ivory simulated leather are available as under:

To take 10×8 print £4.00 ☐
(This is for Frame only)
All prices inclusive of V.A.T. and Postage.

Amount enclosed £_____

RETURN LABEL (do not detach). Set out your NAME and ADDRESS in BLOCK LETTERS.
PLEASE COMPLETE BOTH ADDRESS LABELS

_____

_____

_____

How much does Lenny pay for two photograph frames with photographs to fit?

b) Lenny eats three tablespoons of Vitalp every day.
It keeps his energy up.

(i) How many kilocalories of energy does it give him every day?

(ii) How much fibre does it give him every week?

Vitalp

A marvellous mixture of these ingredients which comes from an old Swiss recipe.

Ingredients
Whole Wheat Flakes
Vine Fruits
Dried Skimmed Milk
Oat Flakes
Soft Brown Sugar
Roasted Hazelnuts
Dried Whey
Malt Extract
Salt

Nutritional Information

Nutrients	per 100 g
Energy	
kilojoules	1537
kilocalories	366
Protein	12.0 g
Fat	5.0 g
Available carbohydrate	65.6 g
Dietary fibre	8.4 g

An average serving of 3 tablespoonfuls is approximately 50 g.

Why Vitalp is not just an ordinary muesli.

Take note

To solve problems we have to
search out the information
we need . . . and then use it.

2 This is the Collins family.
They are going on holiday to France in their motor van.
They will cross the Channel from
Portsmouth to Cherbourg.

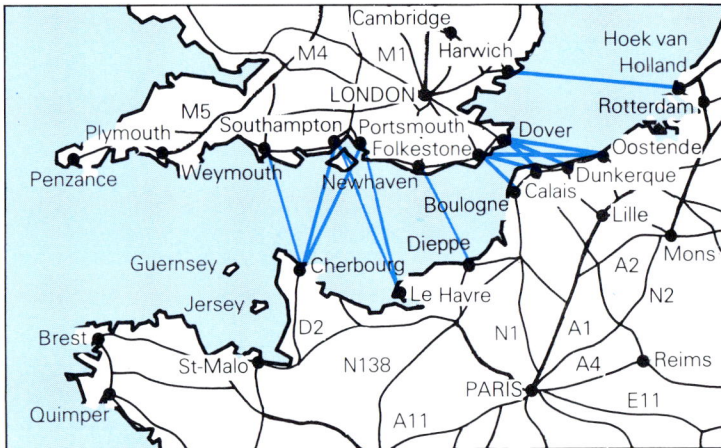

The Collins' motor van
(4.4 m long)

Paul Collins
(3 years)

Ben Collins
(12 years)

Mary Collins
(17 years)

Mr Collins

Mrs Collins

Make out your own
'What will it cost' chart.
Work out how much it will
cost the Collins to cross the Channel.

What will it cost?

Vehicles and Passengers

Your Car/Minibus/Camper

Overall Length not exceeding 4 m (13'1")	£43.00
Overall Length not exceeding 4.5 m (14'9")	£52.00
Overall Length over 4.5 m (14'9" +)	£58.00
Insert Cost as Appropriate	£ .00
Each Adult @ £9.00	£ .00
Each Child (under 16) @ £5.00	£ .00
Each Infant (under 4)	FREE
SINGLE FARE	£ .00

Take note

Don't rush into problems.
Decide what information you need . . . then make a list.

With a friend

3   Meg is using this recipe from *The Pooh Cook Book*.

Honey tart
(Serves 6)

- 180 g plain flour
- pinch of salt
- 100 g butter (or margarine)
  mixed with vegetable fat
  (or shortening)
- 3–4 tablespoons water to mix

For the filling:
- 3 rounded tablespoons honey
  (you can also use golden syrup)
- 60 g fresh white breadcrumbs
- little grated lemon rind
- 1 tablespoon lemon juice

She has to buy all of the ingredients.
The local shop has these.

10p EACH

TABLE SALT 1kg 23p

35p

GOLDEN SYRUP 57p 250g

12p EACH

SELF RAISING FLOUR 42p 1kg

BUTTER 25p 120g

VEGETABLE OIL 75p 500ml

MARGARINE 17p 120g

6p EACH

LEMON JUICE 26p

PLAIN FLOUR 20p ½ kg

CHEESE 7?p

BREAD

HONEY 70p 250g

HONEY 40p 125g

Sea SALT 100g 18p

MARGARINE 3?p 200g

Decide between you what Meg should buy.
Work out the total cost.
Try to keep the cost down.
Each of you make a list.

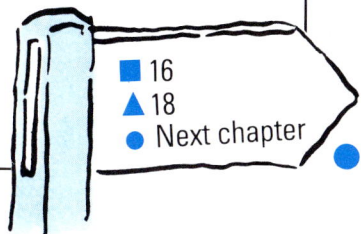

LARD 19p 250g

PLAIN FLOUR 38p 1kg

■ 16
▲ 18
● Next chapter

Paper clips and rock cakes

A 1 Here are two 'paper clip problems'.

Problem A

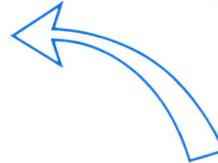

Meg uses 8 paper clips from this box.
How many are left?

Problem B

Each clip in this box is made from
12 cm of wire.
How much wire is used to make
the whole boxful?

This is the missing information
for both problems.

There are 100 paper clips to a box.

a) Answer each problem.

b) Write a 'paper clip problem' of your own which needs this information.
Make it different from problems **A** and **B**.

2 a) Here are two 'baking problems'.
Some more information is needed for each one.
Write down what it might be.

Problem A
Lenny wants to bake 20 rock cakes.
How much flour does he need?

Problem B
Meg weighs out 500 g of flour
to make bread cakes.
How many bread cakes does
she make?

b) Here is another piece of information.
It belongs to both problems.
What is the answer to each problem?

Each cake needs
50 g of flour.

Being choosy

B 1 a) What size sweatshirt would you buy if you joined the Gremlin Gang?

b) How long might you have to wait for it to arrive?

c) How much would it cost you?

2 Lenny has become house-proud! He is papering his bedroom. This is his wallpaper.

7 rolls of Anaglypta – blue and black stripes!

All-purpose wallpaper adhesive.
Makes 12 pints for normal wallpapers.
STICKIT sticks it!

APPLICATION (per sachet)	Quantity of cold water to use	Approximate coverage*
Normal wallpapers	12 pints	8–10 rolls
Washable and Vinyl wallcoverings	10 pints	8 rolls
Novamura	8 pints	6 rolls
Embossed papers (e.g. Anaglypta)	8 pints	3–4 rolls
Heavy embossed papers (e.g. Supaglypta)	7 pints	2–4 rolls
Polystyrene tiles (not veneer)	7 pints	140 sq. ft.
NB: These figures are intended as a guide only.		

a) How many pints of Stickit will he need?

b) How many sachets of Stickit will he need?

3

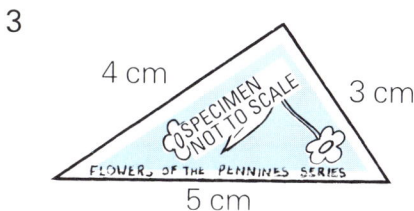

4 cm   3 cm

SPECIMEN NOT TO SCALE

FLOWERS OF THE PENNINES SERIES

5 cm

What area of paper is needed to make 20 of these stamps?

New issue stamps
Flowers of the Pennines

▲ 18
● Next chapter

Collecting your own information

c *With a friend*

1 Sometimes you have to collect information yourself to solve problems.
Discuss with your friend how you would solve each of these.
For each one write down

- what information you would collect
- how you would collect the information
- how you would use the information to solve the problem.

Problem A
Find approximately how quickly the surface water in a river is flowing.

Problem B
Find the approximate length of the groove on one side of a record.

Problem C
Find approximately how many daisies there are on the school playing field.

Think it through

2 Do these two estimations.
You will need to plan and collect information for each one.
Write down how you arrived at each of your estimates.

a)

> Estimate how many kilograms of food you eat each year.

b)

> Estimate how many times your classroom door is opened each year.

3 This is the sketch plan for the foundations of a house. It is not drawn to scale.

a) Would you be able to mark out the foundations using the plan?

b) What is the total distance around the outside of the foundation trench?

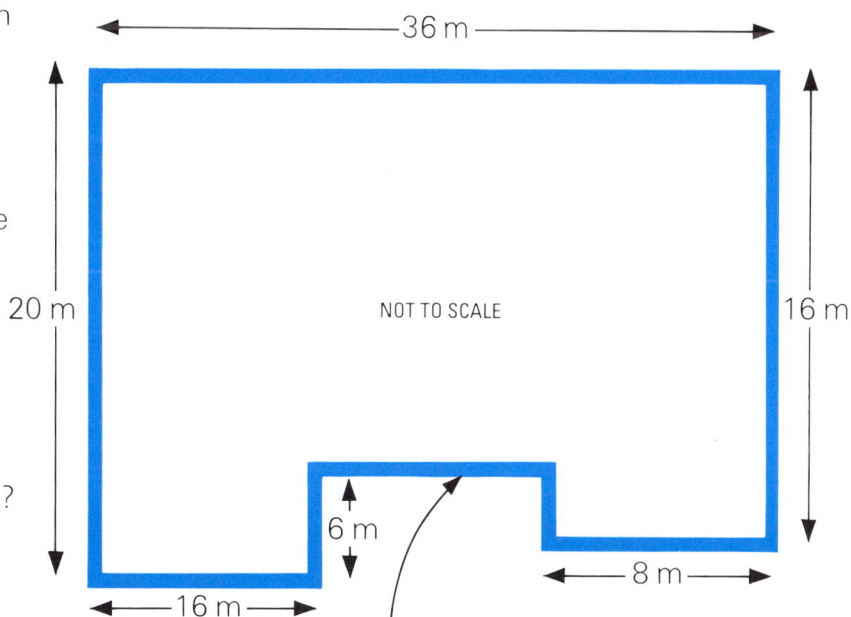

36 m

20 m

NOT TO SCALE

16 m

6 m

16 m

8 m

trench to be ½ m wide, 1 m deep

3 Thinking about symmetry

A

DO YOU REMEMBER...?

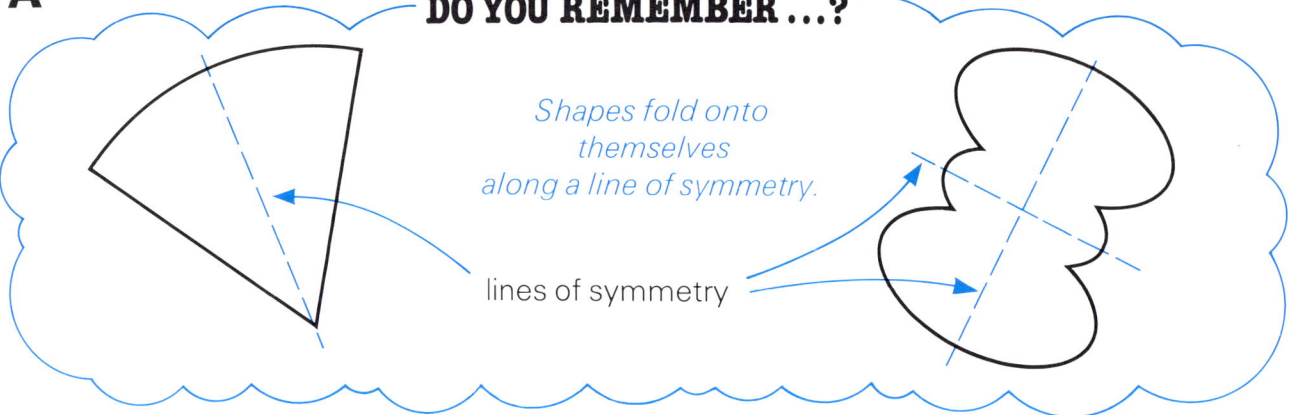

Shapes fold onto themselves along a line of symmetry.

lines of symmetry

1 a) Ben folded a piece of paper in half to make this.

Use squared paper.
Make Ben's starting shape.
Stick it in your book.

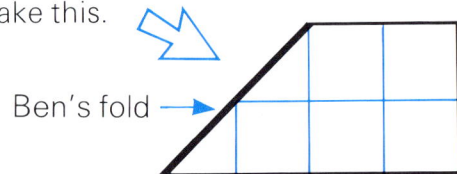

Ben's fold →

b) There are three other starting shapes
you could fold to make this.
Sketch them.
Mark the fold line on each of your shapes.

2 This is one half of a starting shape ...
and part of the other half.
The dotted line is the fold line.
Copy and complete the whole shape
on squared paper.
Cut out your shape and check
if you are correct.

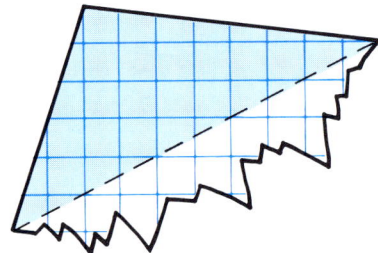

Think it through

3 You need squared paper, scissors and glue.
Ben folded a piece of paper
in half ...
then in half again.

This is the result.

a) Make Ben's starting shape.
Stick it in your book.
Mark the first fold line on
Ben's starting shape.

b) Sketch Ben's shape *after*
he had made the first fold.
Mark the second fold line on your sketch.

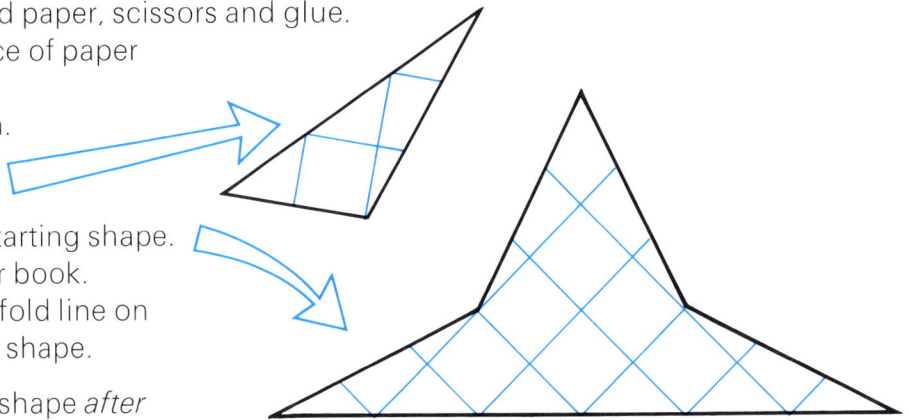

With a friend: *Line symmetry game*

4 Each of you needs a pencil and a rubber.
One of you sketch this.
Use pencil.

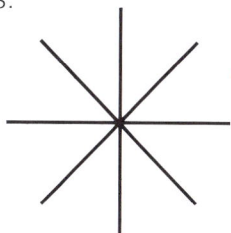

Decide who is player 1 and who is player 2.

Player 1   Add a circle to the end of one arm.

Example:

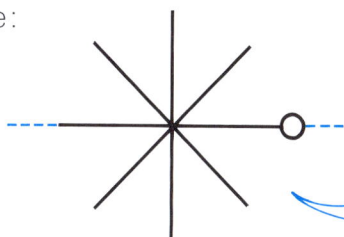

Notice:
It has 8 lines of symmetry.

The shape has
1 line of symmetry.

1 penalty point.

Player 2   Add a circle to any other arm.

Example:

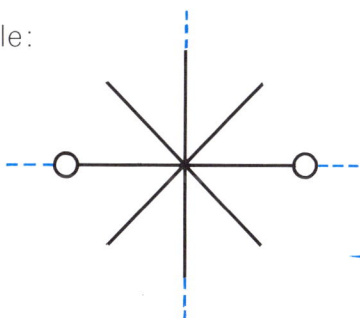

2 lines of symmetry.

2 penalty points.

Player 1   Add a circle to another arm ... and so on,
until there are eight circles, one on each arm.

Then start to rub out circles, one by one.
Player 2 starts with any
circle ...

... until

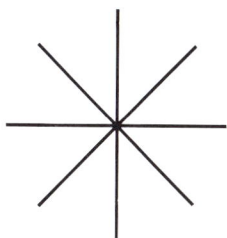

1 line of symmetry.

1 penalty point.

Example:

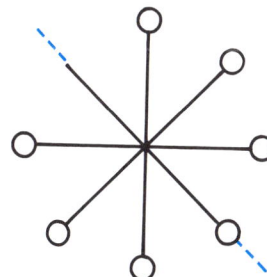

Scoring
The player with more penalty points **loses**.
Play the game four times.
Take turns to start.

Turn symmetry

B 1 Think about the 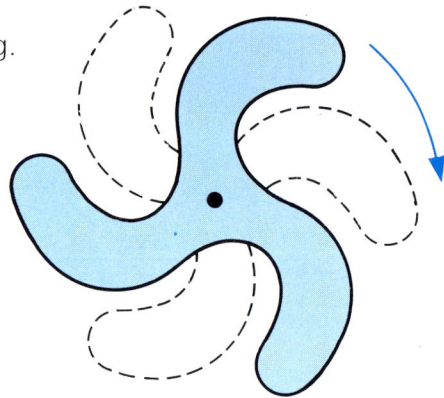 shape turning.

In a full turn, how many times
will it fit the dotted outline?

2 Think about these shapes turning.
In a full turn, how many times will they fit their dotted outlines?

a)

b)
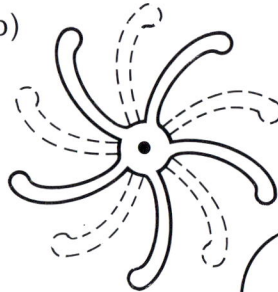

c)
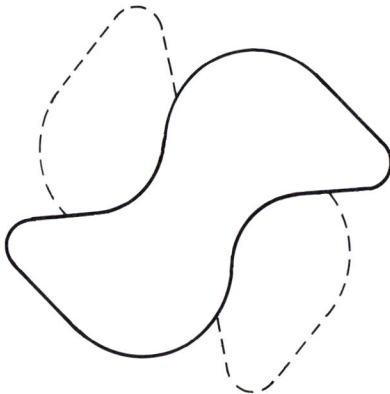

d)
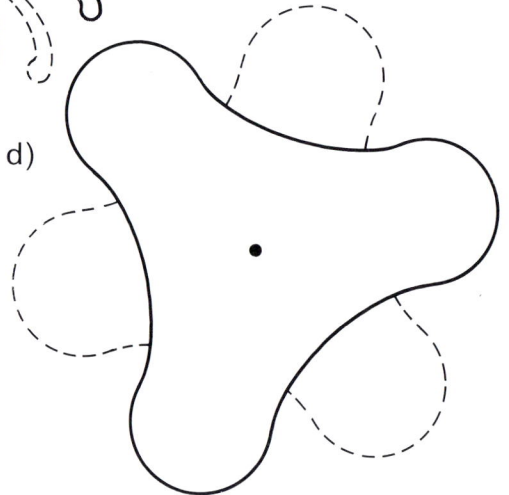

e)
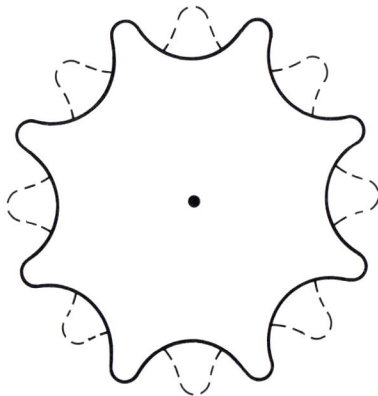

f)
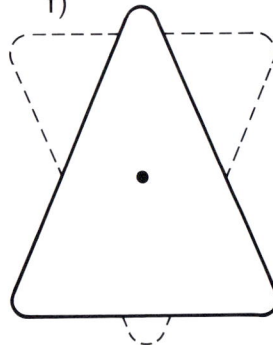

g)
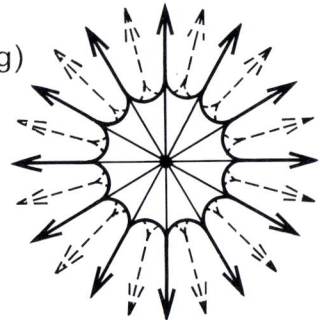

3 Draw your own shape.
In a full turn it must fit its dotted outline six times.

Order of turn symmetry

C

Take note

In a full turn, this shape fits its dotted outline **four** times. We say it has **turn symmetry of order 4.**

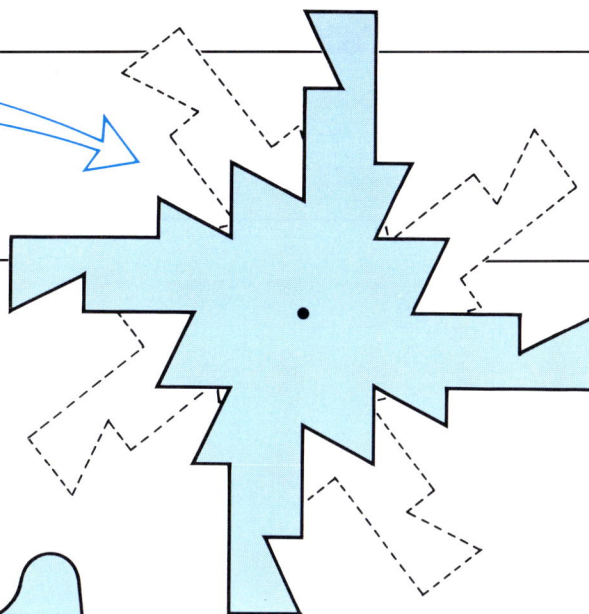

1 How many degrees must it turn between each 'fitting'?

2 This shape has turn symmetry of order 2.
It fits its dotted outline twice in a full turn.
It starts like this.

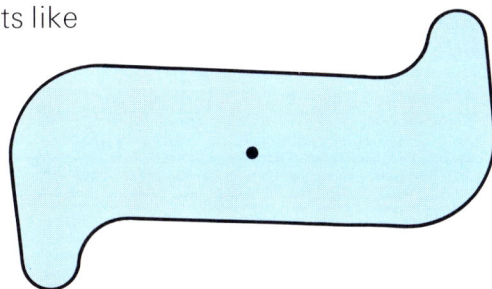

How many degrees must it turn before it fits onto itself again
a) for the first time?
b) for the second time?
c) for the tenth time?

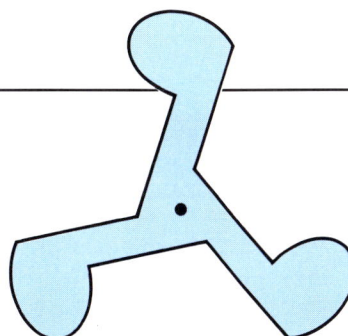

3 This broken shape has turn symmetry of order 2.
Draw the unbroken shape on squared paper.

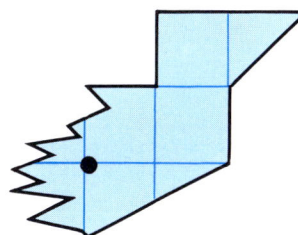

Copy the broken shape, trace your copy, then turn your tracing.

Think it through

4 This shape has turn symmetry of order 3.
How many degrees must it turn before it fits onto itself again
a) for the first time?
b) for the second time?
c) for the tenth time?

Take note

In a full turn, this shape fits its
dotted outline only **once**.
It has turn symmetry of order 1.

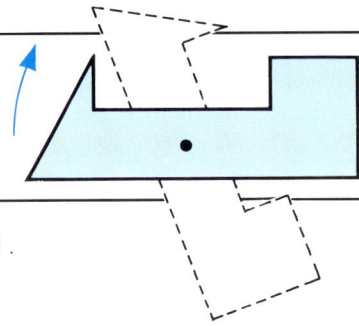

5 Think about shapes with turn symmetry of order 1.
 Draw two examples of your own.

6 These four flag shapes are all different!

A   B   C   D

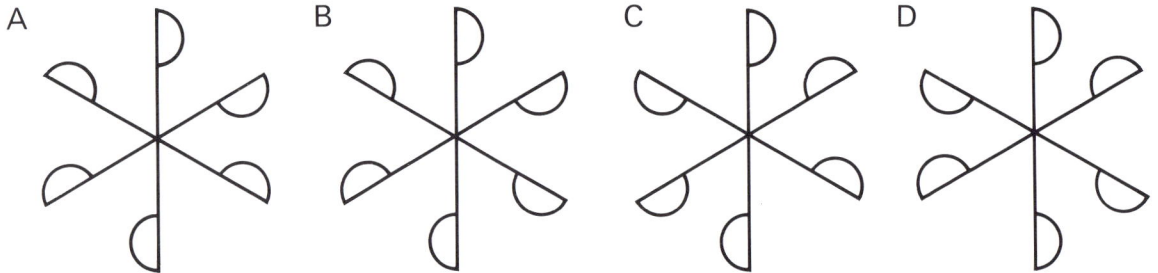

Which one has turn symmetry of order
a) 6 b) 3 c) 2 d) 1?

7 This shape has eight arms.
 It has turn symmetry of order 2.
 Draw your own eight-armed shapes.
 Give them turn symmetry of order
 a) 8 b) 4 c) 2 d) 1.

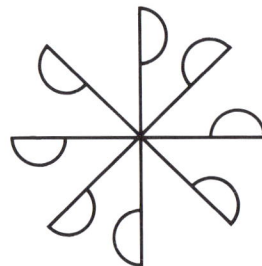

8 Study this shape carefully!
 What is its order of turn symmetry?

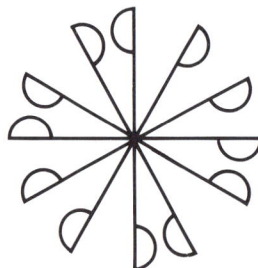

*more than
you can
count*

Challenge

9 a) Draw a shape whose order of turn symmetry is **infinite**.

 b) A shape fits itself again after turning through 72°.
   What is its order of turn symmetry?
   Sketch what the shape might look like.

■ 25
▲ 27
● Next chapter

Symmetry by turning

B **Activity:** *Turning shapes*

1 a) Make two copies of this shape on squared paper.
      Colour one copy.
      Cut out the coloured copy.
      Place it on top of the other.

   b) Do not flip your coloured shape over.
      In how many ways can you fit it on
      top of the other?

   c) Use your pencil point.
      Pin your coloured shape over
      the copy, like this.

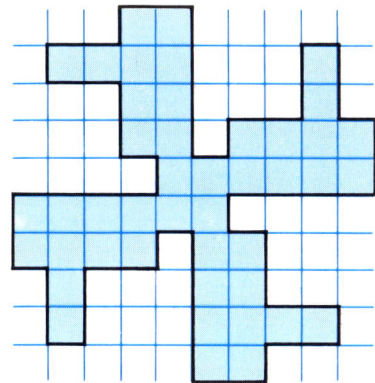

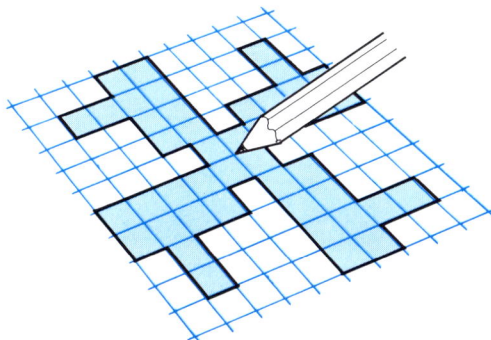

Turn the top shape slowly . . .
until it fits over the bottom
shape again.

How many degrees did you turn it?

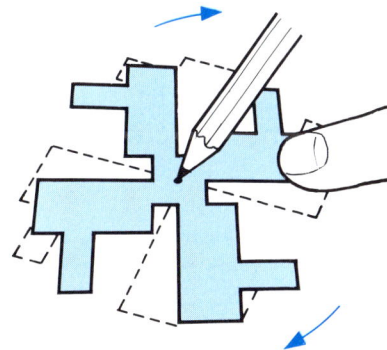

   d) Turn the top shape again . . .
      until it fits over the bottom shape.

      How many degrees has it turned **altogether**?

   e) . . . and how many degrees altogether when it fits again?

   f) . . . and again?

   g) . . . and again?

   h) . . . for the tenth time?

C 1 Which of these shapes have turn symmetry of order
   a) 1? (2 shapes)
   b) 2? (2 shapes)
   c) 3? (2 shapes)
   d) 4? (1 shape)

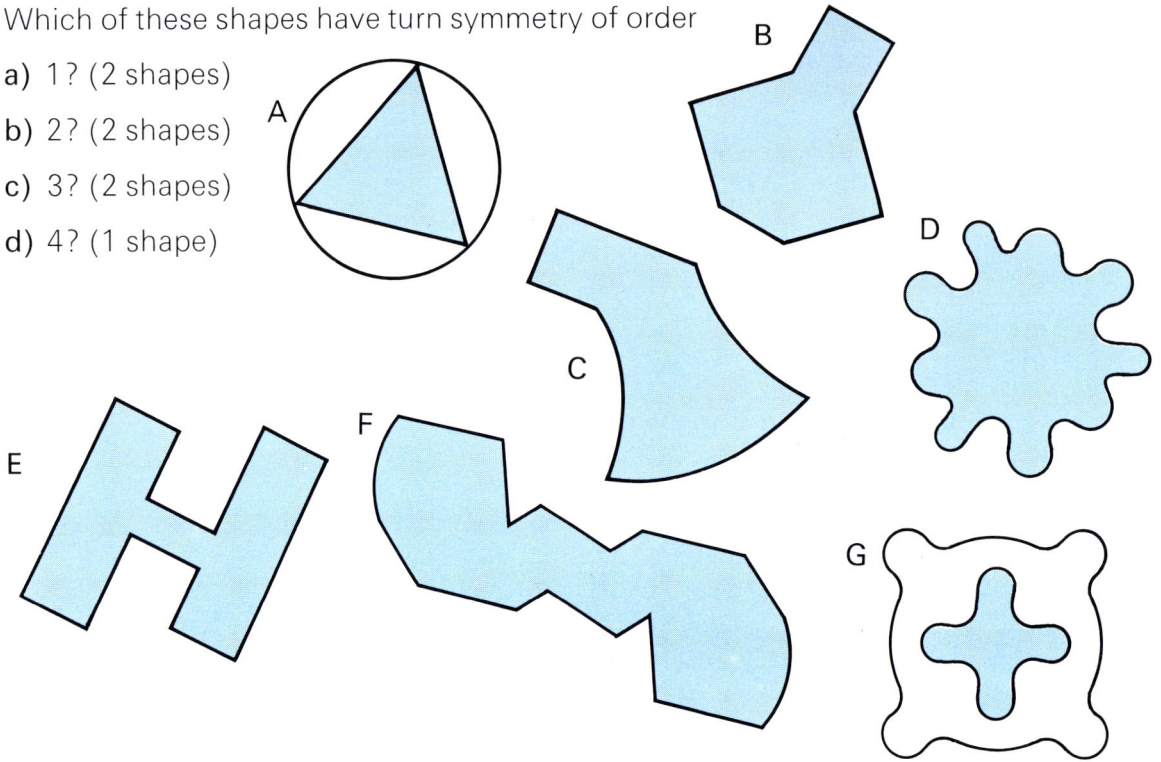

2 You need squared paper and
   tracing paper.
   Only parts of this shape
   have been drawn.
   When finished, it will have
   turn symmetry of order 4.

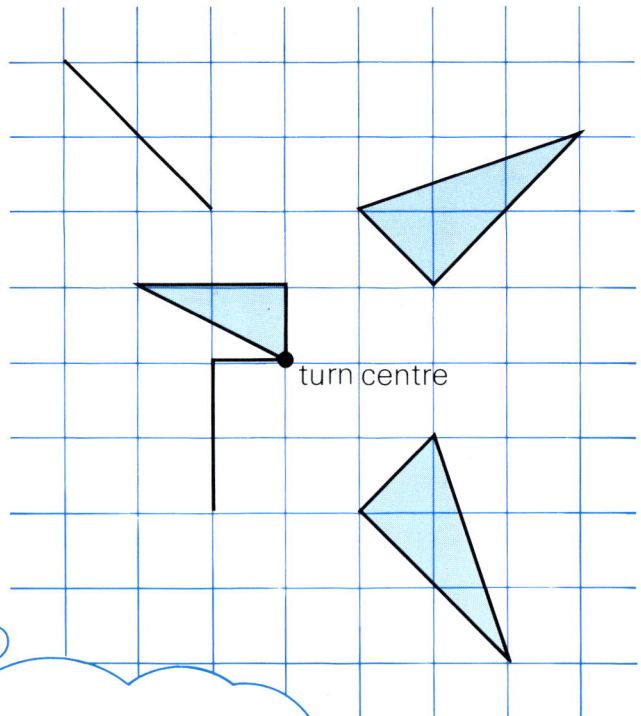

turn centre

Copy and complete it.

Trace your copy, then turn your tracing.

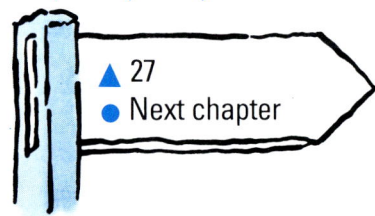

▲ 27
● Next chapter

26

Turn symmetry and line symmetry

D 1 You need squared paper.
   A and **B** are two broken shapes.
   The unbroken shapes both have turn
   symmetry of order 2.

   a) Draw each unbroken shape.

   b) How many lines of symmetry
      does each shape have?

   c) Draw another example of
      a shape which has
      (i) turn symmetry of order 2
           but no lines of symmetry.
      (ii) turn symmetry of order 2
           and two lines of symmetry.

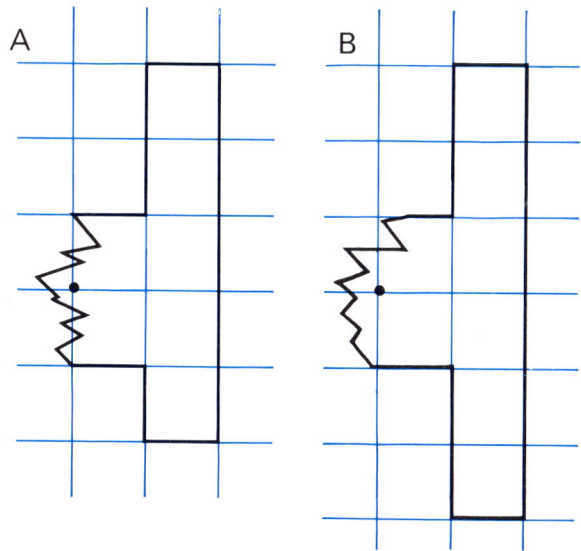

2 a) Shapes **C** and **D** both have turn symmetry of order 3 but . . .
      which one has more lines of symmetry?

   b) How many lines of symmetry does
      each shape have?

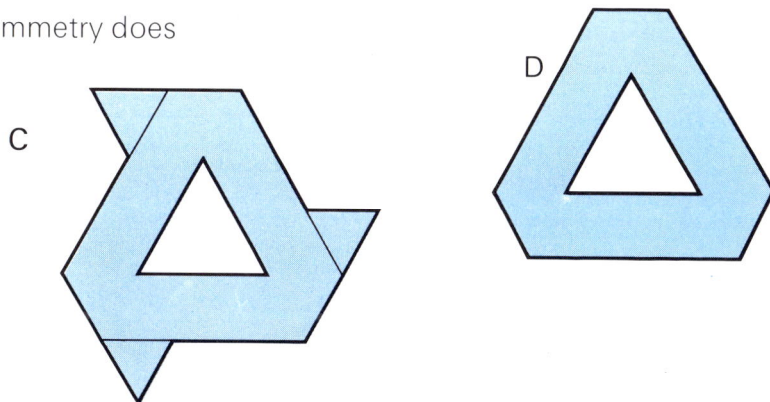

3 a) What is the order of turn symmetry
      of each shape?

   b) Which shape has more lines
      of symmetry?

   c) How many lines of symmetry
      does each shape have?

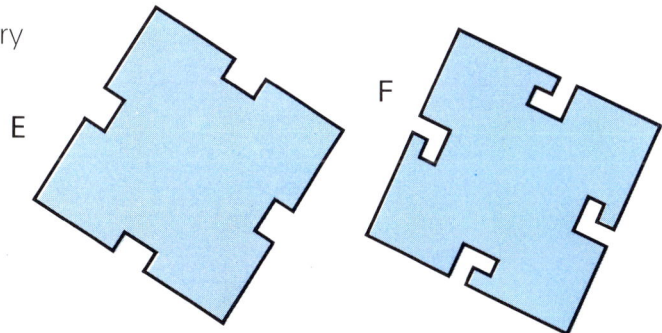

Challenge

4 When this shape is complete,
   it will have two lines of symmetry
   and turn symmetry of order 2.
   Copy and complete it.

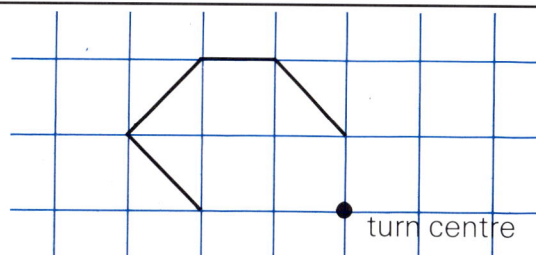

turn centre

▲ 27

Activity: *Drawing shapes*

5 You need dotted isometric paper.
   a) Copy lines **A** to **F**.
      Shapes **A** to **F** are
      well-known shapes.

squares, rectangles, . . .

One side has been drawn for you.
Here is some information
about their symmetries.
Use it to complete a shape
which fits.

A (4 sides)

B (4 sides)

C (4 sides)

D (6 sides)

E (4 sides)

F (3 sides)

Shape **A**
   Turn symmetry of order 2
   and two lines of symmetry.

Shape **B**
   Turn symmetry of order 2
   and no lines of symmetry.

Shape **C**
   Turn symmetry of order 1
   and one line of symmetry.

Shape **D**
   Turn symmetry of order 6
   and six lines of symmetry.

Shape **E**
   Turn symmetry of order 1
   and no lines of symmetry.

Shape **F**
   Turn symmetry of order 3
   and three lines of symmetry.

b) Write the common name
of each shape inside
your drawing.

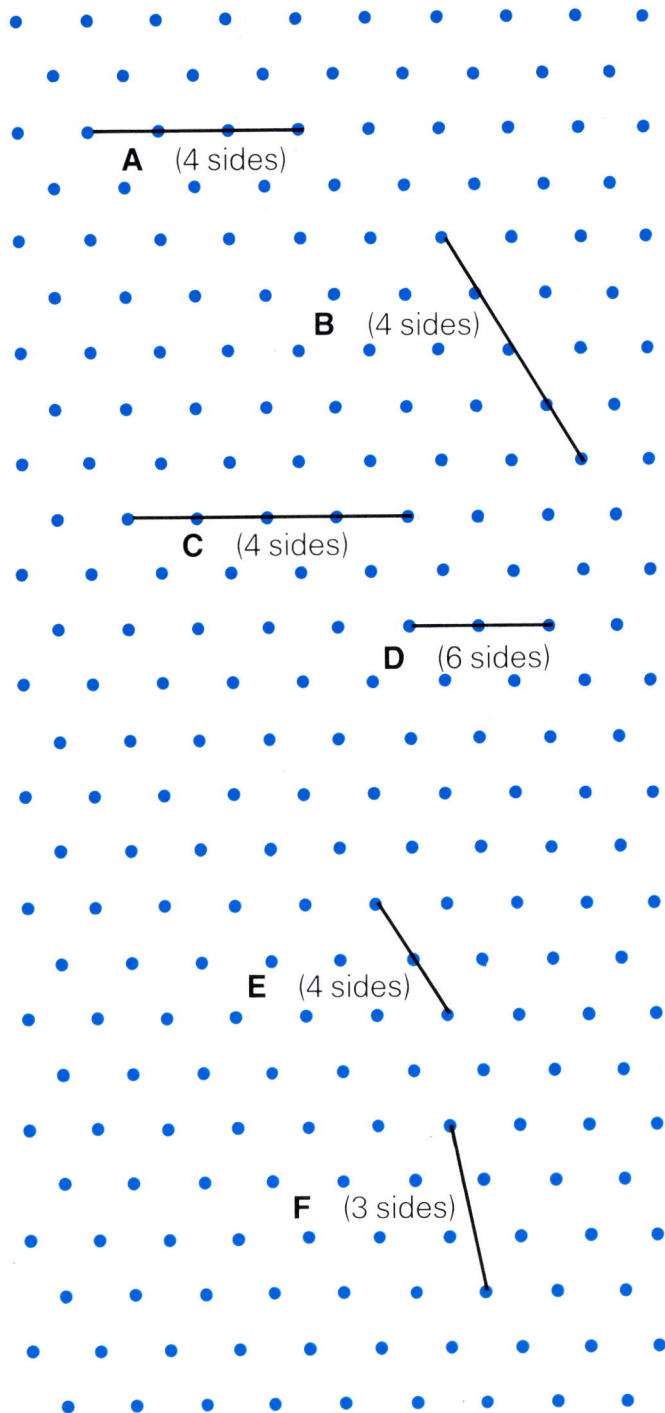

Classifying quadrilaterals

E *With a friend:* Quadrilateral types

1 Each of you copy the table.
Make it large enough to fill a whole page.

Order of turn symmetry

	1	2	3	4
0				
1				
2				
3				
4				

Number of lines of symmetry

Fill as many of the boxes as you can with different types of quadrilaterals.
You might draw more than one quadrilateral in a box.
Mark the equal sides, parallel sides and equal angles of each quadrilateral.

Include these:

'arrowhead' kite   isosceles trapezium   rhombus

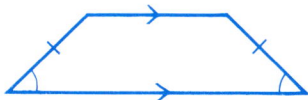

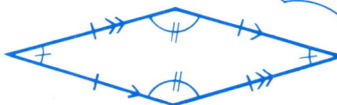

● Next chapter

▲ 29

4 Working with fractions

A ___Activity___

1 You need some 1 cm dotted paper and some 1 cm squared paper.

a) Draw this rectangle.
Write down its area in cm².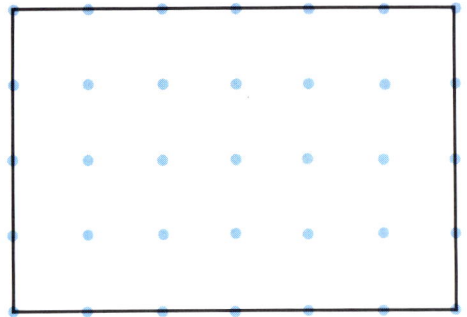

b) Use your squared paper.
Cut out four of these tiles

A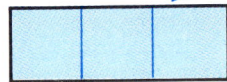

and four of these.
Cover the rectangle with them.

B

c) Which is correct?

is greater than

Area of tile **A** > Area of tile **B**

Area of tile **A** < Area of tile **B**

is less than

Area of tile A = Area of tile B

___Take note___

This is one way in which the rectangle can be covered with **eight** tiles.
The area of each tile is **one-eighth** ($\frac{1}{8}$) of the area of the rectangle.
The area of the coloured tiles is **three-eighths** ($\frac{3}{8}$) of the area of the rectangle.

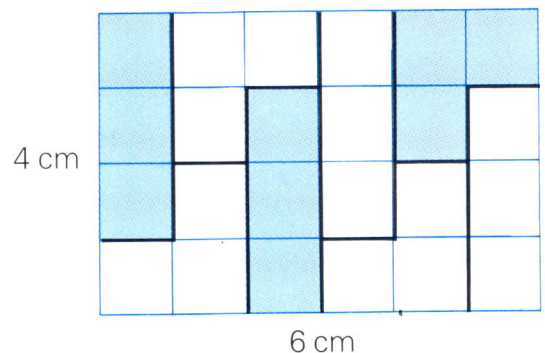

4 cm

6 cm

___Think it through___

2 Copy and complete:

a) $\frac{1}{8}$ of 24 cm² = ☐ cm²

b) $\frac{3}{8}$ of 24 cm² = ☐ cm²

c) $\frac{5}{8}$ of 24 cm² = ☐ cm²

d) $\frac{7}{8}$ of 24 cm² = ☐ cm²

Take note

denominator
(the number of
equal size parts)

3
—
8

a fraction

numerator
(the number of
parts we are
thinking of)

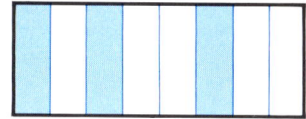

3 a) One of the rectangles has $\frac{5}{8}$ of
its area coloured.
Which is it?

b) Which rectangle has more of
its area coloured than the other two?

c) The area of **A** is 32 cm².
Copy and complete:

(i) $\frac{1}{8}$ of 32 cm² = ☐

(ii) $\frac{3}{8}$ of 32 cm² = ☐

(iii) $\frac{5}{8}$ of 32 cm² = ☐

A
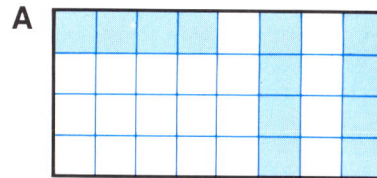

B
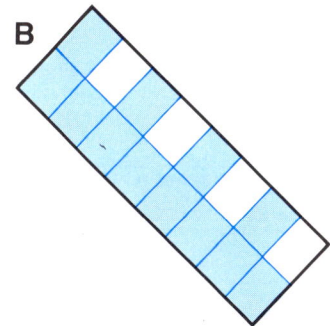

C
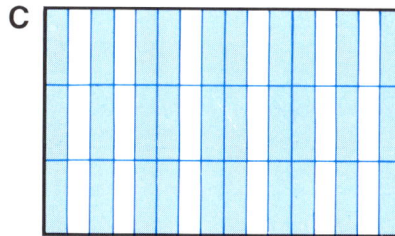

4 a) One of the circles has $\frac{3}{4}$ of its area coloured.
Which is it?

b) Which circle has more of
its area coloured than the
other two?

c) Copy and complete for circle **C**:

Fraction of area coloured $= \dfrac{☐}{☐}$

A
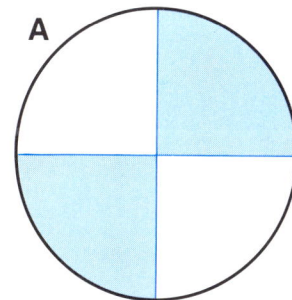

C
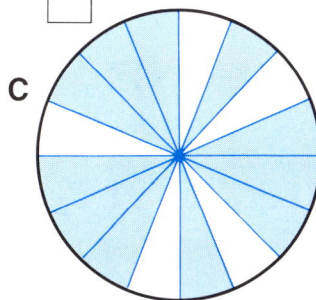

B
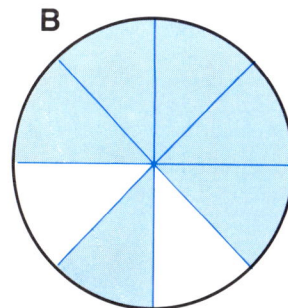

Challenge

5 Draw a circle with radius 3 cm.
Shade in $\frac{4}{5}$ of its area.

6   The rectangle is covered by four tiles.

   a) Which tile covers

      (i) $\frac{3}{10}$ of the total area?

      (ii) $\frac{1}{2}$ of the total area?

   b) What fraction of the total area does tile **D** cover?

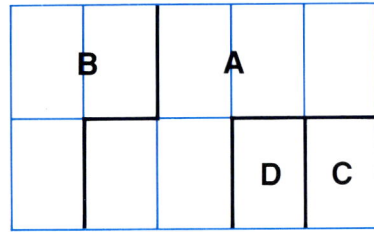

7   This rectangle is also covered by four tiles.

   a) Which tile covers

      (i) $\frac{3}{10}$ of the total area?

      (ii) $\frac{1}{5}$ of the total area?

   b) What fraction does tile **A** cover?

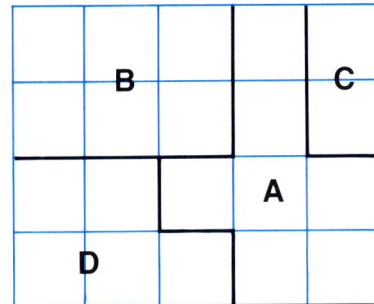

8 a) The area of the circle is 24 cm².

      How many cm² is   (i) coloured?

                         (ii) shaded?

                         (iii) white?

   b) Copy and complete:

      (i) $\frac{3}{12}$ of 24 cm² = ☐ cm²

      (ii) $\frac{5}{12}$ of 24 cm² = ☐ cm²

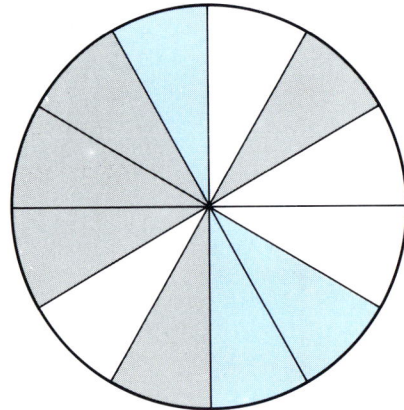

9   Here are four attempts at working out $\frac{5}{9}$ of 54 kg.

   Only two are correct.
   Which two?

   Winston

   C 5 4 ÷ 5 x 9 =

   C 5 4 x 5 ÷ 9 =

   Lenny

   C 5 4 x 9 ÷ 5 =

   C 5 4 ÷ 9 x 5 =

   Fiona

   Maisie

10 Use your calculator to work out these:

a) $\frac{2}{3}$ of £2.52   b) $\frac{3}{11}$ of 143 kg   c) $\frac{7}{8}$ of 72 m

11 How much does each of these cost in the sale?

a)
BMX BIKES
$\frac{3}{4}$ OF THE
USUAL PRICE

Was
£120

b)
Originally
£24

ANORAKS
$\frac{1}{2}$ PRICE
BARGAIN!

c)
TRAINING
SHOES
$\frac{1}{3}$ OFF!

Usually
£18

d)
Normally
£25

JEANS
THREE-FIFTHS
THE USUAL
PRICE

e)
SHIRTS DOWN TO
$\frac{9}{10}$ OF NORMAL PRICE

Originally
£12

f)
Were
£6

TIES DOWN TO
$\frac{3}{10}$ OF
NORMAL PRICE!

Challenge: *Design a flag*

12 Midge is designing a school flag.

She wants to make it $\frac{1}{6}$ red
$\frac{3}{8}$ yellow
$\frac{2}{15}$ green
and the rest white.

Use dotted paper.
Design Midge's flag for her.

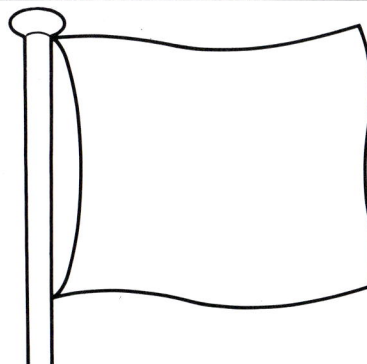

13 a)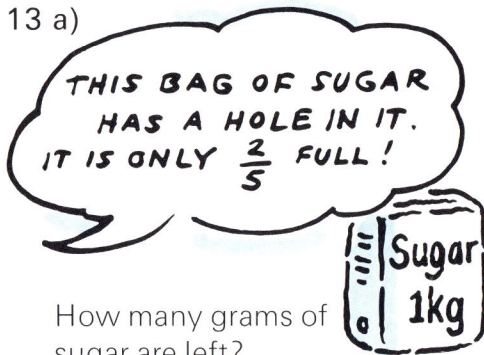

THIS BAG OF SUGAR HAS A HOLE IN IT. IT IS ONLY $\frac{2}{5}$ FULL!

How many grams of sugar are left?

b)

HOW HIGH IS THAT TABLE?

ABOUT $\frac{3}{4}$ OF A METRE.

How many centimetres is this?

c) Copy and complete:

(i) $\frac{1}{5}$ of 1 kg = ☐ g   (ii) $\frac{1}{8}$ of 1 km = ☐ m

d) Write in centimetres: $\frac{3}{10}$ m

$\frac{3}{10}$ of 1 m

e) Write in grams: $\frac{2}{5}$ kg

$\frac{2}{5}$ of 1 kg

Challenge

14 Copy and complete each label:

a)

Rioja WINE 75cl

$\frac{☐}{4}$ ℓ

b)

Brie 625 g

$\frac{☐}{8}$ kg

c)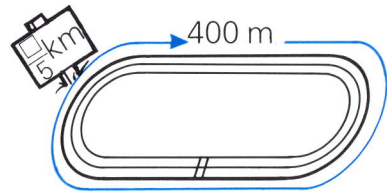

$\frac{☐}{5}$ km   400 m

Activity

15 a) This scale has to show

- whole kilograms
- tenths of a kilogram

and • eighths of a kilogram

up to 2 kg.

Design the scale yourself.

Make it large and easy to read.

b) On your scale, draw arrows pointing to

(i) $\frac{1}{2}$ kg   (ii) $\frac{7}{10}$ kg   (iii) $\frac{2}{5}$ kg   (iv) $\frac{9}{8}$ kg   (v) $\frac{7}{4}$ kg

■ 35
▲ 36
● Next chapter

Fractions of amounts

A 1 Use squared paper.
Make six copies of the rectangle.
Shade these amounts on your rectangles.
Use a new rectangle for each part.

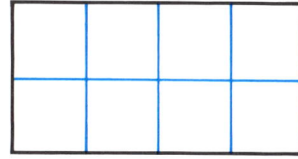

a) $\frac{3}{8}$ of the area   b) $\frac{1}{2}$ of the area

c) $\frac{1}{4}$ of the area   d) $\frac{7}{8}$ of the area

e) $\frac{3}{4}$ of the area   f) $\frac{8}{8}$ of the area

2 Use squared paper.
Make six copies of the letter **L**.
Shade these amounts on your copies.
Use a new copy for each part.

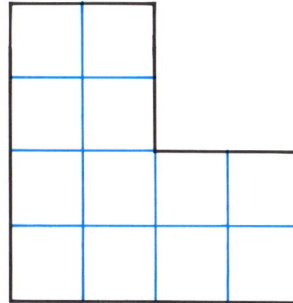

a) $\frac{1}{12}$ of the area   b) $\frac{7}{12}$ of the area

c) $\frac{1}{6}$ of the area   d) $\frac{5}{6}$ of the area

e) $\frac{1}{4}$ of the area   f) $\frac{1}{3}$ of the area

3 Calculate these:

a) $\frac{3}{8}$ of 16 cm²   b) $\frac{7}{8}$ of 24 cm²

c) $\frac{7}{12}$ of 24 cm²   d) $\frac{1}{6}$ of £24

e) $\frac{5}{6}$ of £24   f) $\frac{2}{3}$ of £27

4 Dry sand is only $\frac{7}{10}$ as heavy as wet sand.
The sand in this bag is wet.
What does it weigh when it is dry?

SAND
35 kg

▲ 36
● Next chapter

■ 35

Chocolate biscuits

B — **_Think it through_**

1 This is a gift box of biscuits.
It contains 60 biscuits.

   a) $\frac{2}{5}$ of the total number should be
chocolate biscuits.
How many chocolate biscuits should
there be?

   b) $\frac{3}{10}$ of the total weight of the
biscuits should be
chocolate coated ones.

      What should the total weight of
the chocolate biscuits be,
in grams?

   c) This is a new label for the box.
It is misleading.
Explain why.

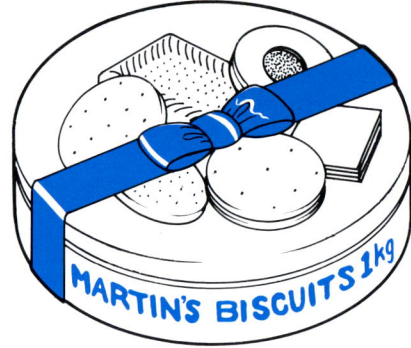

> **GIFT BOX**
>
> $\frac{2}{5}$ *chocolate coated !*
>
> **MARTIN'S BISCUITS 1kg**

   d) The biscuits in the box cannot all be the same weight.
Explain why.

With a friend

2 Work on this problem together.
You have to design a gift box of 12 Mammoth biscuits.
$\frac{2}{3}$ of the total number should be chocolate bicuits.
$\frac{3}{4}$ of the total weight of the biscuits should be chocolate coated ones.
The total weight of biscuits is $\frac{1}{2}$ kg.
Draw 12 Mammoth biscuits for the gift box.
Show which ones are chocolate.
Underneath each biscuit, write its mass in grams.

Three challenges

C Challenge 1

You need some 1 cm dotted paper.

a) Draw two 6 cm by 4 cm rectangles, side by side.

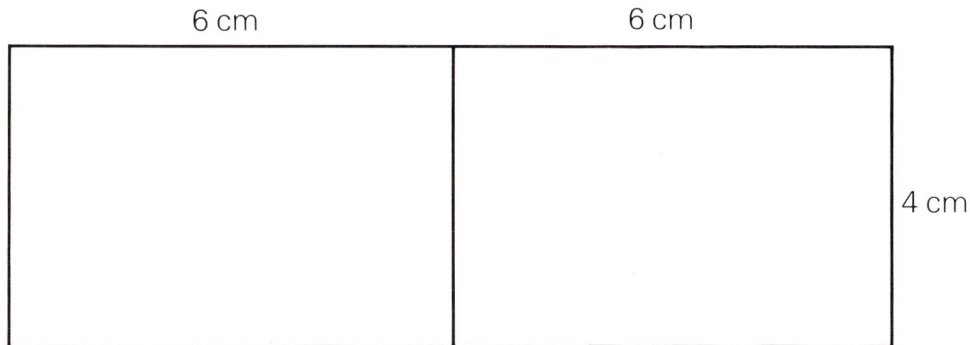

6 cm   6 cm

4 cm

Colour in an area of $\frac{7}{6}$ of 24 cm².

b) Copy the line:

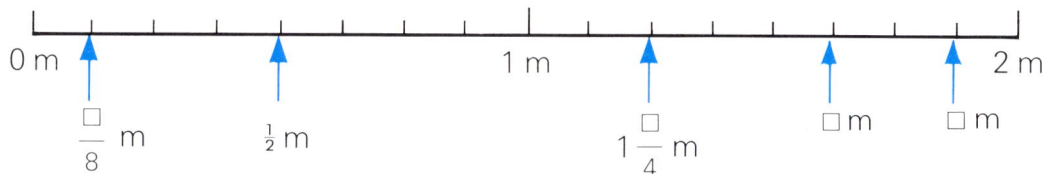

0 m   1 m   2 m

$\frac{\square}{8}$ m $\frac{1}{2}$ m $1\frac{\square}{4}$ m $\square$ m $\square$ m

Complete the distances on your drawing.

Challenge 2

Jake is planting 26 raspberry canes in a row.
From the first cane to the last, the row is 18 m long.
The canes are equally spaced.
Write down how far apart the canes are
a) as a fraction of a metre.
b) in centimetres.

Challenge 3

Lenny writes a note to Fiona on a rectangular sheet of paper.
Then he folds it like this:

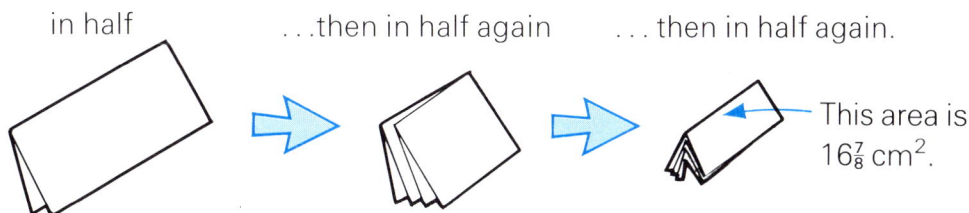

in half   ...then in half again   ...then in half again.

This area is $16\frac{7}{8}$ cm².

Draw what Lenny's notepaper might look like when it is out flat.
Write down the lengths of its sides.

● Next chapter

▲ 37

5 Angles and shapes

A Take note

Angles less than 90° are called **acute** angles.

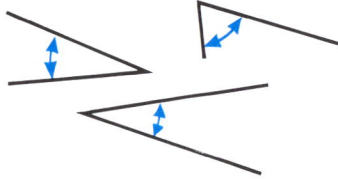

Angles between 90° and 180° are called **obtuse** angles.

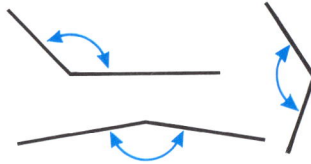

Angles between 180° and 360° are called **reflex** angles.

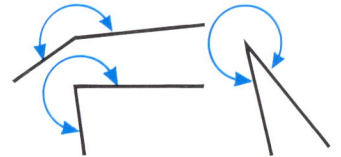

1 Shape **A** has 1 reflex angle and 3 acute angles.
Copy and complete the table
for the five shapes.

Shape	Number of angles which are			
	acute	obtuse	reflex	right
A	3	0	1	0
B				
C				
D				
E				

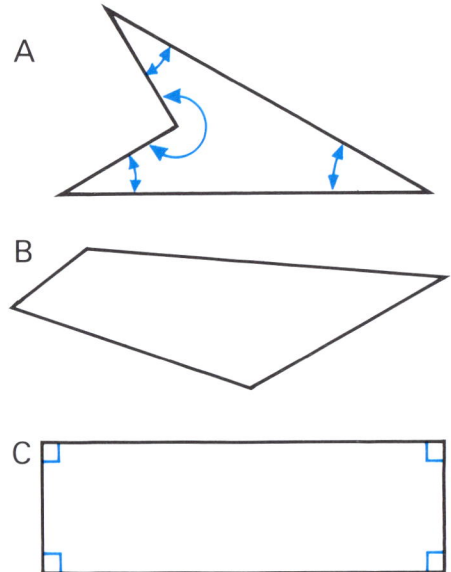

A

B

C

2 Draw a shape

a) with 4 angles:
3 of them acute,
the other one obtuse.

b) with 5 angles:
2 of them right,
2 of them acute,
the other one reflex.

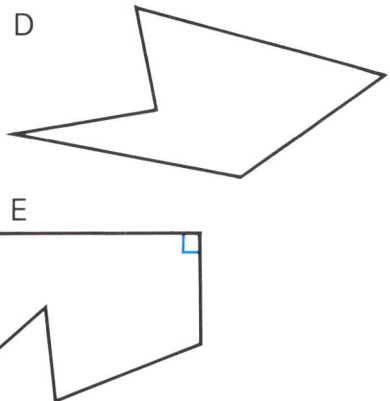

D

E

Think it through

3 A triangle can have 3 acute angles
or 1 right angle and 2 acute angles.

What other possibilities are there?
Sketch a triangle for each possibility.

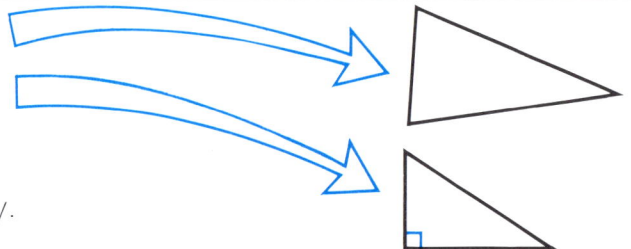

Shapes from isosceles triangles

B   *Activity*

1   You need some card and a protractor.
Draw these isosceles triangles on card and cut them out.

Remember . . . ?
Triangles which have at least
two equal length sides.

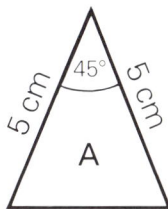

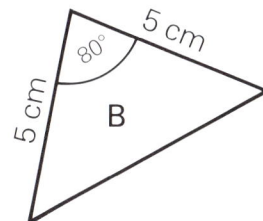

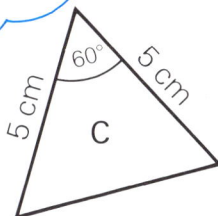

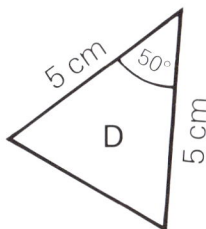

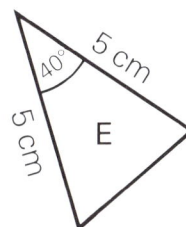

A: 5 cm, 5 cm, 45°

B: 5 cm, 5 cm, 80°

C: 5 cm, 5 cm, 60°

D: 5 cm, 5 cm, 50°

E: 5 cm, 5 cm, 40°

a) Use triangle **A**.
Follow these instructions.

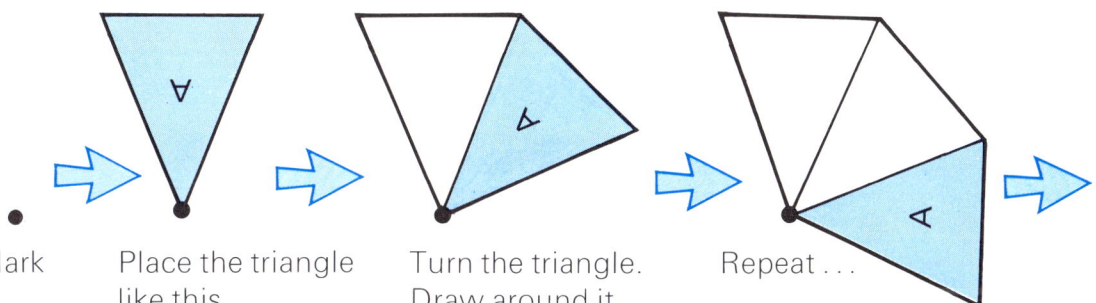

Mark a point.

Place the triangle like this.
Draw around it.

Turn the triangle.
Draw around it.

Repeat . . .

. . . until you have a 'perfect' polygon.

b) Find out which of the other isosceles triangles make perfect polygons.

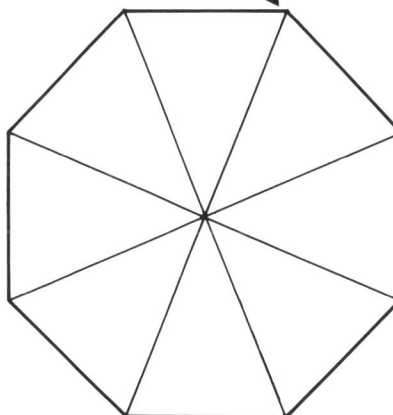

Think it through

2   a) Draw a triangle which makes a perfect polygon with five sides.
Mark the size of each angle of your triangle.

b) Write down what is special about triangles which make perfect polygons.

3 a) Which of these triangles can be used to make a perfect polygon?

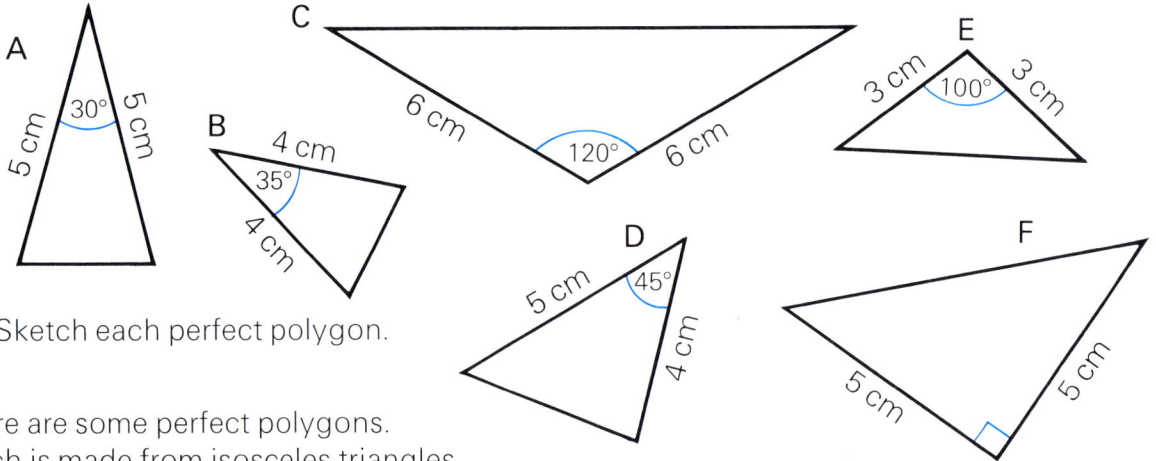

A — 5 cm, 30°, 5 cm

B — 4 cm, 35°, 4 cm

C — 6 cm, 120°, 6 cm

D — 5 cm, 45°, 4 cm

E — 3 cm, 100°, 3 cm

F — 5 cm, 5 cm (right angle)

5

b) Sketch each perfect polygon.

4 Here are some perfect polygons.
Each is made from isosceles triangles.

a) Write down the size of the angle marked ⌐ in each polygon.

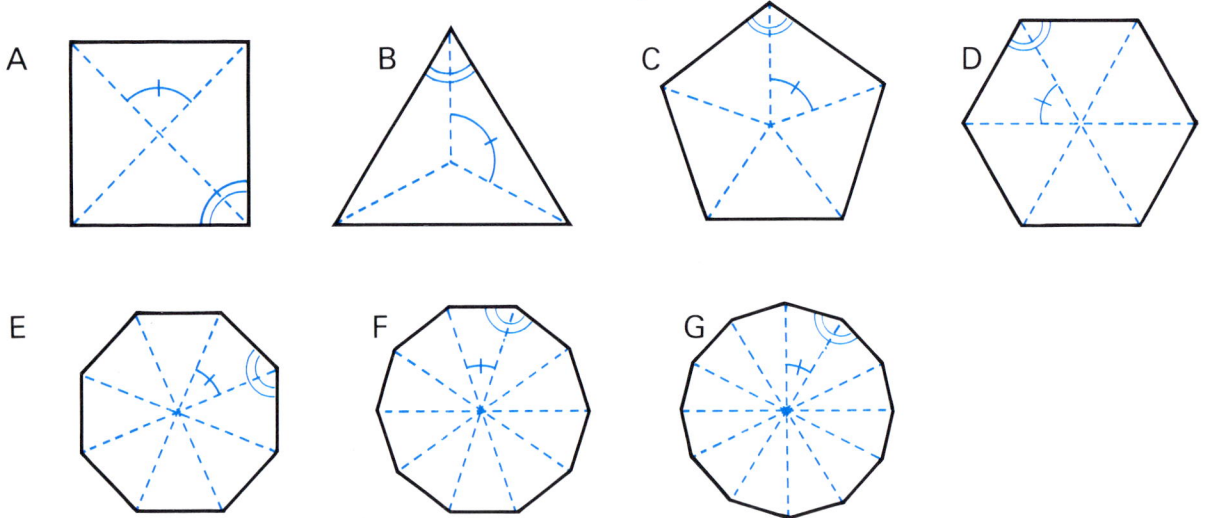

A   B   C   D

E   F   G

b) Write down the size of the angle marked)) in each isosceles triangle.

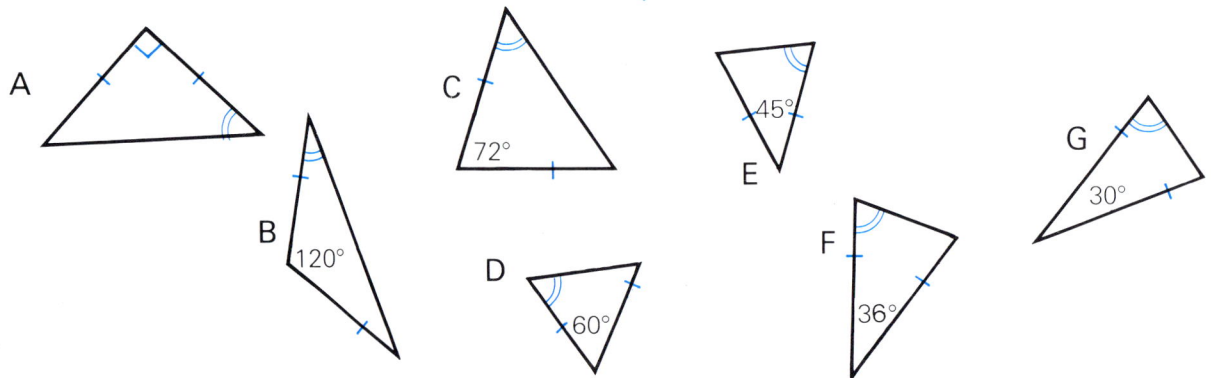

A (right angle)

B 120°

C 72°

D 60°

E 45°

F 36°

G 30°

c) Use your answers to **(b)** to help you.
Write down the size of the angle marked)) in each polygon from **(a)**.

Regular polygons

5

C

Take note

Polygons which have equal length sides **and** equal size angles are called **regular** polygons.

regular octagon

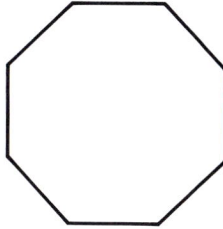

regular quadrilateral (square)

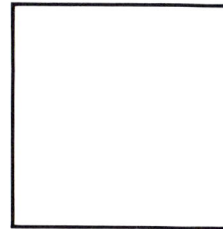

1   Which of these are **regular** polygons?
    Write **yes** or **no** for each one.
    If you write **no**, explain why.

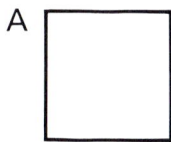

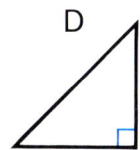

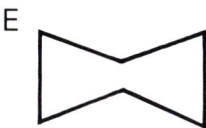

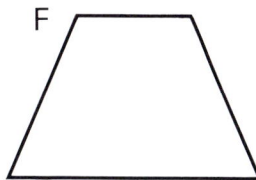

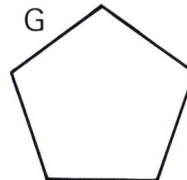

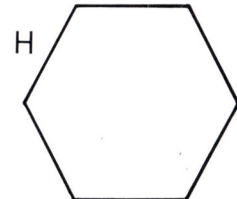

A

B

C

D

E

F

G

H

Think it through

2   A regular polygon has 12 sides.

    a) What is the size of each interior angle?

    b) What is the size of each exterior angle?

interior angle   exterior angle

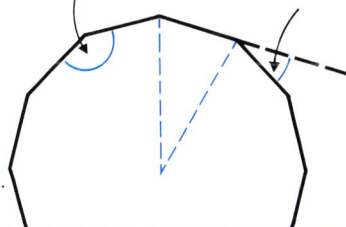

3   Copy and complete this table for regular polygons.

Number of sides	3	4	5	6	8	9	10	12	15	18	20
Size of interior angle											
Size of exterior angle											

■ 42
▲ 43
● Next chapter

Drawing regular polygons

B
C

Activity

1 You need a protractor.

a) Follow the instructions.

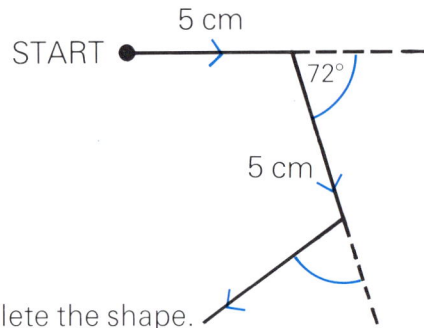

5 cm

START

72°

5 cm

Complete the shape.

Move forward 5 cm.
Turn 72° clockwise.
Repeat until you
return to START.

72°

b) Your shape in part **(a)**
should look like this.

How many degrees is the angle marked ⟨⟨ ?

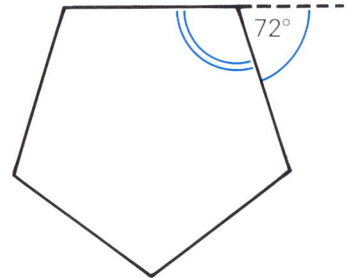

Take note

Your shape is called a regular pentagon.
It has equal length sides
and equal size interior angles.
Each interior angle measures 108°.

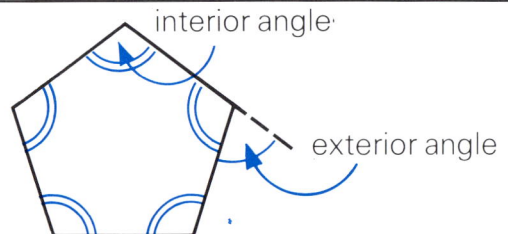

interior angle

exterior angle

c) Follow these instructions.

Move forward 4 cm.
Turn 45° clockwise.
Repeat until you
return to START.

d) Your shape in
part **(c)** should
look like this.
How many degrees
is each angle
marked ⟨⟨ ?

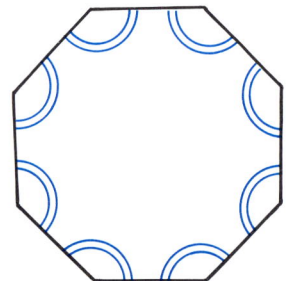

Take note

Shapes which have
equal length sides and equal size angles
are called **regular** polygons.

▲ 43
● Next chapter

Designing fishponds

Think it through

1 a) This is a ten-sided fishpond.
      There is one like it at the
      Briggens House Hotel, near Harlow.
      The ten edging stones look like
      this.

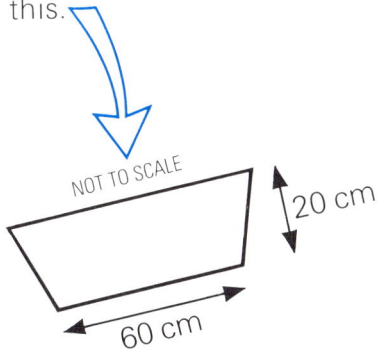

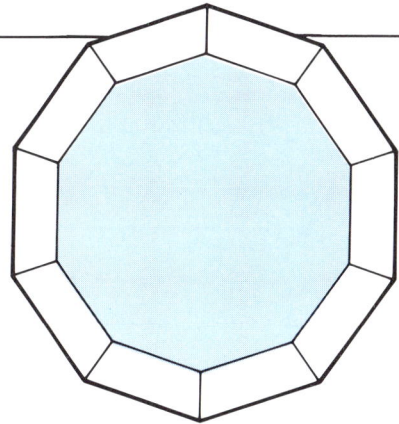

NOT TO SCALE

20 cm

60 cm

Make your own scale drawing of an edging stone.
You will have to calculate the
angles for yourself.

*Think about joining
each corner to the
centre of the pond.*

b) Stonemasons draw patterns for
   edging stones.
   This pattern is for this pond.

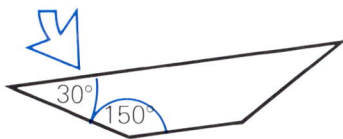

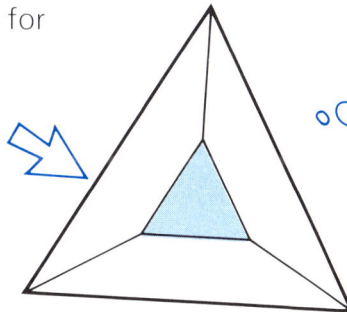

30°   150°

*Equilateral
triangle*

This pattern is for square ponds.
What are *x* and *y*?

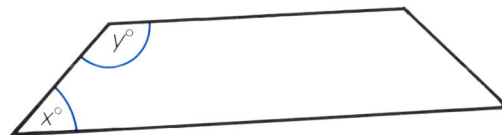

y°

x°

c) Draw your own patterns for
   ponds with these numbers of sides.
   In each one, mark the sizes of
   the two angles.
   (i)   5 sides
   (ii) 6 sides
   (iii) 8 sides
   (iv) 12 sides

*The shape of each
pond is a regular
polygon.*

5

▲ 43

Stars or polygons?

EXPLORATION

1 You need a protractor.
   a) Follow these instructions.

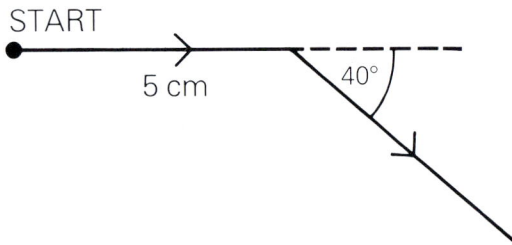

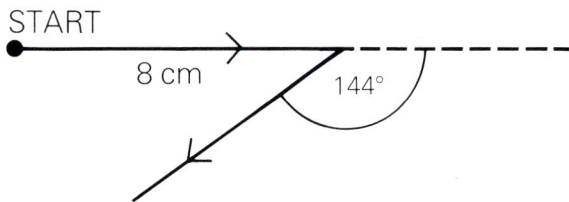

START

5 cm   40°

Move forward 5 cm.
Turn 40° clockwise.
Repeat until you
return to START.

5

   b) Your finished drawing is a **regular polygon**.
       How many sides does it have?

   c) Follow these instructions.

START

8 cm   144°

Move forward 8 cm.
Turn 144° clockwise.
Repeat until you
return to START.

You should draw a **star**.
Write down how many points it has.

2 Look at these instructions.

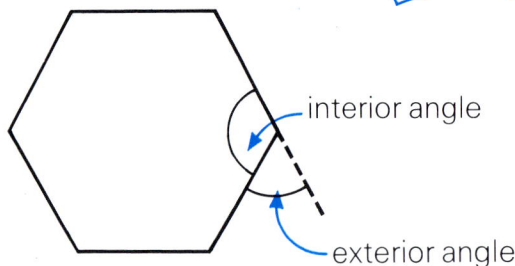

interior angle

exterior angle

Move forward 8 cm.
Turn ⬚° clockwise.
Repeat until you
return to START.

   a) Find **five** whole number replacements for ⬚
       which lead to **regular polygons**.
       Sketch each polygon.
       On each one mark the sizes of the
       interior and exterior angles.

interior angle

exterior angle

   b) Find **five** whole number replacements for ⬚
       which lead to **stars**.
       Sketch each star.
       On each one mark the sizes of the interior and exterior angles.

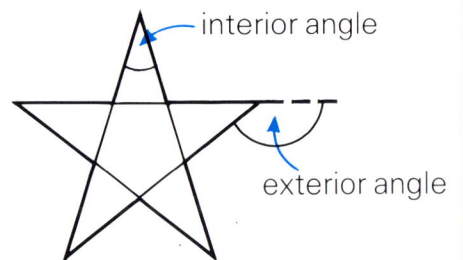

Tiling patterns

EXPLORATION

1 Regular pentagons do not make a tiling pattern.

equilateral triangles

Regular triangles do make a tiling pattern.

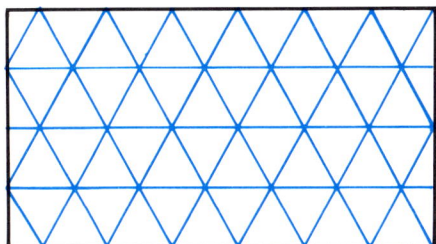

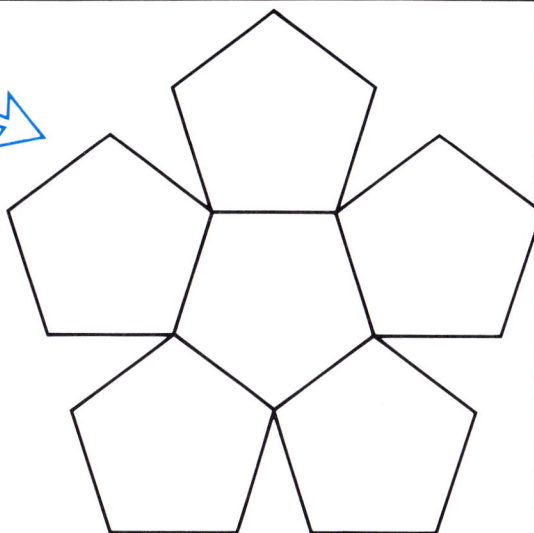

a) Copy and complete this table for regular polygons.

Number of sides	3	4	5	6	7	8	9	10	11	12
Size of interior angle	60°	90°								
Size of exterior angle	120°									
Make a tiling pattern?	√									

b) List all the regular polygons which make a tiling pattern. Start with equilateral triangles.

c) Sketch a tiling pattern for each polygon you ticked in the table.

d) You are told the size of the interior angle of a regular polygon. How can you tell if it will make a tiling pattern?

without drawing!

2 Regular octagons have 135° interior angles. Squares have 90° interior angles.

a) Together they make a tiling pattern. This has something to do with the sentence:

$$2 \times 135 + 90 = 360$$

Explain what it is.

b) Find other tiling patterns made from regular polygons. Sketch each pattern.

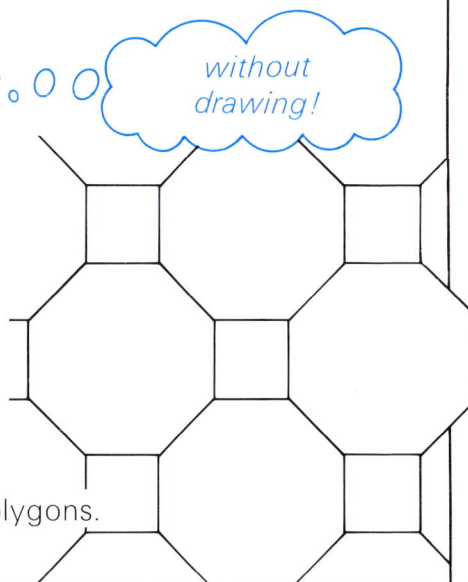

Angles and symmetry

G 1 In each drawing the coloured lines are lines of symmetry.
Calculate x and y.

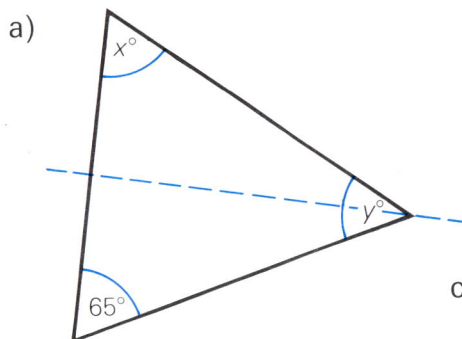

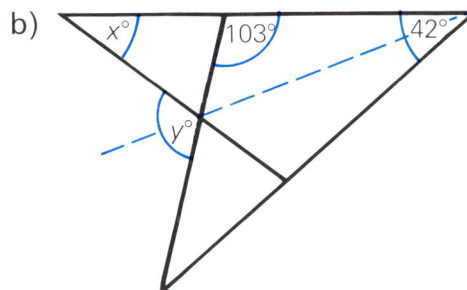

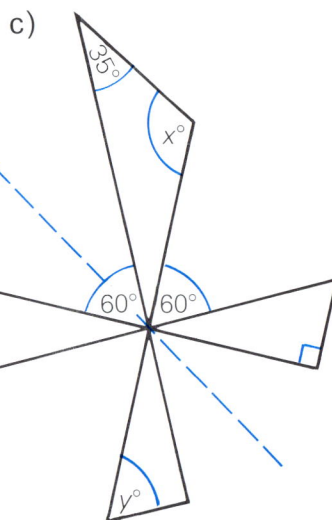

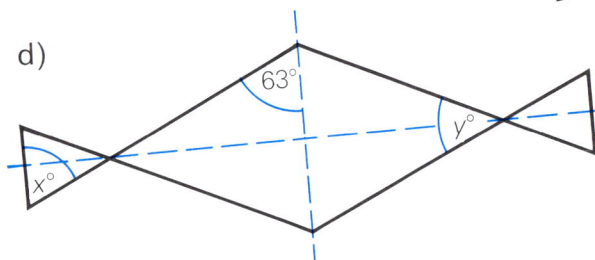

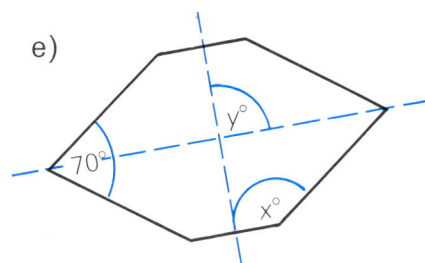

a)

$x°$

$y°$

$65°$

b)

$x°$ $103°$ $42°$

$y°$

c)

$35°$

$x°$

$60°$ $60°$

$y°$

d)

$63°$

$y°$

$x°$

e)

$y°$

$70°$

$x°$

Challenge

2 This silver star has three lines of symmetry.
Calculate x.

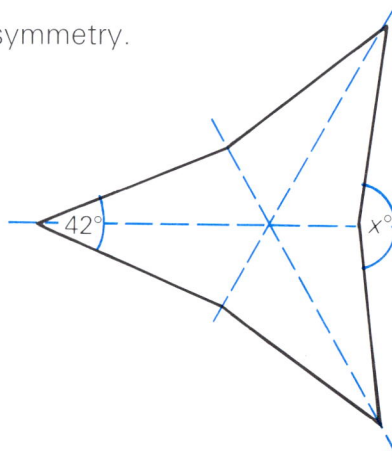

$42°$

$x°$

● Next chapter

46

6 Moving shapes

A **Activity 1:** *Making a border pattern*

1 You need a piece of card about 4 cm × 4 cm,
a pair of scissors, and
two long strips of paper, each about 4 cm wide
and 30 cm long.

a) Cut a shape from your card.
Make sure it is **not** symmetrical.

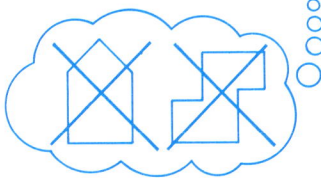

Here are some examples:

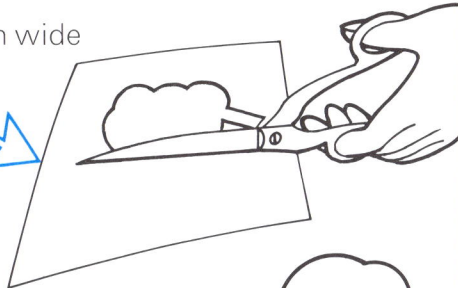

b) Draw around your shape.
Make two border patterns like **A** and **B**.
Colour them.

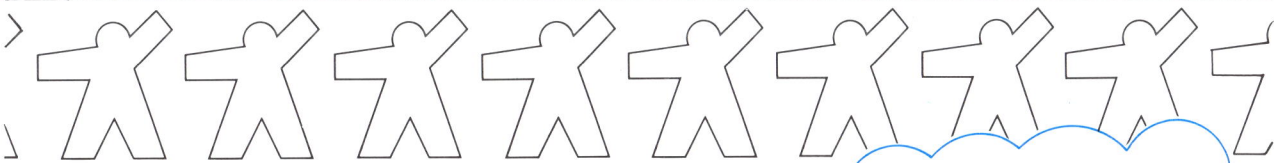

A A border pattern made by
sliding a template

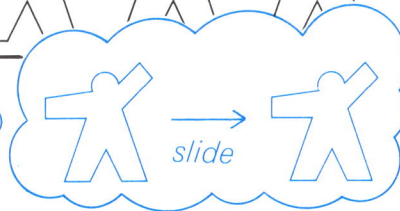

slide

B A border pattern made by sliding
and turning a template

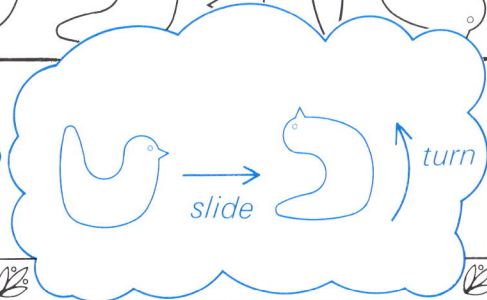

slide   *turn*

Activity 2: *A different kind of border pattern*

2 Use your template from Activity 1 ... or borrow a friend's.
Make two border patterns like these.

A border pattern made by
sliding and flipping a template.

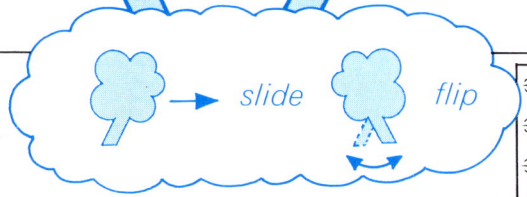

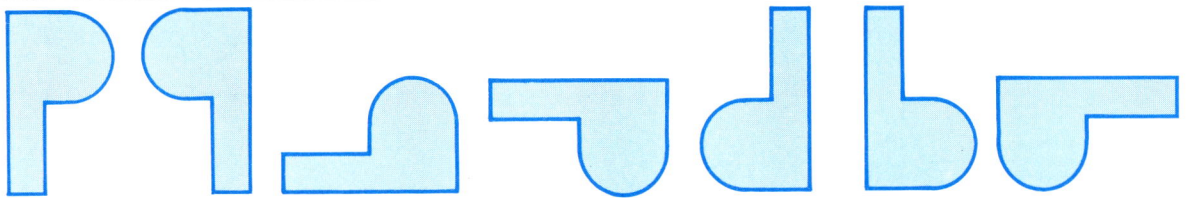

slide *flip*

A border pattern made by sliding and flipping,
then sliding and turning a template.

ASSIGNMENT:

3 Collect some patterns like these.

Look in newspapers, magazines, ...
Find examples of wallpaper patterns,
border patterns,
wrapping paper patterns, ...

Make a display.
Next to each pattern, write down
how it can be made.

*This pattern can be made
by sliding a stencil.*

*This pattern can be made
by sliding and flipping a stencil.*

Activity

4 You need scissors and 1 cm squared paper.
   Cut this shape from the squared paper.

Place it here.

Move it to the new positions.
Each time, write down
what you had to do.

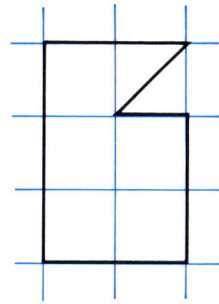

● *slide only*
or ● *slide and turn*
or ● *slide and flip*
or ● *slide, flip and turn*

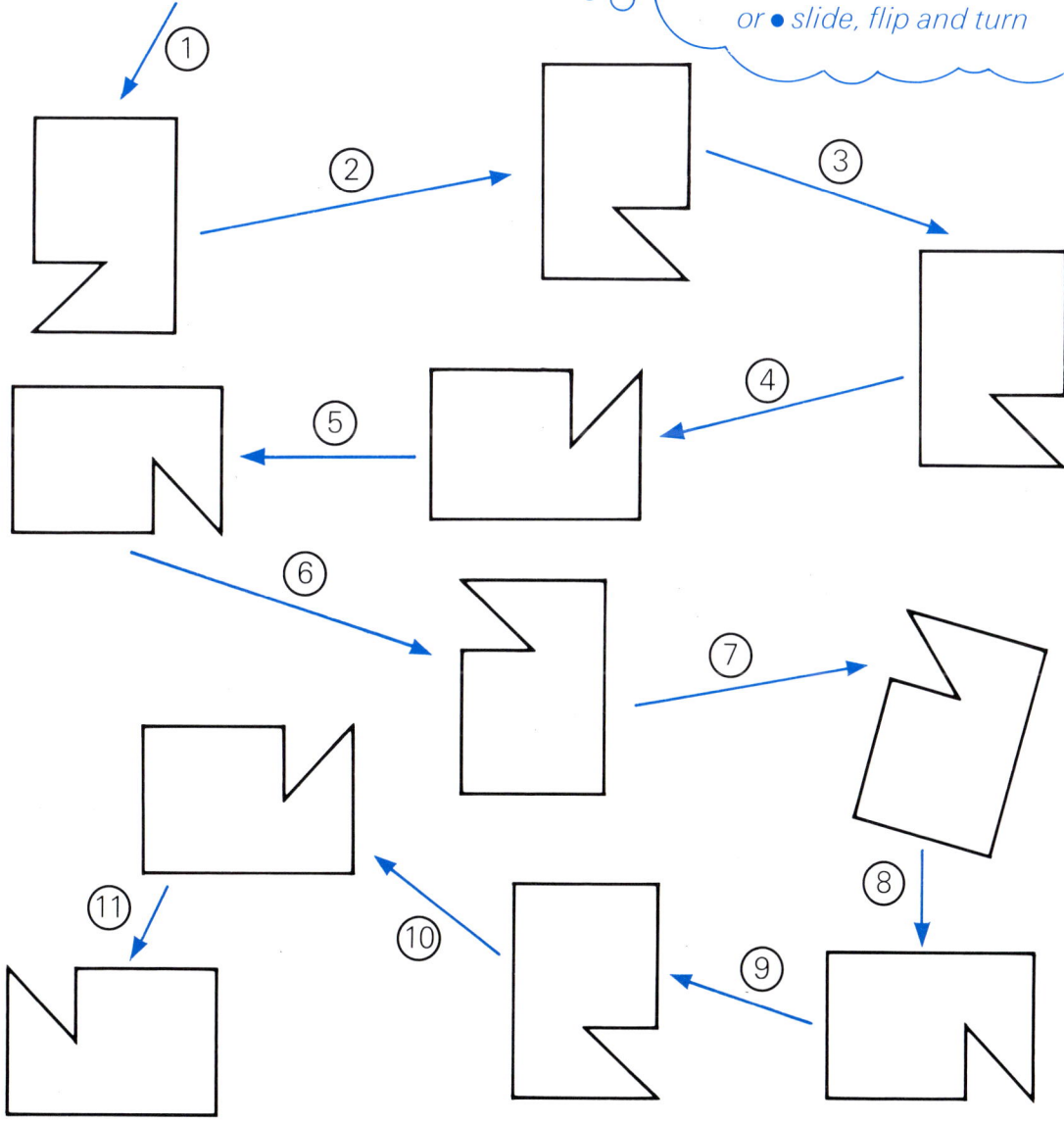

① ② ③ ④ ⑤ ⑥ ⑦ ⑧ ⑨ ⑩ ⑪

6

5

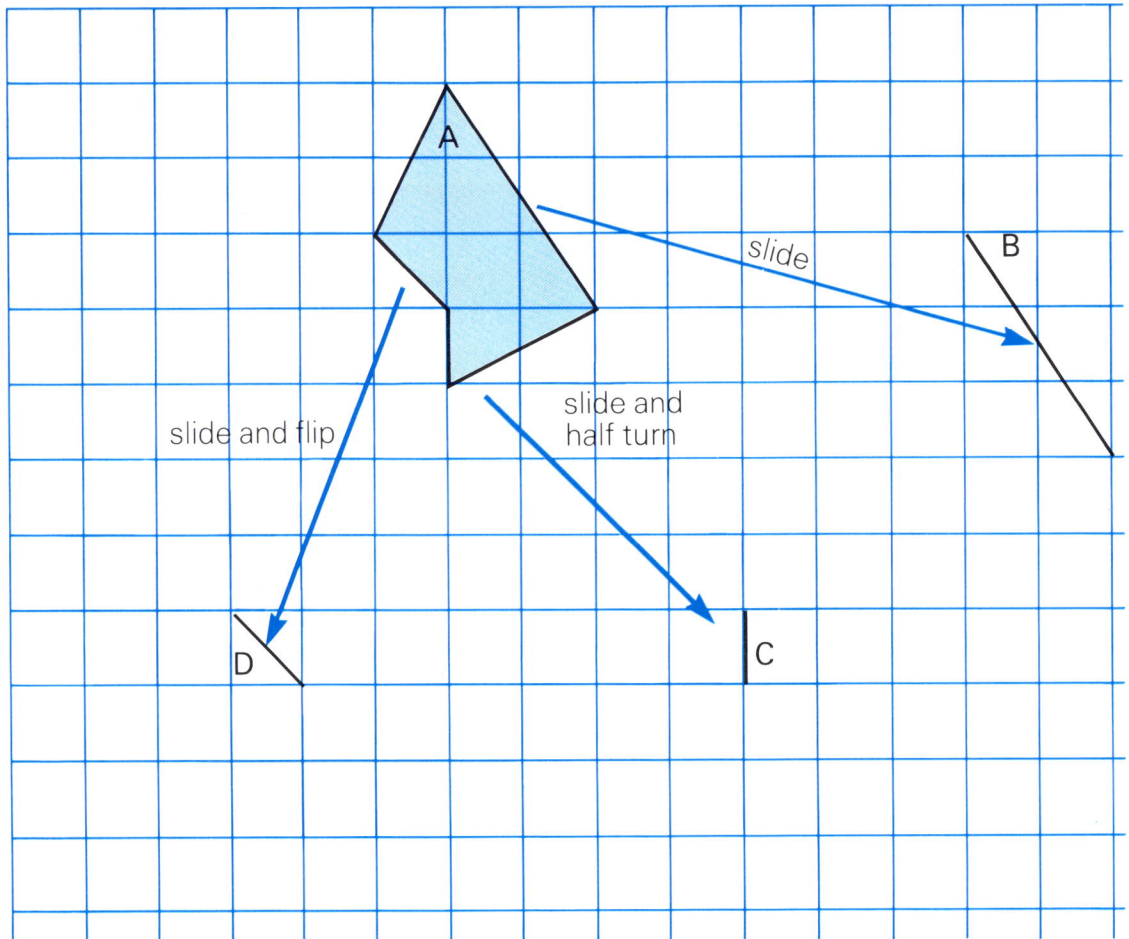

You need squared paper.

a) Copy the shape onto your squared paper.
The shape starts at **A** and slides to **B**.
One side has been drawn in its new position.
Complete the shape on your squared paper.

b) Start again at **A**.
The shape slides to **C** and does a half turn.
One side has been drawn in its new position.
Complete the shape on your squared paper.

c) Start again at **A**.
The shape slides to **D** and flips.
One side has been drawn in its new position.
There are **two** possible positions for the shape.
Draw them both.
Use a different colour for each position.

6

Flipping symmetrical shapes

B 1 Horace chooses one of these templates.

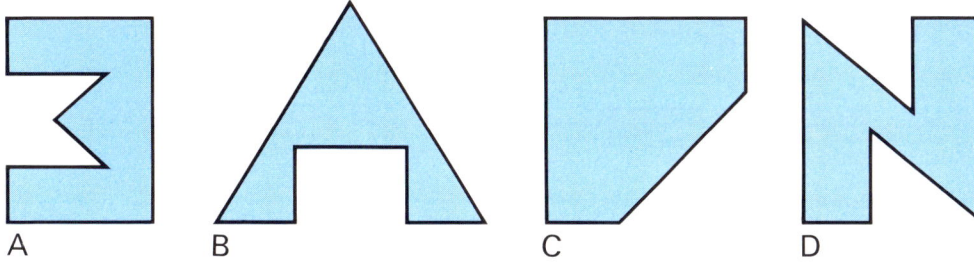

A   B   C   D

a) He **flips** it left to right.
   It looks exactly the same as before!
   Which template did he choose?

b) He chooses another.
   He gives it a **half turn**.
   It looks exactly the same as before!
   Which template did he choose?

c) He chooses another.
   He **flips** it top to bottom.
   It looks exactly the same as before!
   Which template did he choose?

d) Horace flips template **C**.
   It looks exactly the same as before!
   Draw a diagram to show how he flipped it.

6

2 Look at the quadrilateral templates.
   Which of them will fit exactly the same position when they are

a) flipped about a diagonal? (4 shapes)

b) flipped about either one of the diagonals? (2 shapes)

c) given a half turn? (4 shapes)

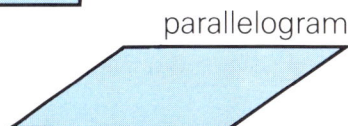

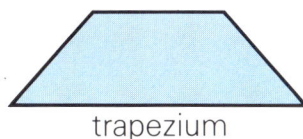

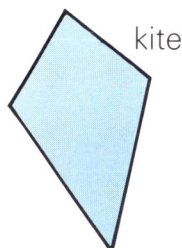

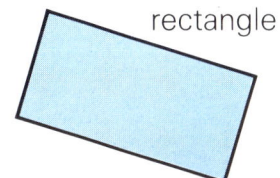

It must fit no matter which diagonal you choose.

square

parallelogram

kite

rectangle

'rhombus

trapezium

'arrowhead' kite

Making patterns

EXPLORATION

1 **a)** This template 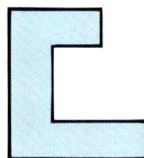 is used to make this border pattern.

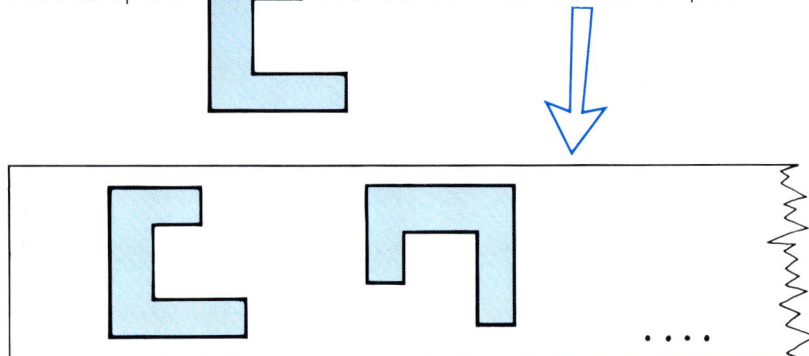

From one position to the next it **slides**,

then **turns** 90°

then **flips** bottom to top.

Draw the next three positions.

b) Start again with

This time the template **slides**,

then **flips** bottom to top,

then **turns** 90°

Does this give the same pattern as in **(a)** or a different one?

c) Investigate the patterns for other combinations of **slide**,

turn 90°

and **flip** bottom to top.

How many different patterns do you get?

d) Investigate the patterns you get for different combinations of

slide,

half turn,

and **flip** left to right.

Write down what you notice about the patterns.

c ■ 53
▲ 55
c ● Next chapter

Sliding, turning and flipping

A 1   a) **Two** of these border patterns were made by sliding this template.
Which ones?

b) **Two** of the patterns were made by sliding
and turning the template.
Which ones?

template

c) For **two** of the patterns the template had to turn, slide and flip from shape to shape.
Which patterns are they?

d) **One** pattern was made by sliding and flipping the template, left to right.
Which one?

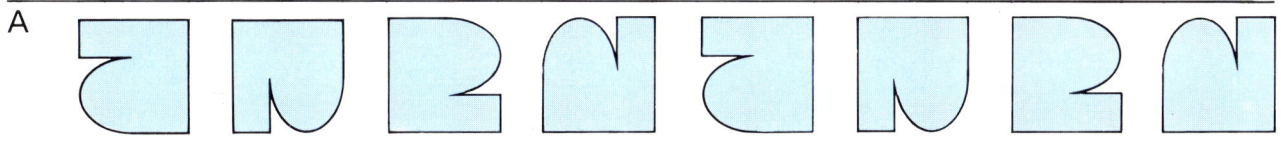

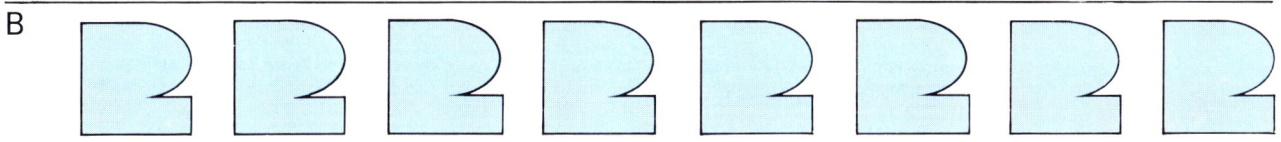

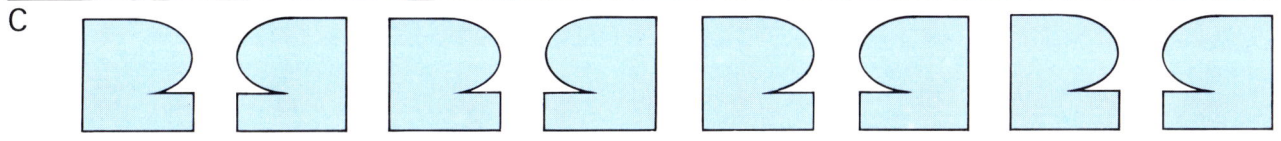

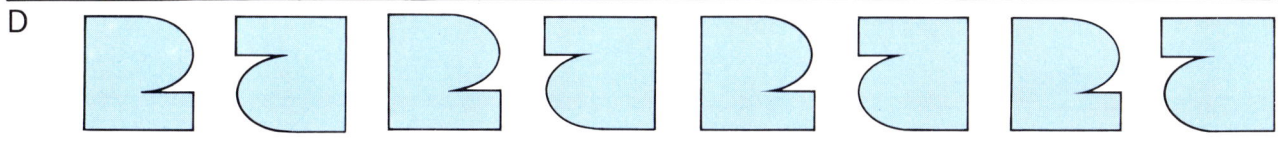

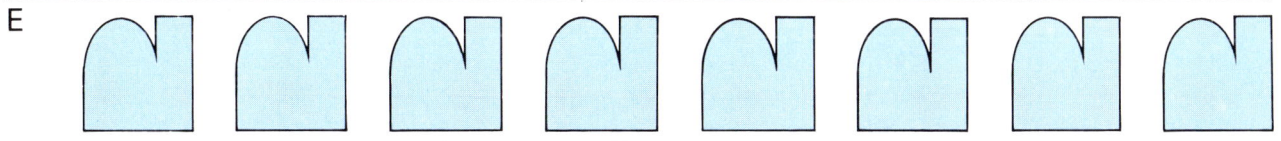

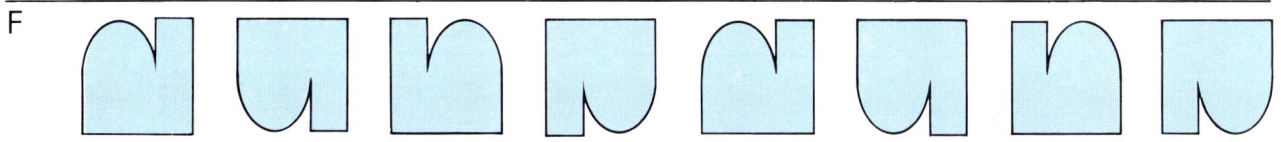

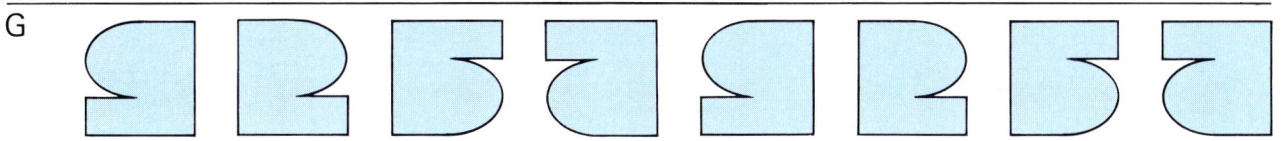

A

B

C

D

E

F

G

2 You need squared paper.

a) Copy the shape.
It slides to position ①.
Draw the shape in its new position.
One side of the shape has been drawn for you.

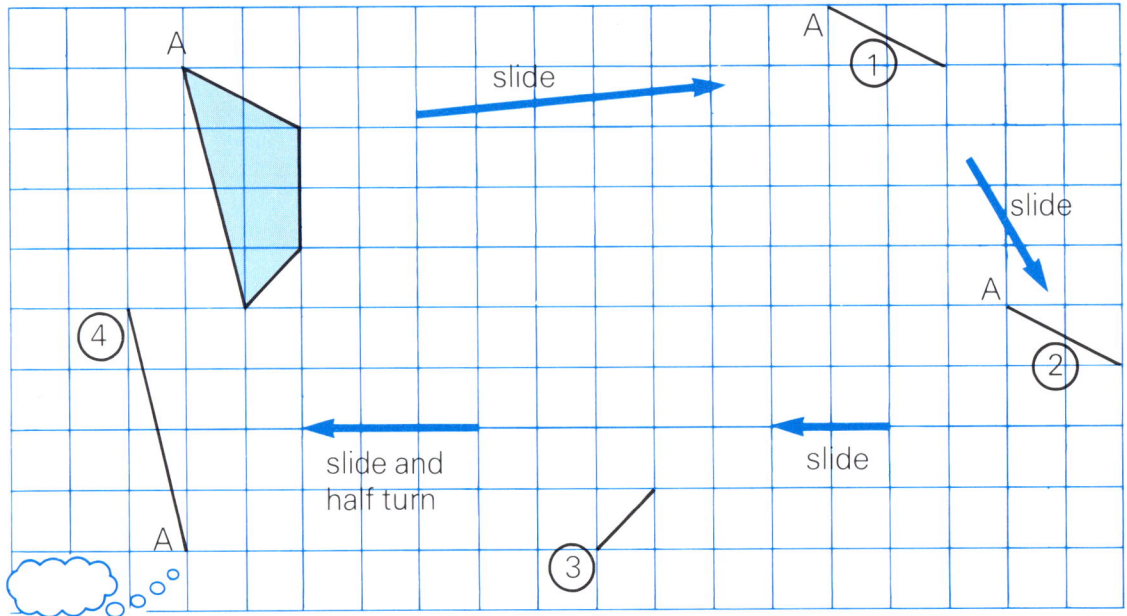

b) The shape slides to position ② and then to ③.
Draw it in these new positions.

c) The shape slides and turns.
Draw it at position ④.

Challenge

3

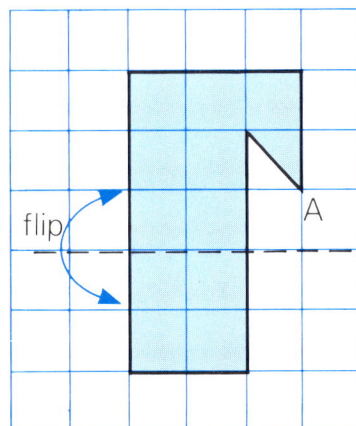

The shape flips over about
the dotted line.
Then it turns 90° clockwise.
Finally it slides so that
point A finishes here.

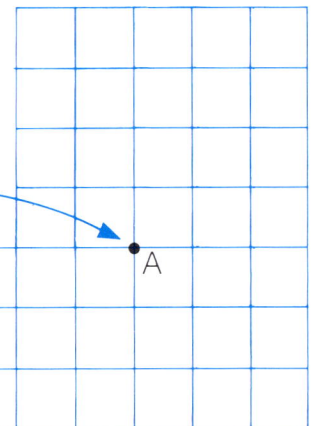

Copy the shape on squared paper.
Draw it in its final position.

▲ 55
● Next chapter

Recognising slides

6

D ___**Challenge**_____

1 You need a ruler.
   Three of the positions numbered ① to ⑧ can be reached just
   by sliding the stencil **A**.
   To reach the other five positions the stencil has also to be
   turned a little.

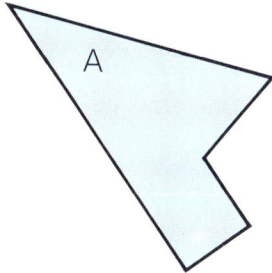
A

   a) **You are allowed only
      to use a ruler.
      You are not allowed
      to make a stencil.**
      Decide which three
      positions can be reached
      by sliding only.

   b) Explain how you decided.

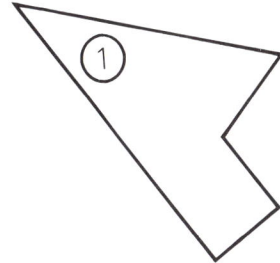

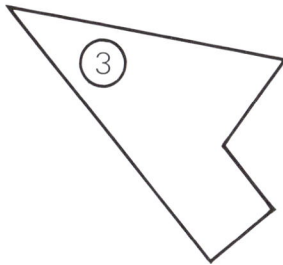

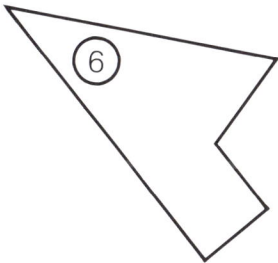

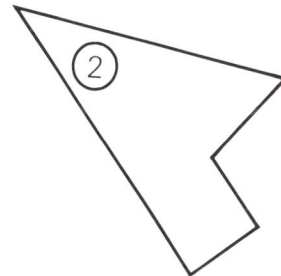

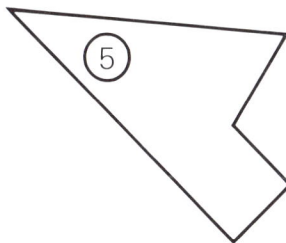

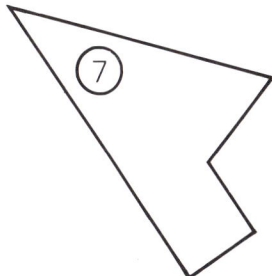

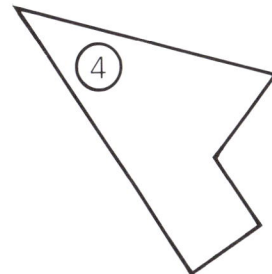

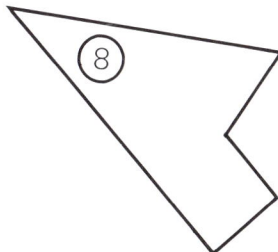

Combining moves

E 1 When this template is flipped from left to right . . .

it does not look the same.
When it is flipped from top to bottom . . .
it does look the same.
Draw your own template which looks the same

a) only when it is flipped from top to bottom.

b) only when it is flipped from left to right.

c) when it is flipped from top to bottom and when
it is flipped from left to right.

2 Jamal cuts out a template.
He draws around it.
He flips the template from left to right . . .
then from top to bottom.
It fits exactly into the shape he drew.
What is the least number of lines of symmetry the template must have?

*Try sketching some templates
to help you find out.*

3 Jamal cuts out a quadrilateral.
He draws around it.
He gives it a half turn.
It fits into the shape he drew.
He flips it about a diagonal.
It fits again.
He gives it a 90° turn, clockwise.
It does not fit.
Draw a quadrilateral which Jamal might have cut out.

4 Glenda cuts out a quadrilateral.
She draws around it.
She flips the quadrilateral about a diagonal.
It fits into the shape she drew.
She gives the quadrilateral a half turn.
It does not fit the shape she drew.
Draw two different types of quadrilateral which
Glenda might have cut out.

5 Jamal cuts out this template.

He starts with it like this. ➡ He flips it left to right. ➡ Then he flips it top to bottom.

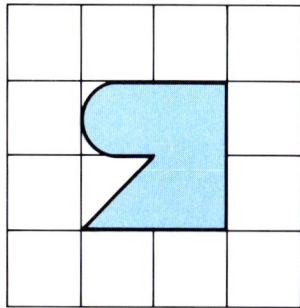

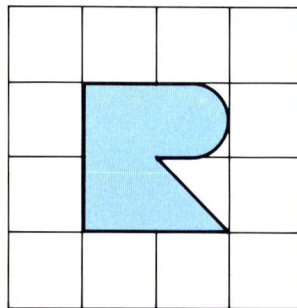

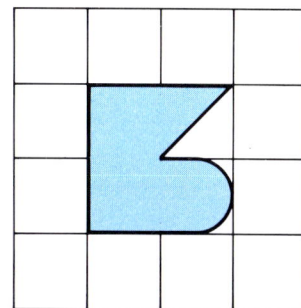

Glenda cuts out a template the same shape.

She starts with it like this. ➡ She gives it a half turn.

Jamal and Glenda write this sentence to describe what they have discovered:

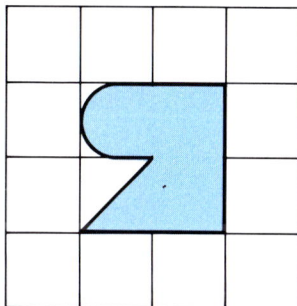

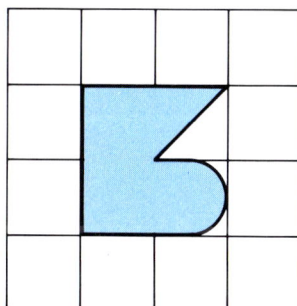

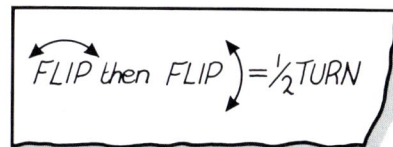

FLIP then FLIP = ½ TURN

a) Later, they write these sentences:

A Flip then ½ turn = flip

B 90° then 90° = flip

C ½ turn then ¼ turn = ¼ turn

D ¼ turn then flip = ¾ turn

Discuss with your friend what the sentences mean.
Write down the ones you would say are correct.

b) Each of you, copy and complete these sentences:

(i) flip then ½ turn = ⬚

(ii) ½ turn then flip = flip then ⬚

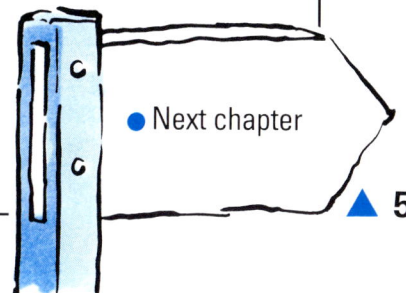

● Next chapter

▲ 57

7 Solving problems

A To do a jigsaw puzzle . . .
- we think about the shapes we have
- we look for clues
- we pick up a piece
- and we test to see if it fits.

1 Which piece fits?

To solve a problem . . .
- we study the problem carefully
- we decide on a possible solution
- and we check to see if it fits.

2 Ben and Glenda want to make
a flagpole 5 m long.
They have two poles,
each 3 m long.

Check if Glenda or Ben is correct.

3 Check if Midge or Lenny is correct.
If they are both wrong, write down
how many goldfish **you** think Meg has.

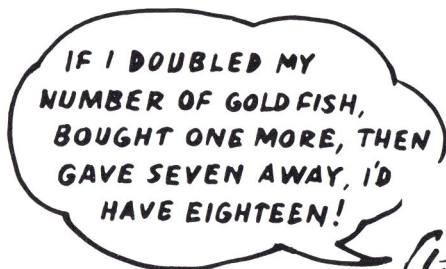

WE HAVE TO OVERLAP THE POLES BY 2m.

YOU MEAN BY 1m, DON'T YOU?

Glenda

Ben

IF I DOUBLED MY NUMBER OF GOLDFISH, BOUGHT ONE MORE, THEN GAVE SEVEN AWAY, I'D HAVE EIGHTEEN!

MEG OWNS SIX GOLDFISH.

NO. SHE HAS THIRTEEN.

Meg

Midge   Lenny

Challenge

4 Which number fits?

?

Double it, divide by 3,
subtract 10, and . . .
the result is 10.

With a friend

5 Work together on these problems.
   Think and test until you find the solutions.

a) Midge doubled a number.
   She then subtracted three.
   The result was larger than her
   starting number by 13.
   What was her starting
   number?

b) Laura Bates has two
   darts left.
   She must score 61
   to win the Championship.
   Her **last** dart must
   score a double or a
   bull (worth 50).
   Write down four ways
   in which she can score
   61 and win.

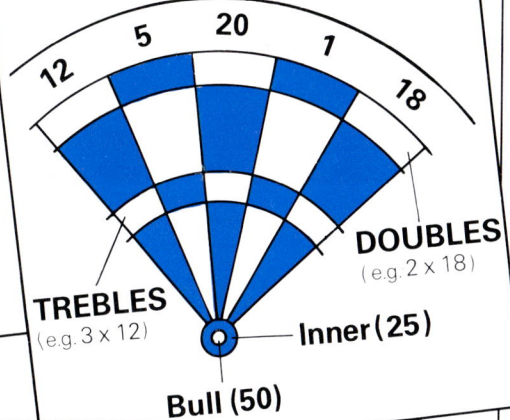

12 5 20 1 18

DOUBLES
(e.g. 2 x 18)

TREBLES
(e.g. 3 x 12)

Inner (25)

Bull (50)

c) Pepper is posting a parcel.
   The postage is 85p.
   She has 20p and 15p stamps.
   Write down how she can make
   up the postage.
   Find more than one
   solution, if possible.

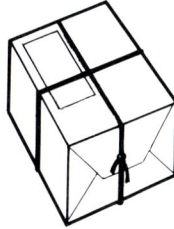

d) Horace is designing a postcard.
   He wants it to be 4 cm longer
   than it is wide.
   He wants its area to be 192 cm².
   How long and how wide should he make it?

Dear Teacher
Wish you were
here
Luv,
Horace

e) Jane Carpenter is making this roof frame.
   She has 29 metres
   of timber.
   How long should she
   make each beam?
       This beam has to be
       3 m longer than this beam,
       but 5 m shorter
       than this one.

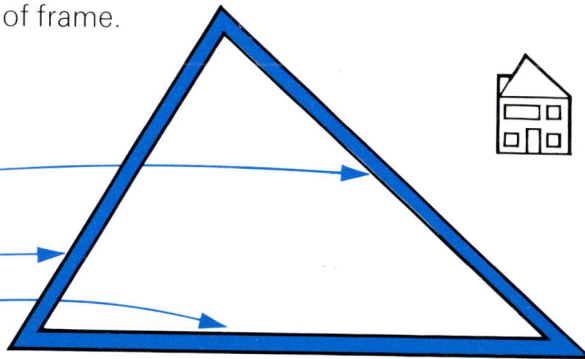

Take note

We can solve problems by thinking and testing.

Number sentences

B 1 Jamal is trying to find the number which fits this sentence.

$$4 \times \, ? + 12 = 36$$

He tries 2.

$$4 \times 2 + 12$$
$$\downarrow$$
$$20$$

a) Is 2 correct?

b) Is 5 correct?

c) What is the correct number?

2 Rupinder is trying to find a number which fits this sentence.

$$? - 8 \times 7 = 56$$

She tries 10 . . .

$$10 - 8 \times 7$$
$$\downarrow$$
$$14$$

. . . then she tries 18.

$$18 - 8 \times 7$$
$$\downarrow$$
$$70$$

a) Which of these do you think is true?
The missing number is

    A less than 10.
    B between 10 and 18, but nearer 10.
    C between 10 and 18, but nearer 18.
    D greater than 18.

b) Find the number yourself.

Challenge

3 Test numbers for **?** until you find the correct one.

    a) $36 \div \, ? + 3 = 12$ b) $20 - ? \times 2 = 6$

Take note

We can find numbers missing from sentences by thinking and testing.

Writing sentences

DO YOU REMEMBER ...?

We can write sentences to help us solve problems.

PROBLEM

An engine is pulling five carriages.
Each carriage is 8 m long.
The whole train is 46 m long.
How long is the engine?

NUMBER SENTENCE

$$5 \times 8 + ? = 46$$

5 carriages

carriage length (in m)

engine length (in m)

train length (in m)

a) How long is the engine?

b) Check that your answer fits the problem **and** the sentence.

2 For each problem:
   (i) complete the number sentence
   (ii) write down the answer to the problem
   (iii) check that your answer fits the problem **and** the sentence

a) **PROBLEM**

IF I DOUBLED MY NUMBER OF GOLDFISH, THEN BOUGHT FIVE MORE, I WOULD HAVE 21. HOW MANY DO I HAVE?

Meg

NUMBER SENTENCE

$$? \times \square + \square = \square$$

the number of goldfish Meg has

b) **PROBLEM**

Midge added another £7 to her savings.
She spent half of what she now had on clothes.
She had £5 left.
How much did she have originally?

NUMBER SENTENCE

$$? + \square \div \square = \square$$

Midge's original savings (in £)

● 61

Challenge

3 a) Write your own number sentence for this problem.

Phil added five more stamps to his collection.
He then gave 20 to Ben.
He now had 17 stamps.
How many did he have originally?

Use ? to stand for the number of stamps he had originally.

b) Check that 32 fits your number sentence.

c) Check that 32 fits the problem.

4 a) Write your own number sentence for this story.

Ben had 32 stamps.
Meg gave him some more.
A week later he sold half
of his stamp collection.
He had 40 stamps left.
How many did Meg give him?

Use ? to stand for the number of stamps Meg gave him.

b) One of these numbers should fit the story.
Write down which one.

52   48   128

c) Check that the number also fits your sentence.
If it does not, check your sentence.

5 a) Write your own stamps story for this sentence.

$$\underline{60 - ?} \div 2 = 25$$

Start your story like this:
Glenda had 60 stamps.
She lost some of them.
Then she . . .

b) Write down how many stamps Glenda lost.

c) Check that your answer fits your story
and your sentence.

Take note

Always check that your solution fits the sentence **and** the story.

63
65
Next chapter

Problems and number sentences

B 1   You are given some numbers in a cloud and a number sentence.
Test each number in the sentence.
Find the one that fits.

a)

4   6
    7

$\underline{? \times 2} - 4 = 10$

b)

$2\frac{1}{2}$ 12
   22

$27 - \underline{2 \times ?} = 3$

c)

21   0   24
    9

$4 + \underline{? \div 3} = 11$

d)

0   2
   8

$\underline{2 \times ?} + \underline{3 - ?} = 3$

Each ? replaces the same number.

7

2   For each sentence,
find the number which fits it.

a)

$\underline{? \times 2} - 1 = 9$

b)

$24 - \underline{? \times 2} = 10$

c)

$15 \div \underline{? + 1} = 3$

d)

$\underline{2 \times ?} + \underline{? \times 5} = 21$

C 1   For each problem:
(i) complete the number sentence
(ii) write down the answer to the problem

a)   **PROBLEM**

Vince cuts 8 cm from a ribbon.
He cuts the remainder in half.
Each half is 12 cm long.
How long was the ribbon?

NUMBER SENTENCE

$\underline{? - \square} \div \square = \square$

*Original length of the
ribbon (in cm)*

b) PROBLEM

At the first station the number
of passengers on the train doubled.
Twelve got off at the next
station, but none got on.
Now there were 48 passengers.
How many were there originally?

NUMBER SENTENCE

$$? \times \boxed{} - \boxed{} = \boxed{}$$

*original number
of passengers*

Challenge

2 a) Write your own number sentence for this problem.
(Use **?** m for the length Jane cut off.)

> Jane cut down an 18 m tall palm tree.
> Then she cut off the top.
> She cut the remainder in half.
> Each half measured $7\frac{1}{2}$ m.
> What length did Jane cut off the top?

b) One of these is the number of metres Jane cut off the top.
Write down which one

9   3   5

c) Check that the number also fits your number sentence.
If it does not, check your sentence again.

3 a) Write your own carpenter story for this sentence.

$$\underline{? - 1} \div 2 = 12$$

Start your story like this:

> Jane cut 1 m off a plank.
> Then she . . .

b) Write down how long the plank
was before Jane cut it.

▲ 65
● Next chapter

7

Writing sentences for problems

D 1 For each problem:
 (i) write down a number sentence
 (ii) write down the answer to the problem

a) Midge added another £7 to her savings.
 She spent half of what she now had
 on clothes.
 She had £4.50 left.
 How much did she have originally?
 (Use £? to stand for the original
 amount of her savings.)

b) The outer diameter of
 this washer is 7 cm.
 What is the radius
 marked '? cm'?

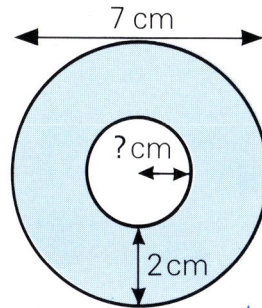

7 cm

? cm

2 cm

c) Henry Plumber welded two
 equal length pipes together.
 He then sawed off 2 m.
 The length of the pipe was now 11 m.
 How long was each of the two pipes?
 (Use ? m for the length of each pipe.)

d) This angle
 is 25° larger
 than this angle.
 How many
 degrees is each
 angle?
 (Use ?° for
 the size of the
 first angle
 mentioned.)

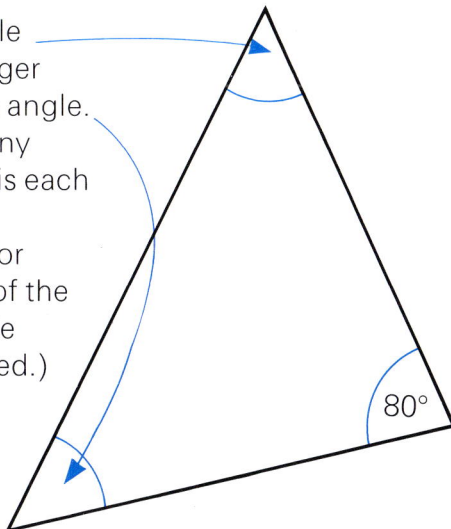

80°

e) The smaller jar holds
 200 g less honey than
 the larger jar.

HONEY HONEY

 Two small jars and one
 large jar together hold
 1400 g of honey.
 How many grams does
 a small jar hold?
 (Use ? g to represent
 the amount of honey a
 small jar holds.)

2 Midge and Phil are doing this problem.

> Ben had 32 stamps.
> Meg gave him some more.
> A week later he sold half his collection.
> He had 40 stamps left.

They both use **?** for the number of stamps Meg gave Ben.

Midge wrote this sentence:

$$32 + ? \div 2 = 40$$

Phil wrote this:

$$40 \times 2 = 32 + ?$$

a) Whose sentence is correct?

b) In each sentence find the number which replaces **?**.

c) How many stamps did Meg give Ben?

3 Now Ben and Meg are doing this problem.

> Phil bought five more stamps for his collection.
> Later in the week he gave 20 to Midge.
> He now had 17 stamps.
> How many did he have originally?

They both use **?** for the number of stamps Phil had originally.

Ben wrote this sentence:

$$? + 5 - 20 = 17$$

Meg wrote this:

$$17 + 20 = ? + 5$$

a) Whose sentence is correct?

b) Find the number which fits each sentence.

c) How many stamps did Phil have originally?

Take note

Sentences for problems can be written in different ways.

Challenge

4 a) Write a sentence using + and × for this story. Then write another sentence using − and ÷.

> Ron cycled 18 km on Monday.
> On Tuesday he cycled twice as far as he cycled on Sunday and Monday together.
> He cycled 39 km on Tuesday.
> How far did he cycle on Sunday?

b) Check that $1\frac{1}{2}$ fits each of your sentences.

● Next chapter

7

8 Thinking about circles

Activity

1   You need a pair of compasses.

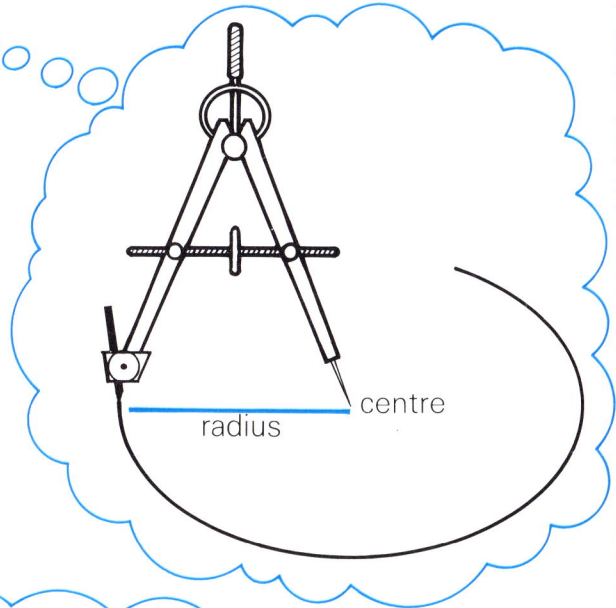

centre

radius

a) Use your compasses to draw
the tennis ball, full size.

*Plan your drawing carefully
before you begin.*

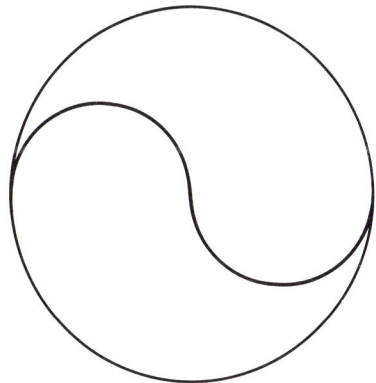

b) Use your compasses to draw this flowerhead design.
Make your own drawing twice the size.

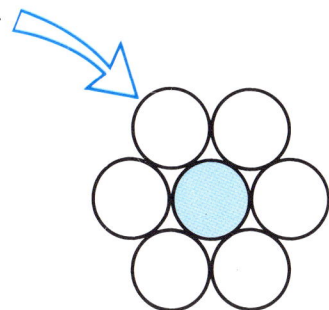

With a friend

2 The tips of the hands of a clock move in circles. Make a list of **six** other things which move in circles. Each of you make sketches to show the circles.

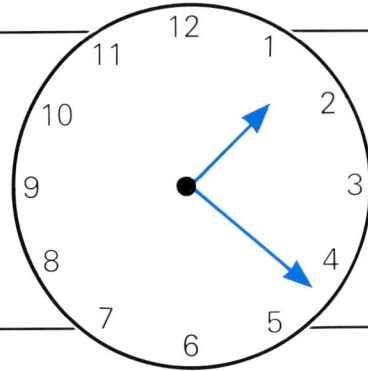

3 a) Draw a circle with radius 3 cm.

b) What is its diameter?

Take note

diameter   radius

centre

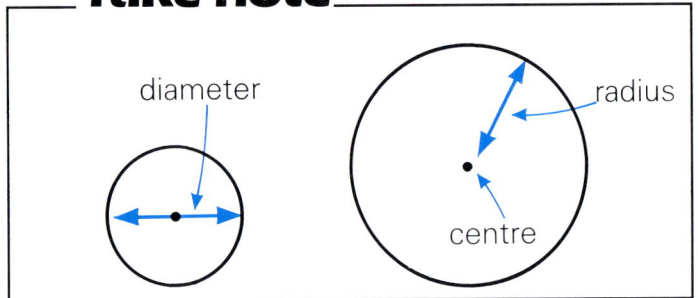

4 Approximately how many millimetres is the radius of

a) the largest button on your clothes?

b) the smallest button on your clothes?

5 Write down (i) the radius and (ii) the diameter of each circle.

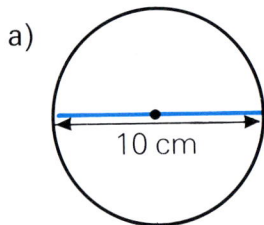

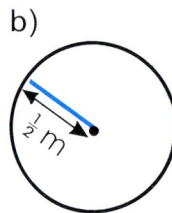

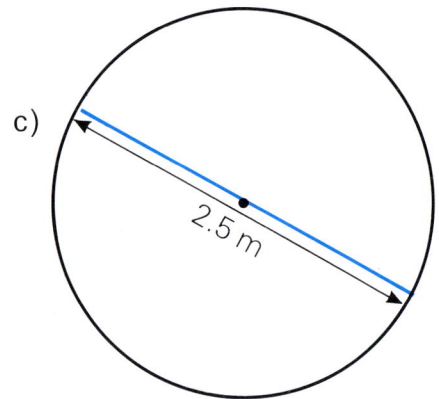

a)
10 cm

b)
½ m

c)
2.5 m

Think it through

6 a) Draw this pattern of circles. Start with the smallest. The dots are the centres of the circles.

b) What is the radius of the first circle? *the smallest circle*

c) What is the radius of the third circle?

d) Imagine more circles have been drawn. What is the radius of the tenth circle?

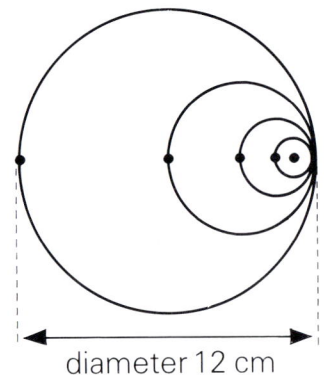

diameter 12 cm

68

Activity

7 Make a list of the diameter and radius
   of each of the circular British coins.   *1p, 2p, 5p, . . .*

8 Cut a disc like this out of paper.
   Fold your disc in half,
   then in half again,
   and so on . . . until you have made this.

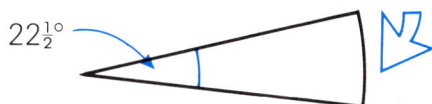

   $22\frac{1}{2}°$

   a) Write down how many times you folded your disc.

   b) How many times altogether must you fold your disc to get an
      angle of less than 5°?

      less than 5°

9 You need tracing paper.
   Mark any two points on your tracing paper.

   a) **Without measuring,** draw
      (i) any circle which passes through both points.
      (ii) the **smallest** circle which passes
            through both points.

   b) Stick your finished drawings in your book.
      Underneath each drawing write down
      (i) the diameter of the circle in millimetres.
      (ii) the distance between the two points in millimetres.

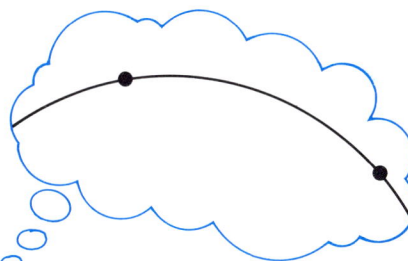

10 You need tracing paper and a can.
   Place your object on the tracing paper.
   Draw around it.

      or a cup, plate, mug, . . .

   a) **Without measuring,** find the centre
      of the circle you have
      drawn.

   b) Write down what you did to find it.

8

Circle problems

B 1   This is the Town Hall clock.
The minute hand is 3 m long.
At 12 o'clock its tip is 62 m
above the ground.
How far above the ground is it at
a) 12:15   b) 12:30?

2   The truck is carrying four rolls of paper.
The largest roll has a radius of 1.6 m.
Each roll has a radius twice as big
as the roll behind it.

    a) What is the radius of each of
the other three rolls?

    b) About how many metres long is
the back of the truck?

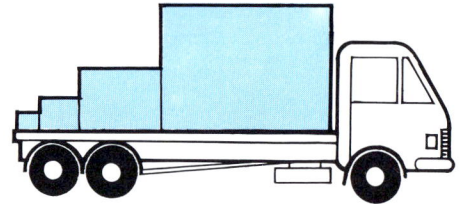

3   a) How many lines of symmetry does a circle have?

    b) The **order of turn symmetry** of this
shape is 4.

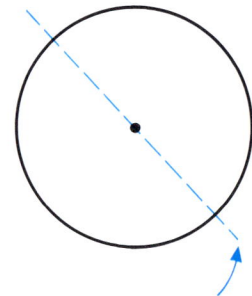

line of symmetry

     What is the order of turn symmetry of a circle?

4   This is the end-view of a water pipe.
What is

    a) its external radius?

    b) its internal radius?

    c) its thickness?

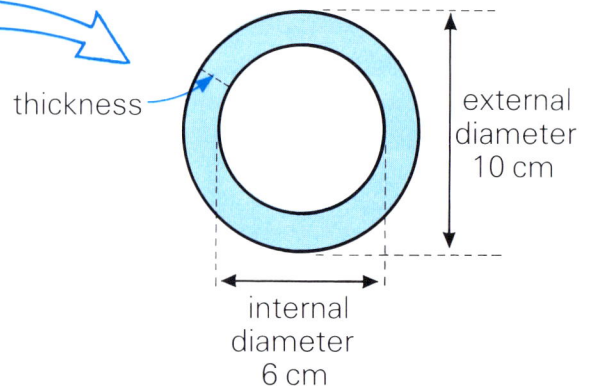

thickness

external
diameter
10 cm

internal
diameter
6 cm

Challenge

5   You need some string.
The distance around a circle is 22 cm.
What is its radius to the nearest centimetre?

Take note

The distance around
a circle is called
its **circumference**.

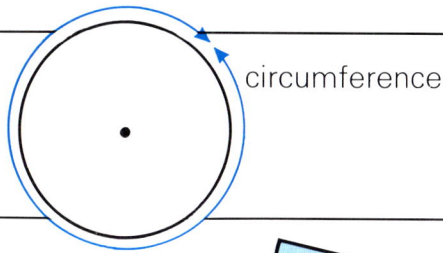

circumference

6 You need a strip of paper about 20 cm long.

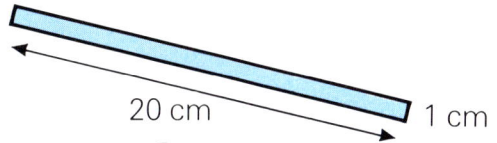

20 cm   1 cm

a) Use it to measure the
   distance around your wrist.
   Write down your result
   to the nearest centimetre.

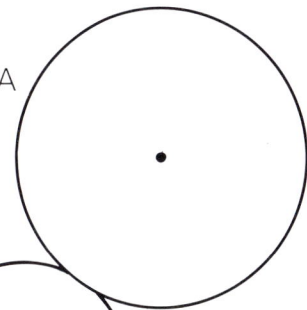

b) Copy and complete the table.

   (Measurements to the nearest centimetre)

Circle	Radius	Diameter	Circumference
A			
B			
C			
D			
E			

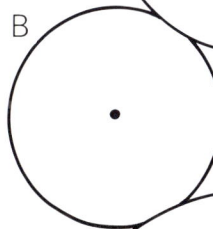

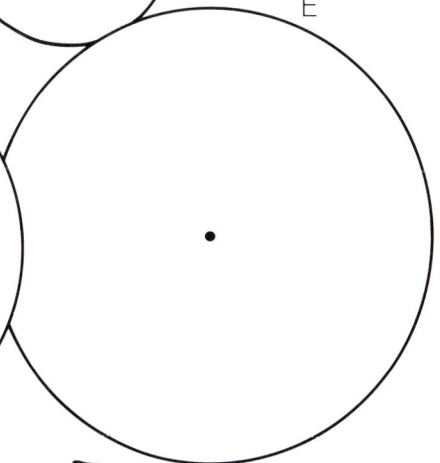

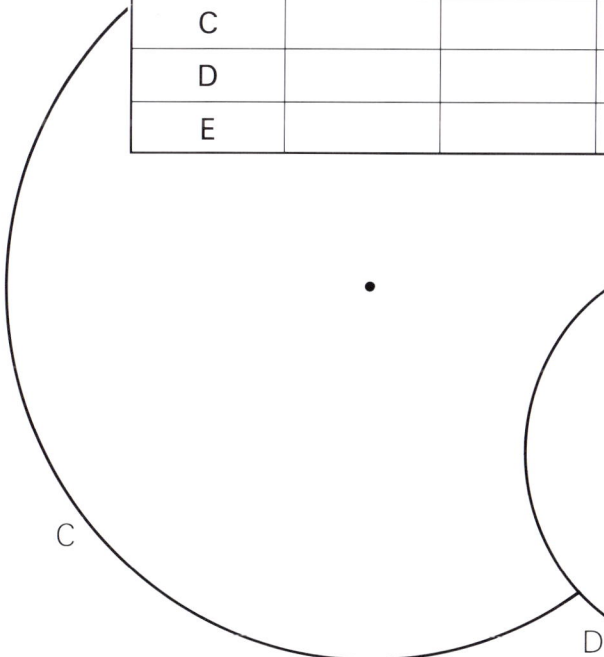

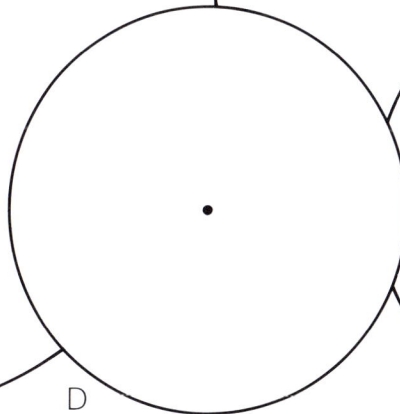

A

B

E

C

D

Challenge

7 You need a pair of compasses and some string.
   An oil drum has a circumference of 500 cm.
   Approximately what is its radius?

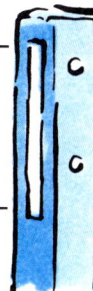

■ 72
▲ 73
● Next chapter

8

Radius, diameter and circumference

A
B

Activity

1   Use compasses to draw this design.
    Make your drawing twice the size.

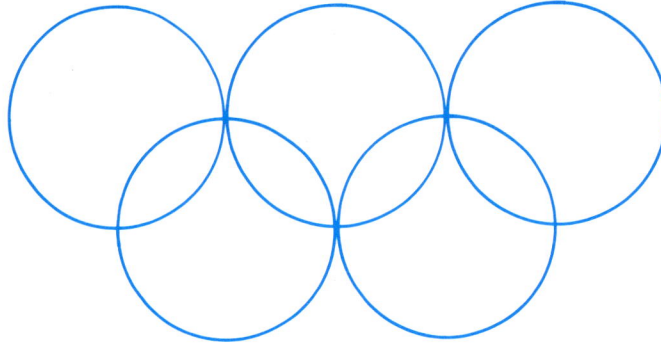

With a friend

8

2   Use string.
    Measure the circumference of
    a) each other's head.
    b) each other's neck.

    Each of you copy and complete this table.

	Circumference (to nearest cm)	
	Head	Neck
Me		
My friend		

3   This is a full-size drawing
    of a metal washer.
    What is

    a) its internal radius?
    b) its external radius?
    c) its width?

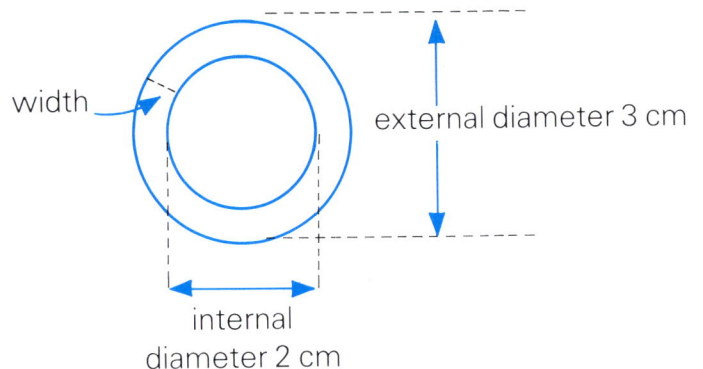

width

external diameter 3 cm

internal diameter 2 cm

4   A washer is 1 cm wide.
    Its internal diameter is 3 cm.

    a) Draw the washer.
    b) What is its external radius?

5   Draw these washers on squared paper.

    a) internal diameter 2 cm, width $\frac{1}{2}$ cm
    b) external diameter 5 cm, width 1 cm

▲ 73
● Next chapter

Area and circumference

C Challenges

1 Use 1 cm squared paper.
   a) Draw a circle whose area is
      less than 24 cm²
      but more than 16 cm².

   b) Estimate the area of
      your circle in cm²
      by counting squares.

2 You need 1 cm squared paper.
   The circumference of a circle is 36 cm.
   Estimate its area in cm²
   by counting squares.

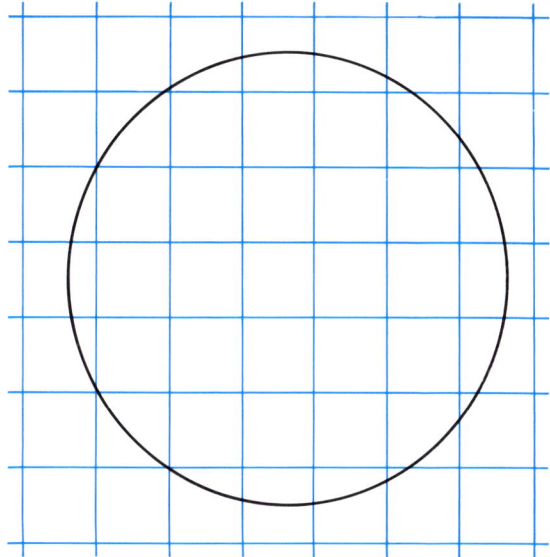

3 A ten-pin bowling ball has a radius of 10 cm.
   It travels 8 m along the gutter.
   Estimate how many full turns it makes.

4 a) Draw ten circles with different radii.
      Use string to measure their circumferences.
      Make a table like this.

   b) Use your table.
      The diameter of a circular pond is 50 m.
      Approximately what is its circumference?

   c) The diameter of the moon is 3475 km.
      An astronaut drives in a circle on the moon.
      Estimate the greatest distance
      she could travel.

Diameter	Circumference (to nearest cm)
1 cm	3 cm
2 cm	
3 cm	
4 cm	
⋮	

Drawing with circles

Activity

1 You need a pair of compasses.
Make an accurate copy of this question mark.

2 a) In the middle of a new page draw this circle.
We will call it the **starting circle**.

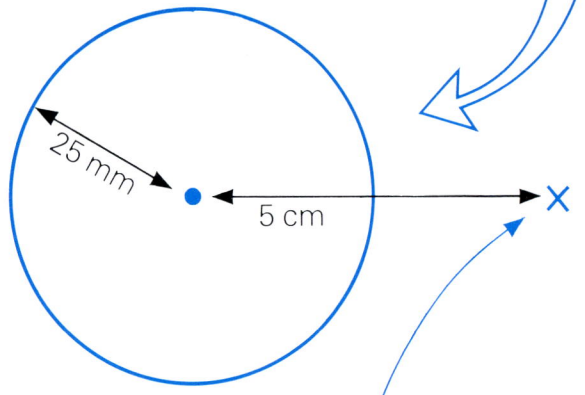

25 mm

5 cm

Mark the ✗ 5 cm to the right of the centre.
Draw another circle.
Its centre must be on the starting circle.
It must pass through the ✗ .

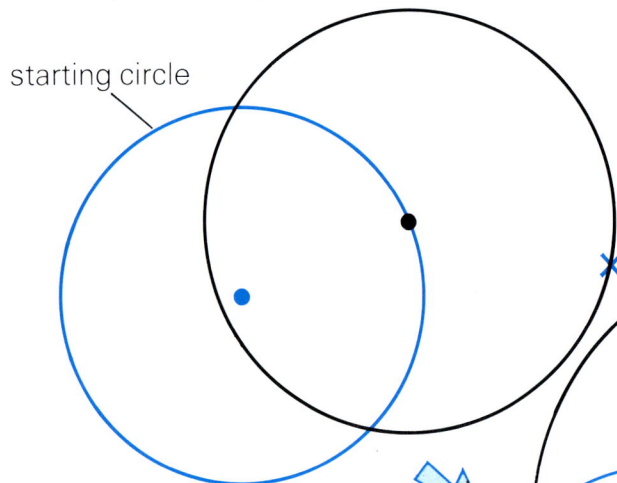

starting circle

Draw another . . .
and so on

Draw at least 30 circles

b) Write a sentence to say
what your drawing looks like.
c) Draw the circle in (a) with the
smallest possible radius.
  (i) Colour it in.
  (ii) Write down its radius.
d) Draw the circle in (a) with the
largest possible radius.
  (i) Shade it in.
  (ii) Write down its radius.

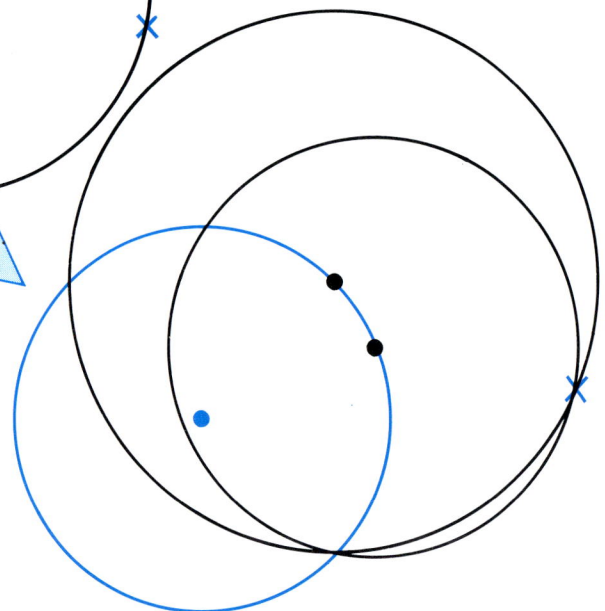

8

Activity

3 **a)** In the middle of a new page mark a dot A.
Draw a circle with centre A and radius 4 cm.

Mark any dot B on the circle.
Join A to B.

Draw a circle with
centre B and radius 4 cm.
Mark any dot C on
the second circle.
Join B to C.

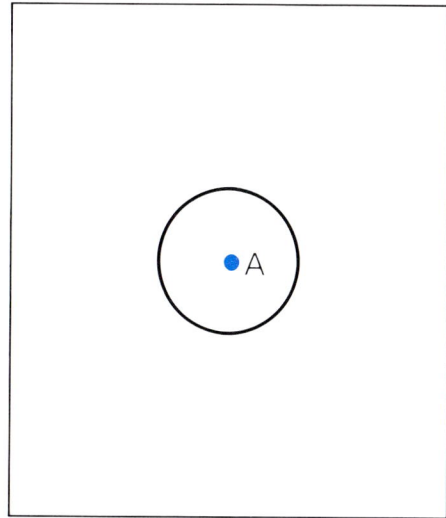

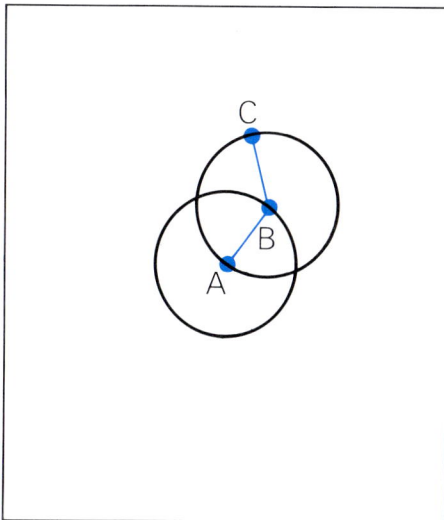

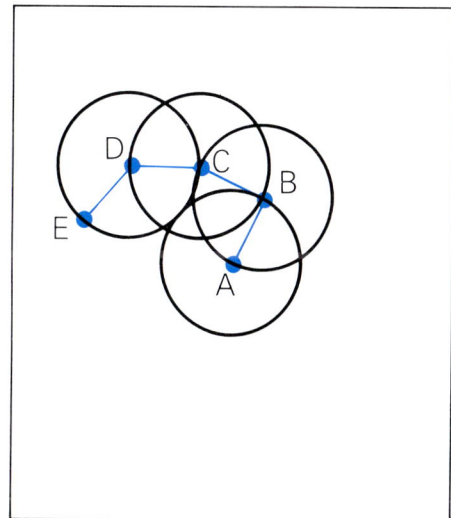

Carry on in the same way,
drawing dots and circles.
Try to make your sixth circle
go through A.
Join F to A.

b) Draw three shapes
like these.

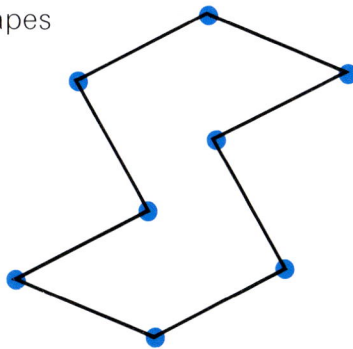

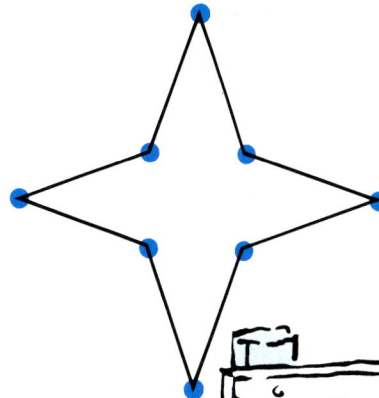

Each shape has eight sides, each 4 cm long.
Use the 'circles method' from **(a)**.

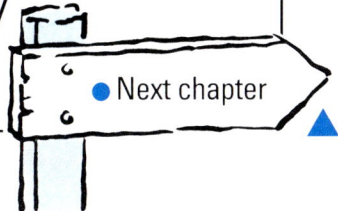

● Next chapter

9 Measures and decimals

A

With a friend

1 Which is heavier:
   a) 7.5 kg or 7.350 kg? b) 9.8 kg or 9.830 kg?

Example:
D A drinking straw 18.3 cm long.

2 Rupinder wrote this list of amounts.
  Decide between you what each amount might represent.
  Write down what you decide.

```
A  £3.27
B  4.12 m
C  7.5 kg
D  18.3 cm
E  30.5 tonne
F  1.75 l
```

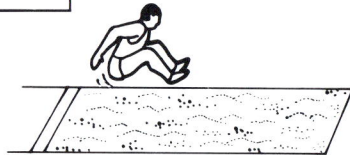

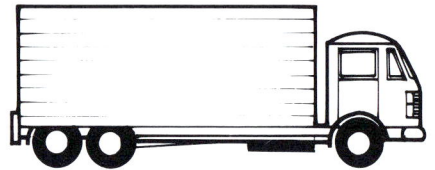

3 a) In **£3.27** the **2** means 20p.
     What does the **7** mean?

   b) What does the **2** mean in **8.25 m**?
      Write your answer
      (i) in centimetres.
      (ii) in millimetres.

Remember...?

```
0m                    1m
                  100 cm
                  1000 mm
```

£3.27

8.25 m

   c) What does the **5**
      mean in **7.5 kg**?
      Write your answer
      in grams.

Remember...?

1kg   1000g

Remember...?

1 tonne   1000 kg

POTATOES
7·5 kg

   d) What does the **4** mean in **2.4 tonne**?
      Write your answer in kilograms.

   e) What does the **2**
      mean in **1.2 l**?
      Write your answer
      (i) in centilitres.
      (ii) in millilitres.

Remember...?

1000 ml
100 cl

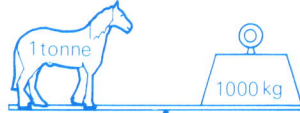

2.4 tonne

1.2 l

4 a) Two of these are correct.
Write down which two.

A $1p = £\frac{1}{10}$ B $10p = £\frac{1}{10}$ C $1p = £\frac{1}{100}$ D $10p = £\frac{1}{100}$

b) Copy and complete these column headings for prices in pounds and pence.

£1	.	10p	□p
3	.	7	5

£1	.	$£\frac{□}{□}$	$£\frac{1}{100}$
3	.	7	5

c) How many pence is £3.75?

5 a) Three of these are correct.
Write down which three.

A $1g = \frac{1}{10}kg$ B $1g = \frac{1}{100}kg$ C $10g = \frac{1}{10}kg$ D $100g = \frac{1}{10}kg$

E $1g = \frac{1}{1000}kg$ F $10g = \frac{1}{100}kg$ G $100g = \frac{1}{100}kg$

b) Copy and complete these column headings for amounts in kilograms and grams.

1 kg	.	□ g	10 g	□ g
3	.	1	7	5

1 kg	.	$\frac{1}{10}kg$	□	$\frac{1}{1000}kg$
3	.	1	7	5

c) How many grams is 3.175 kg?

6 a) Two of these are correct.
Write down which two.

A $1cm = \frac{1}{10}m$ B $10cm = \frac{1}{10}m$ C $1cm = \frac{1}{100}m$ D $10cm = \frac{1}{100}m$

b) Copy and complete these column headings for distances in metres and centimetres.

1 m	.	10 cm	□ cm
7	.	6	5

1 m	.	□ m	.	□ m
7	.	6		5

c) How many centimetres is 7.65 m?

7 a) Three of these are correct.
Write down which three.

A $1\,mm = \frac{1}{10}\,m$

B $10\,mm = \frac{1}{10}\,m$

C $100\,mm = \frac{1}{10}\,m$

D $1\,mm = \frac{1}{100}\,m$

E $10\,mm = \frac{1}{100}\,m$

F $100\,mm = \frac{1}{100}\,m$

G $1\,mm = \frac{1}{1000}\,m$

b) Copy and complete these column headings for distances in metres and millimetres.

1 m .	□ mm	□ mm	1 mm
3 .	1	2	5

1 m .	$\frac{1}{10}$ m	□ m	□ m
3 .	1	2	5

9

c) How many millimetres is 3.125 m?

8 a) Three of these are correct.
Write down which three.

A $1\,ml = \frac{1}{100}\,l$

B $10\,ml = \frac{1}{100}\,l$

C $100\,ml = \frac{1}{100}\,l$

D $1\,ml = \frac{1}{1000}\,l$

E $10\,ml = \frac{1}{10}\,l$

F $100\,ml = \frac{1}{10}\,l$

b) Copy and complete these column headings for amounts in litres and millilitres.

c) How many millilitres is 2.615 l?

1 l .	□ ml	10 ml	□ ml
2 .	6	1	5

1 l .	□ l	$\frac{1}{100}$ l	□ l
2 .	6	1	5

9 a) Copy and complete these headings for amounts in litres and centilitres.

b) How many centilitres is 0.75 l?

1 l .	□ cl	□ cl
0 .	7	5

1 l .	□ l	□ l
0 .	7	5

Writing amounts in different ways

B *With a friend*

1 In each part, one person has made a mistake.
  Decide between you who it is.
  Write down what you decide.

a)

CHAMPION PARSNIP 1.78kg

Alice: 1 kg 78g! THAT'S QUITE A PARSNIP!

Jim: IF YOU ASK ME, IT'S $1\frac{78}{100}$ kg.

Lenny: ACTUALLY, IT WEIGHS 1780g.

b)

CRUSHED ORANGE CONCENTRATE 675ml

Horace: THIS IS GOOD VALUE! YOU GET 6.75 ℓ.

Fiona: NO YOU DON'T. IT'S ONLY 0.675 ℓ.

Jim: IF YOU ASK ME, IT'S $\frac{675}{1000}$ ℓ.

c)

DOORS 2.5m HIGH

Horace: TWO AND A HALF METRES. JUST WHAT I NEED!

Fiona: 2m 5cm. MY DAD WOULD BANG HIS HEAD.

Lenny: 2500mm. JUST RIGHT!

2 This is a Death's Head Hawkmoth.
  It is the largest moth found in Britain.
  Its wingspan is 145 mm.

  a) Draw it, full size.

  b) Find **eight** different ways to
     write its wingspan.

145 mm

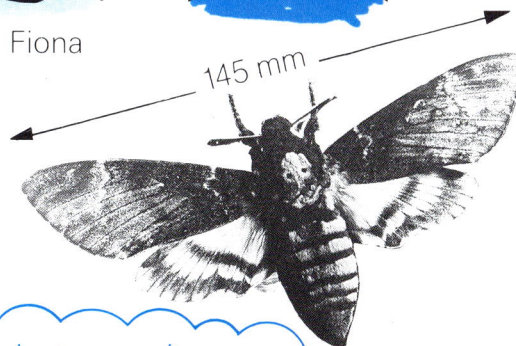

Think in millimetres, centimetres, metres, kilometres.

Use just one unit, mixed units, fractions of a unit, the decimal point.

Each of you write down all the ways you find.

9

■ 81
▲ Next chapter

Fractions and the decimal point

A
B

1   Write the length of each piece of wood
   (i)   in metres, using the decimal point.

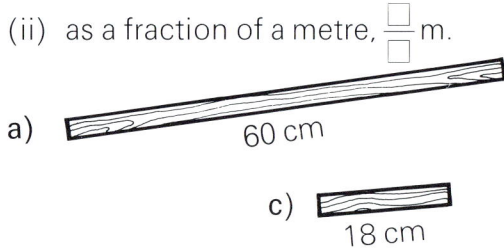

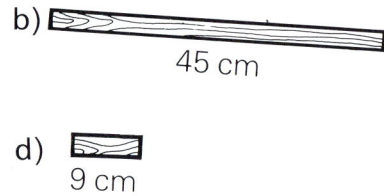

   (ii) as a fraction of a metre, $\frac{\square}{\square}$ m.

b) 45 cm

a) 60 cm

d) 9 cm

c) 18 cm

2   Write the amount of milk
   shown on each container
   (i)   in millilitres.
   (ii) as a fraction
      of a litre, $\square \frac{\square}{\square}$ l.

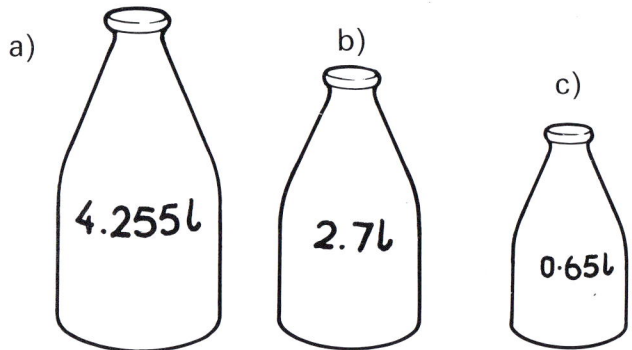

a) 4.255 l

b) 2.7 l

c) 0·65 l

9

3   Write down how much breakfast cereal
   each box holds
   (i)   in grams.
   (ii)   in kilograms, using
        the decimal point.
   (iii) as a fraction of a

        kilogram, $\frac{\square}{\square}$ kg.

a) SUNRISE FLAKES 0.74 g

b) CRUSTY CRUNCHIES 0.45 kg

c) Happy ☺ Mornings 520 g

4   There are two young elephants in a zoo, Jumbo and Petal.
      Jumbo weighs 200 kg
and Petal weighs 184 kg.

Write down the mass of each elephant
   (i)   in tonnes, using the decimal point.

   (ii) as a fraction of a tonne, $\frac{\square}{\square}$ tonne.

Think it through

5   A tree in Jake's garden is 0.0153 km tall.
   How many centimetres is this?

6 a) Weigh yourself (or estimate your mass) in kilograms.

   b) Write your mass
      (i)   in tonnes, using the decimal point.
      (ii) as a fraction of a tonne, $\frac{\square}{\square}$ tonne.

7 a) Measure (or estimate) your height in centimetres.

210 cm

200 cm

190 cm

180 cm

170 cm

160 cm

150 cm

140 cm

9

   b) Write your height
      (i)   in metres, using the decimal point.
      (ii) in kilometres, using the decimal point.
      (iii) as a fraction of a metre, $\frac{\square}{\square}$ m.
      (iv) as a fraction of a kilometre, $\frac{\square}{\square}$ km.

8 Cecil wrote this.

1 min 23 s = 1.23 min

   a) Is he correct?
   b) Explain your answer to (a).

▲ 82
● Next chapter

Working with the decimal point

C 1 A 2.56 km stretch of motorway is being resurfaced. 375 m has been completed.

Write down the distance yet to be completed
a) in metres.

b) in kilometres, using the decimal point.

2 Glenda is flying to the Isle of Man.
She is allowed 20 kg of hand luggage.
She wants to take these items.

clothes 6.75 kg   books $7\frac{1}{2}$ kg   presents 4.4 kg   food $2\frac{1}{4}$ kg

a) Can she take them all?

b) If you say **no**, write down how many kilograms too heavy they are.

3 a) Jake's lawn mower holds 1.75 l of petrol. How many times can he fill it completely from the 5 l can?

b) Jake has a full 5 l can.
He fills the tank of the lawn mower.
Write down how much petrol is left in the can
(i) in litres, using the decimal point.
(ii) in centilitres.
(iii) as a fraction of a litre, $\boxed{}\frac{\boxed{}}{\boxed{}}$ l.

4 A carton of ice cream holds 3 litres of ice cream.
The ice cream weighs 3.6 kg.

a) How many 100 g scoops is this?

b) How many 15 cl tubs can be filled from the carton?

• Next chapter

10 Thinking about solids

A 1 Imagine a perfectly smooth cannonball. It is a solid **sphere**. Write your own example of

   a) a solid sphere.

   b) a hollow sphere.

2 Imagine a perfectly smooth stick of rock. It is a solid **cylinder**. Write your own example of

   a) a solid cylinder.

   b) a hollow cylinder.

   Your example may or may not have a lid or a base.

3 Imagine filling a traffic cone with concrete. When the concrete is perfectly smooth, its shape is a solid **cone**. Write your own example of

   a) a solid cone.

   b) a hollow cone.

   Your example may or may not have a lid or a base.

4 Imagine the Egyptian Pyramids . . . polished perfectly smooth. These are solid pyramids. Write your own example of

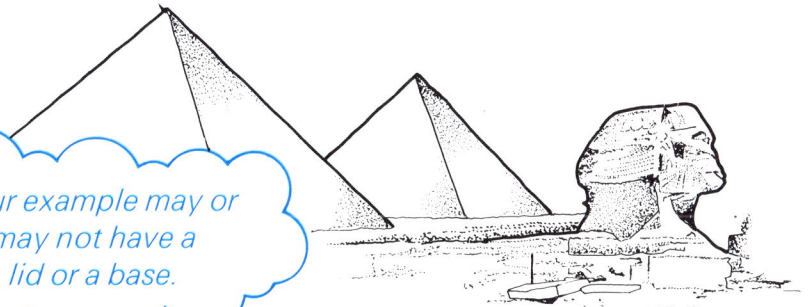

   a) a solid pyramid.

   b) a hollow pyramid.

   Your example may or may not have a lid or a base.

   These are all pyramids.

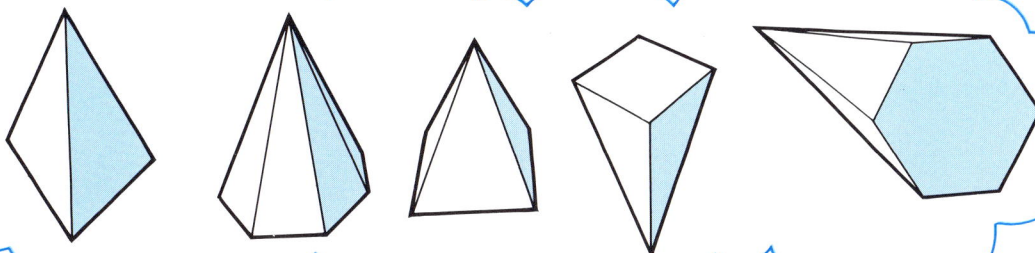

10

Different views

B 1   Write down the name of each of these solid shapes.

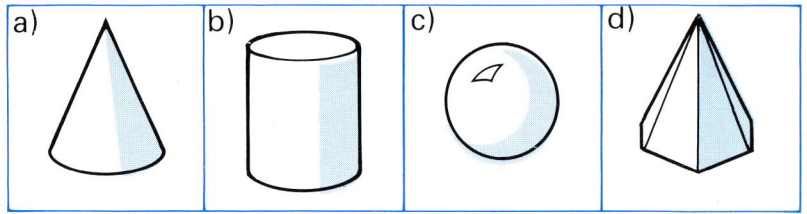

a)	b)	c)	d)

2   These are the same four solids. Write down the name of each one.

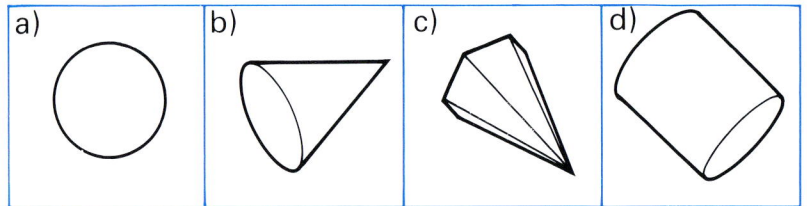

a)	b)	c)	d)

3   These are the same solids again. Name each one.

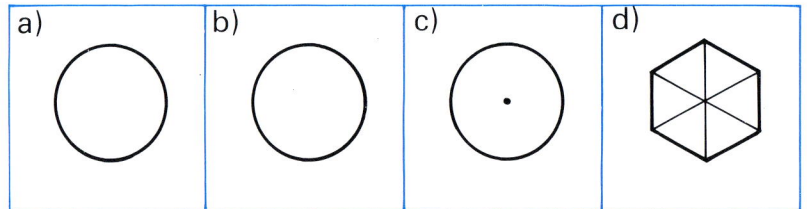

a)	b)	c)	d)

4   Here they are once more. Which one is which?

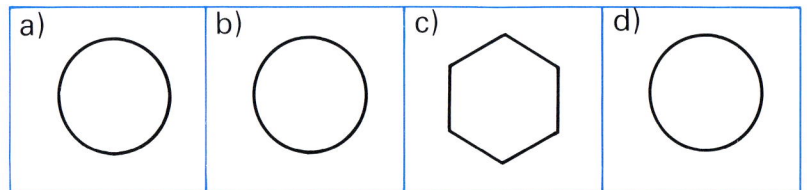

a)	b)	c)	d)

5   These are still the same four solids. Name each one.

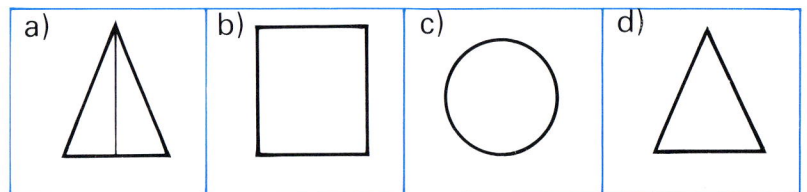

a)	b)	c)	d)

6

*We say that this surface is **curved***

*... and this surface is **plane** (flat).*

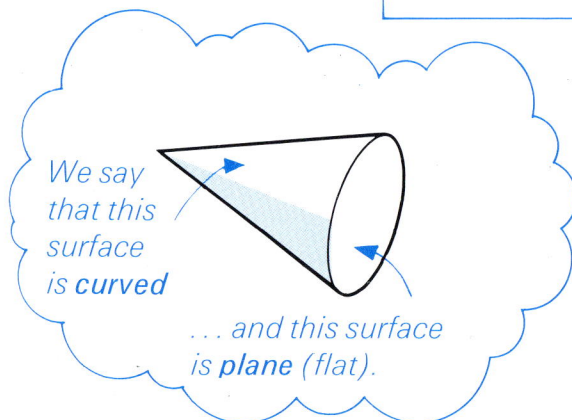

Look back at the four solid shapes. Write down which one has

a) **no** curved surface.

b) **no** plane surface.

c) **one** curved and **one** plane surface.

d) **one** curved and **two** plane surfaces.

Paperweights

C 1 Bernie makes glass paperweights.
Here is one of her **cube** paperweights.

a) This is an **edge**.
How many edges are
there altogether?

b) This is a **vertex**.
How many vertices are
there altogether?

*Read this as VER-tiss-eeze.
For more than one vertex,
we use the word **vertices**.*

c) This is a **plane face**.
How many plane faces are there altogether?

d) How many **curved surfaces** does a cube have?

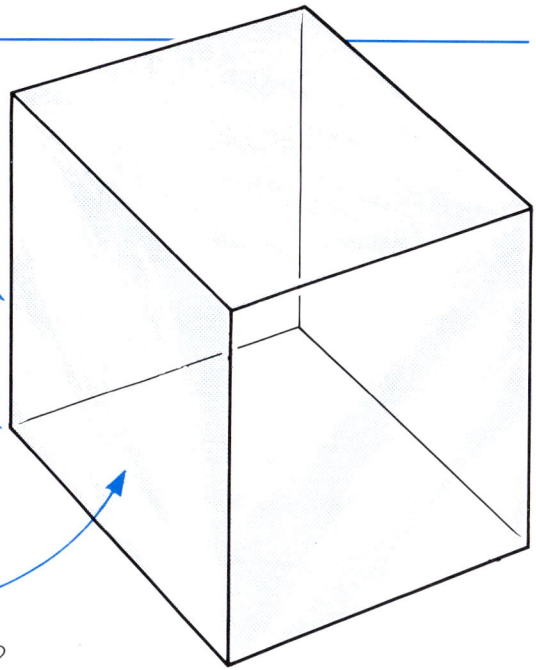

2 This is one of Bernie's **cone** paperweights.
It has just one edge.
Write down how many of these there are:
a) vertices   b) plane faces   c) curved surfaces

3 Here are some more of Bernie's paperweights.

a) solid cylinder   b) solid sphere   c) solid pyramid

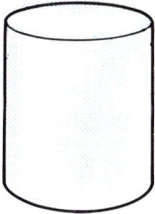

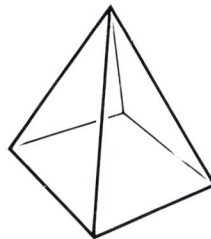

Copy and complete this table for them.

Solid	Number of			
	edges	vertices	plane faces	curved surfaces
Cylinder				
Sphere				
Pyramid				

Think it through *flat*

4 Think about solids with only plane faces.
Draw one of the solids with the smallest possible number of faces.

10

Leaning cones

D 1   Bernie made all of these paperweights.
She put felt on the **base** of each one.
Describe what shape the felt is for each paperweight.

a) cube   b) cylinder   c) cone   d) pyramid   e) pyramid

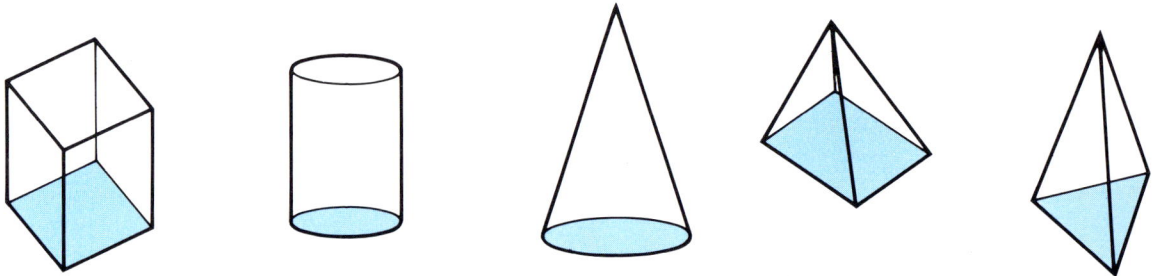

2   Bernie made these cone-shaped paperweights.
Some are slanting.
But they all have the same **height**.

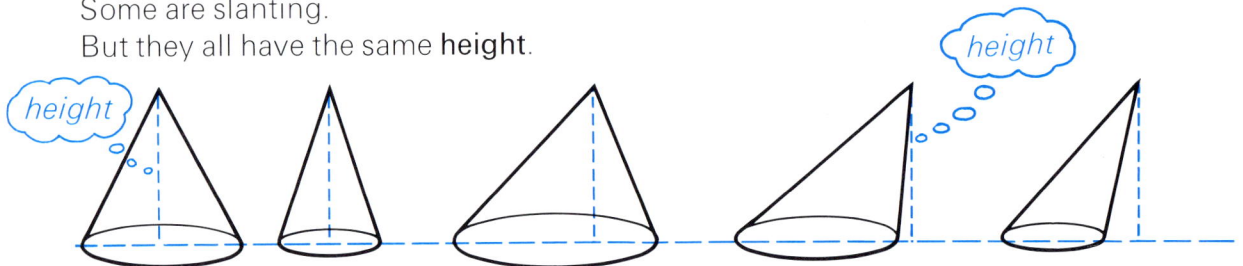

height   *height*

a)   Sketch these front views of the paperweights.

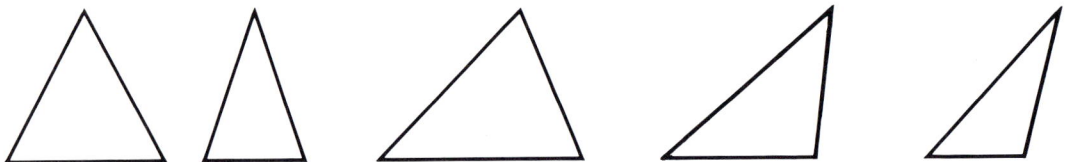

b)   Draw a dotted line to show the height of each one.

c)   How many centimetres high is each one?

Think it through

3   You need a protractor.

Bernie made these
cylinder-shaped paperweights.
In real life the cylinders
are 8 cm high.
Cylinder **C** leans over
at 35° to the vertical.
It measures 5 cm across.

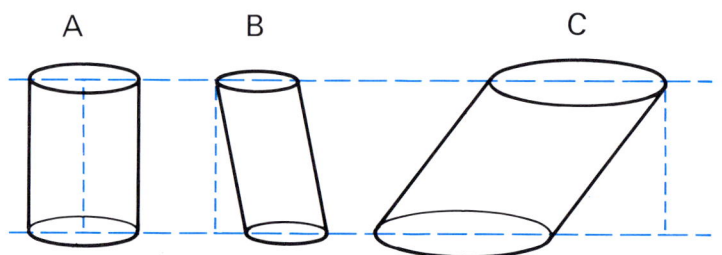

A   B   C

a)   Draw its front view, full size.

b)   Draw it, full size, when you are looking from the left.

10

Some solid shapes

E 1 Copy the *Take note*.
In it, draw a different example of a pyramid.

Take note

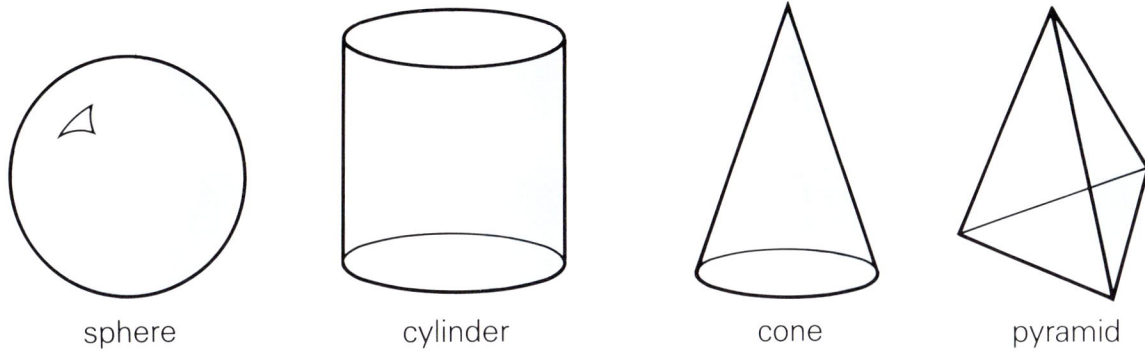

sphere   cylinder   cone   pyramid

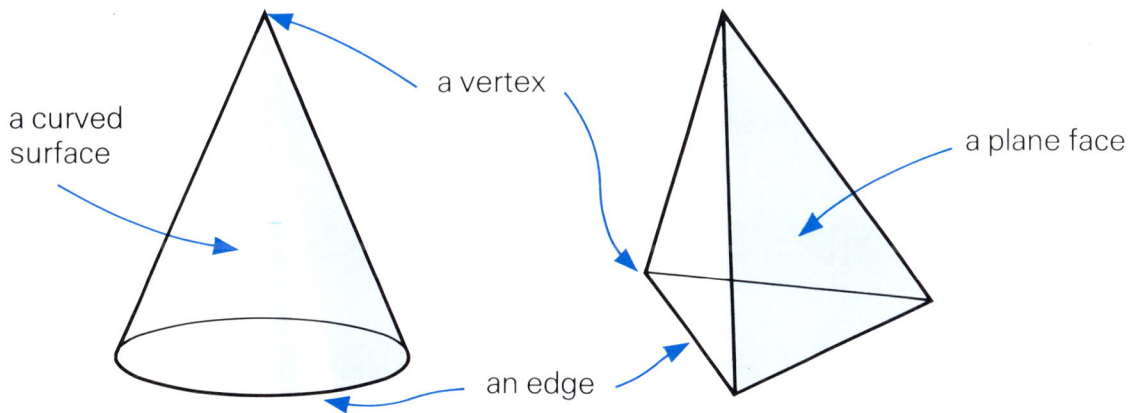

a vertex

a curved surface

a plane face

an edge

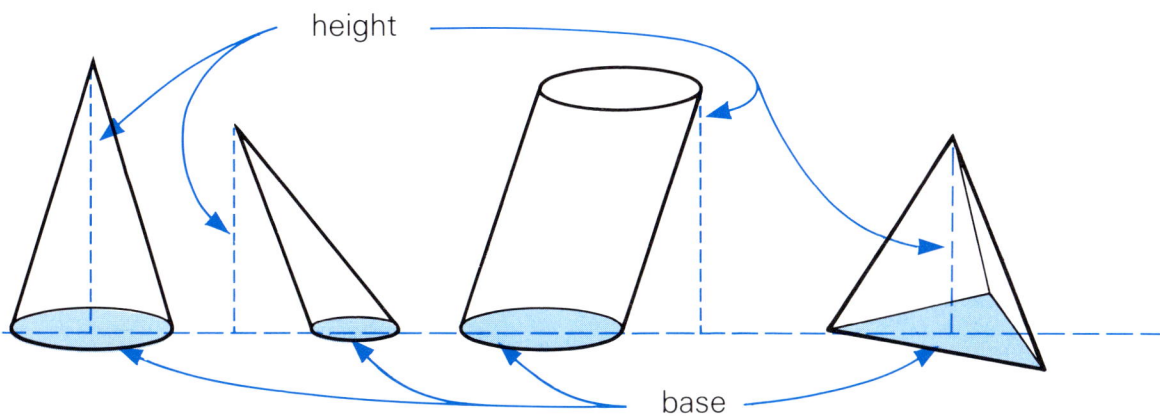

height

base

Slices

F 1 Victoria is slicing a carrot.

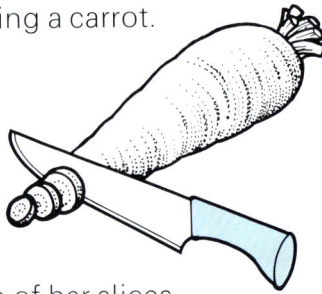

Here are some of her slices.
Draw three more.

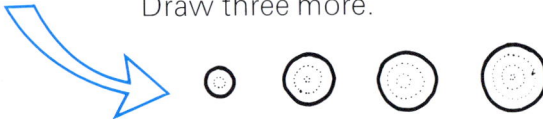

2 Ken is also slicing a carrot.

Here are some of his slices.
Draw three more.

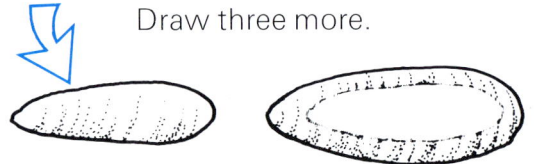

3 Victoria always slices 'crossways';
Ken always slices 'lengthways'.
  a) They both slice a hard-boiled egg.
  Here are the outlines of some of their slices.
  Draw two more of each person's slices.

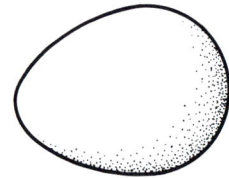

Victoria    Ken

b) Victoria and Ken slice these foods.
  Draw the outline of two slices for each person.

(i) pear   (ii) salami   (iii) orange   (iv) loaf of bread

   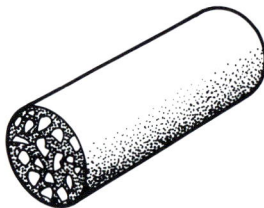   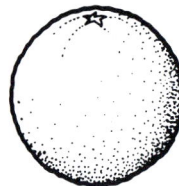   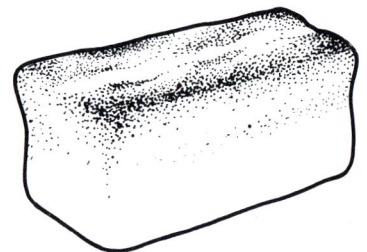

Think it through

4 Darren comes along to help.
Here, he is slicing a carrot.
Darren slices 'diagonally'.
Draw two of his
  a) carrot slices.
  b) salami slices.
  c) pear slices.
  d) orange slices.

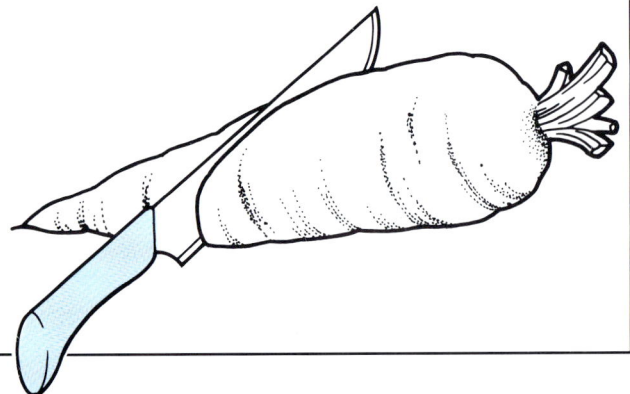

10

Different views

G 1 These are Bernie's working sketches for some paperweights.
For each one

(i) write down the name of the solid.
(ii) sketch the missing view.

a)

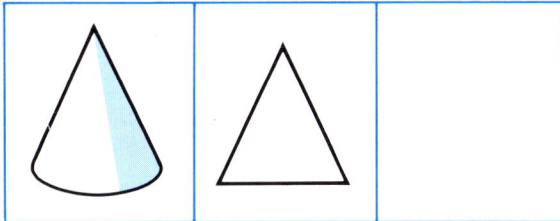

3D view   front view   top view

This means 'three-dimensional'

b)

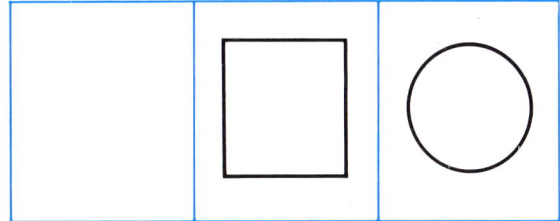

3D view   front view   top view

c)

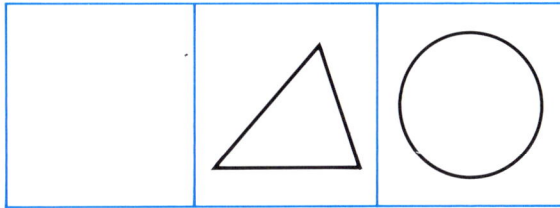

3D view   front view   top view

d)

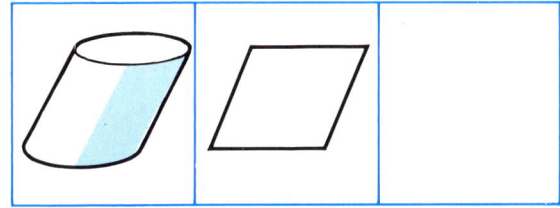

3D view   front view   top view

e)

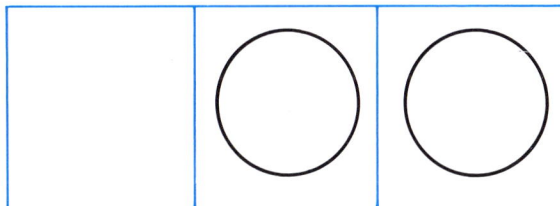

3D view   front view   top view

f)

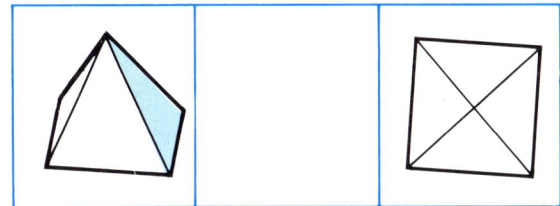

3D view   front view   top view

g)

3D view   front view   top view

h)

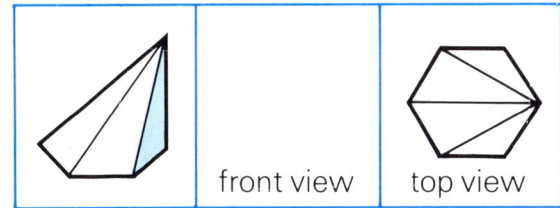

3D view   front view   top view

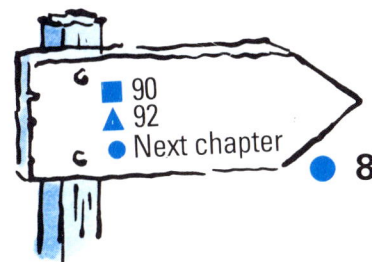

■ 90
▲ 92
● Next chapter

Faces, edges and vertices

A
E

1   Look at these.
   a) Write down which one is a cylinder.
   b) Write down how many
      (i)   curved surfaces it has.
      (ii) plane surfaces it has.
      (iii) edges it has.
      (iv) vertices it has.

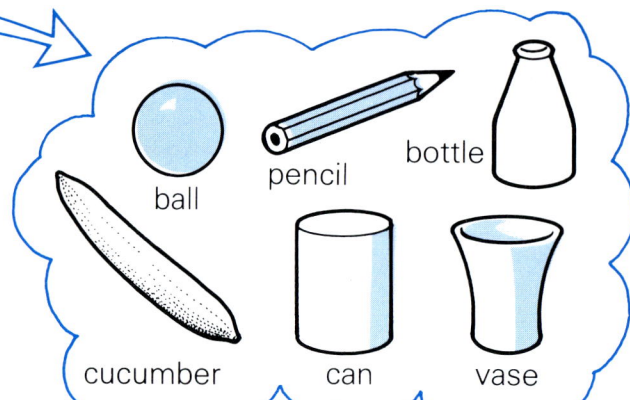

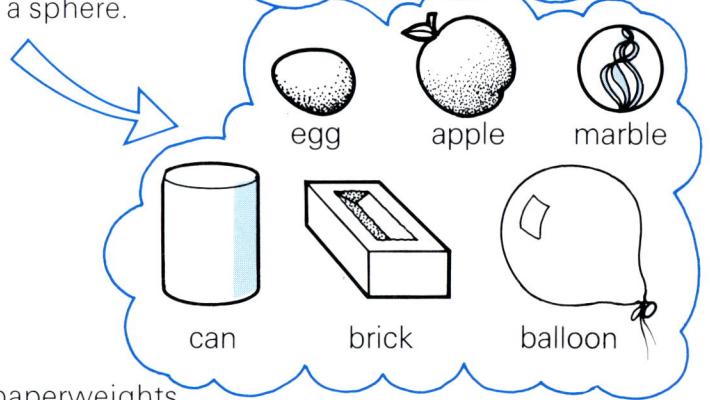

ball   pencil   bottle

cucumber   can   vase

2   a) Write down which of these is a sphere.
   b) Write down how many
      (i)   curved surfaces it has.
      (ii) plane surfaces it has.
      (iii) edges it has.
      (iv) vertices it has.

egg   apple   marble

can   brick   balloon

3   Here are some more of Bernie's paperweights.
   a) Which one is a cone?
   b) Which one is a pyramid?

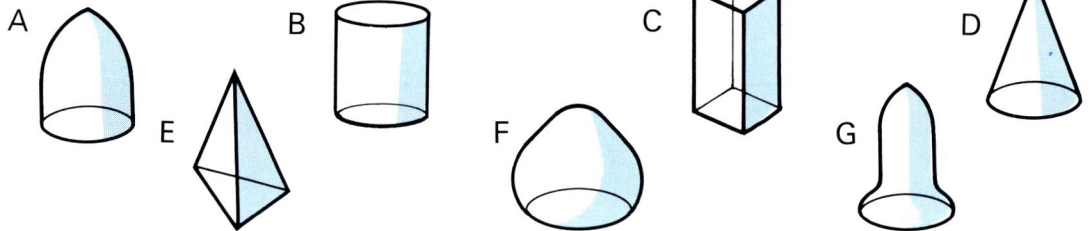

A   B   C   D

E   F   G

F
G

Think it through

1   Vince is sawing a cone and a cylinder into slices.

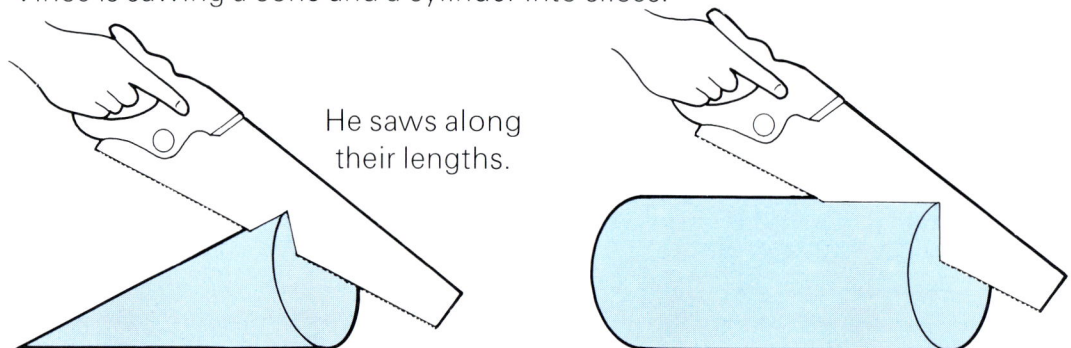

He saws along
their lengths.

For each solid, draw two of the new faces he makes.

10

90

2 Ken has made these objects in his woodwork class.

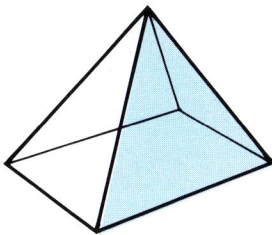 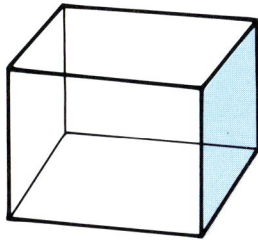 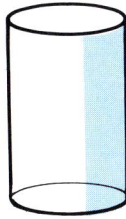 

pyramid   cuboid   cylinder   cone   sphere

Vince slices them horizontally.
These are some of the slices.

a)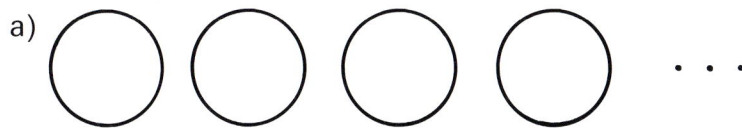

b)

c)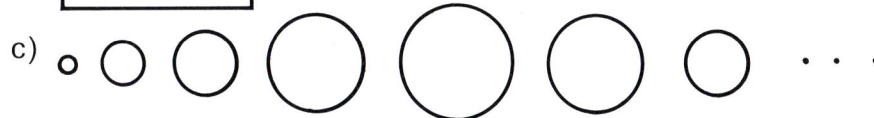

d)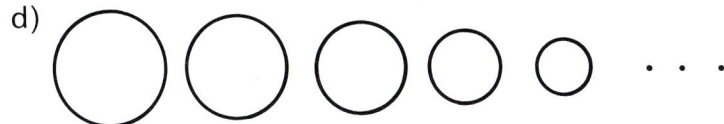

e)

Write down which slices belong to which solid.

Challenge

3 Vince cuts this pyramid exactly in half.
The two halves have the same shape and size.
a) What kind of solid is each half?
b) Sketch one half, to show how Vince
might have made the cut.

He can do this in lots of ways.

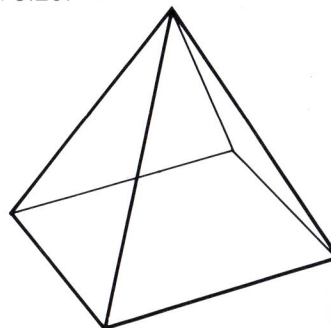

▲ 92
● Next chapter

10

Prisms

H 1 Here is a piece of wood moulding.
The two end faces
look like this:

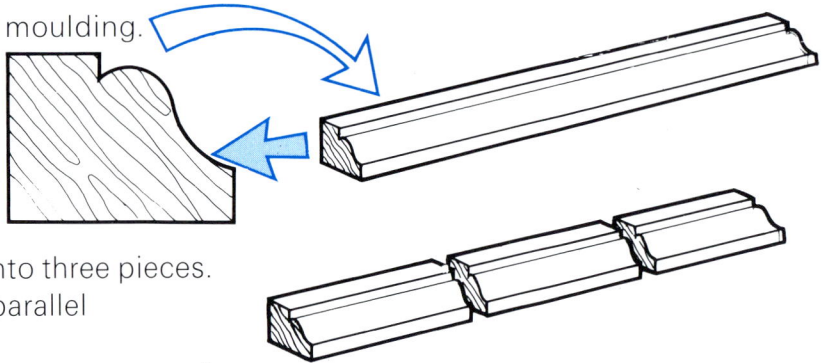

Vince saws the wood into three pieces.
The cuts he makes are parallel
to the end faces.
Think about the new faces his cuts make.
What can you say about their shape?

2 Vince's piece of wood was
a **prism**.
Roughly speaking,
all but one of these
are prisms.
Which one is not?

Take note

Prisms have
parallel end faces.
Cuts parallel to them have
the same **shape** and **size**
all the way through.

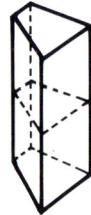

A
a packet
of mints

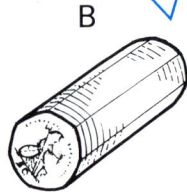

B
a packet
of 50p coins

C
a chocolate
packet

D
a cream
carton

E
a milk
carton

3 Here are two pieces of wood.

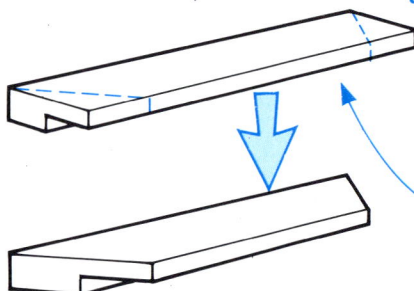

They are both prisms.

Vince cuts the
ends off.
He saws along the
dotted lines.

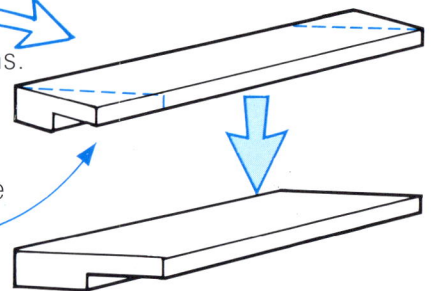

a) Why is this no longer a prism?

b) Why is this still a prism?

Take note

These are prisms.

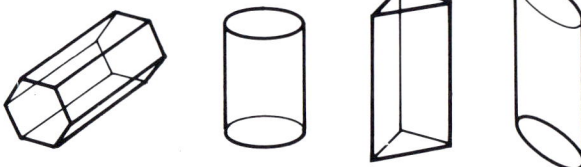

These are not prisms.

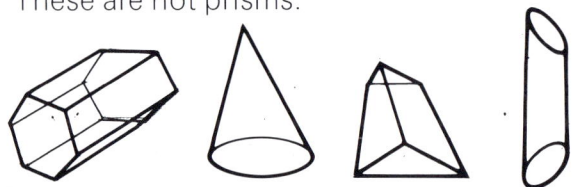

Think it through

4 You need some squared paper.

a) Vince is making a picture frame from this piece of wood.

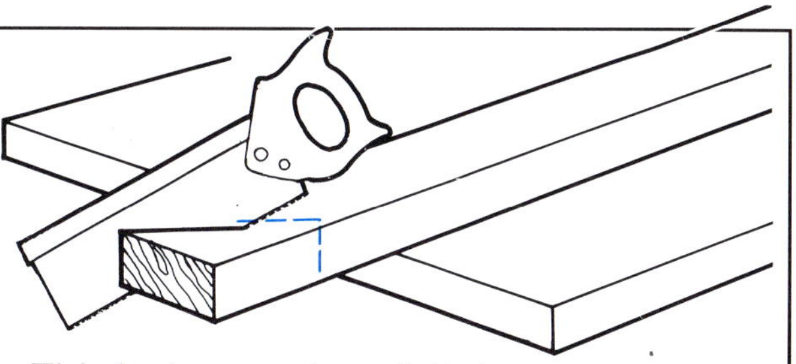

This is the end view of the wood, full size.

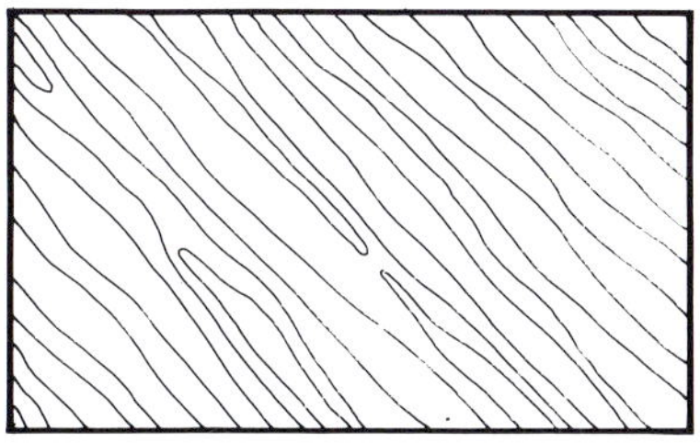

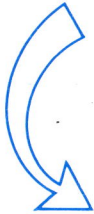

This is the top view, full size.

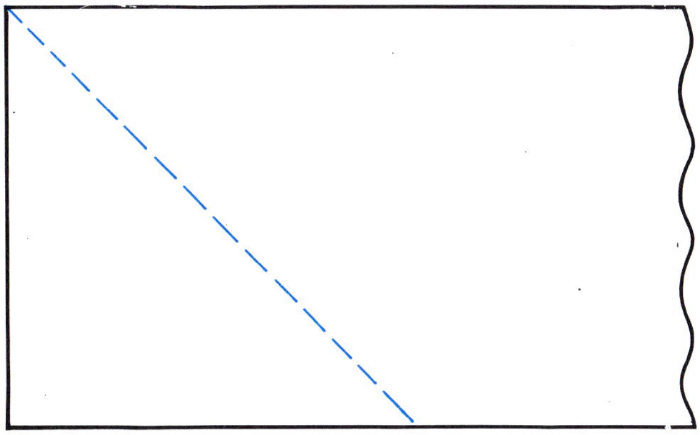

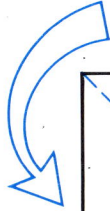

Vince is making a vertical cut along the dotted line.

Make an accurate, full-size drawing of the new face his cut will make.

b) Vince now makes a picture frame from this piece of wood moulding.

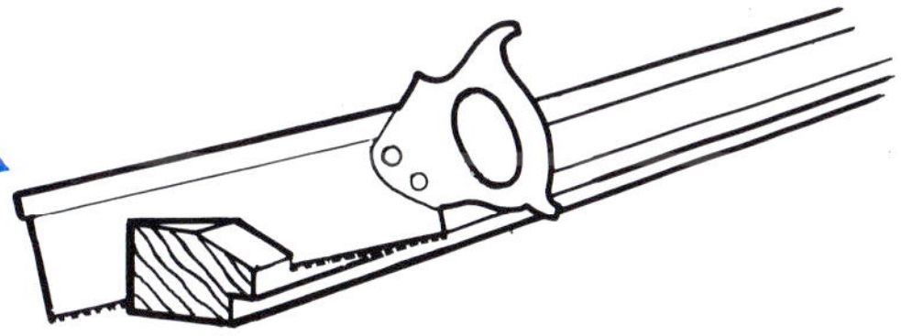

The end face of the wood is this size and shape.

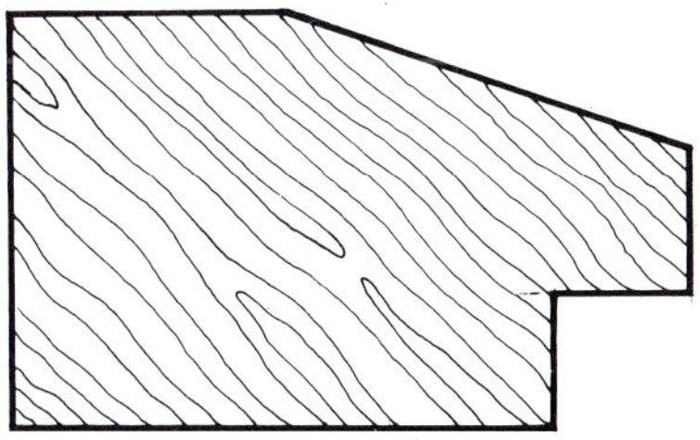

The top view looks like this.

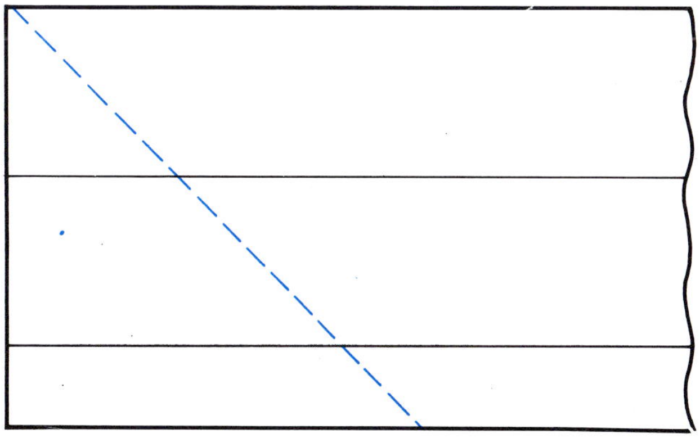

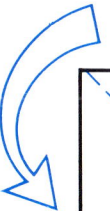

Vince makes a vertical cut along the dotted line.

Make an accurate, full-size drawing of the new face his cut will make.

10

Cutting butter

Activity: Cutting butter

1 You need some squared paper.

Victoria, Ken and Darren each have a wrapped pack of butter.

They cut them in half, like this:

Victoria 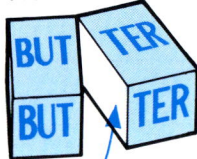  Ken

Darren

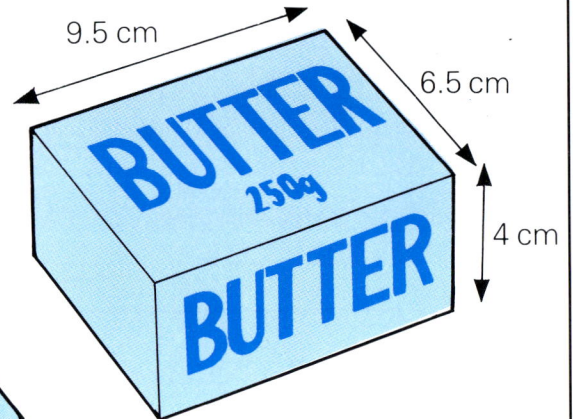

9.5 cm
6.5 cm
4 cm

Each new face is a rectangle.

a) Victoria's rectangle is 4 cm high and 6.5 cm wide. What size is Ken's rectangle?

b) Draw Darren's rectangle accurately, full size.

2 Darren cuts a new pack of butter like this.

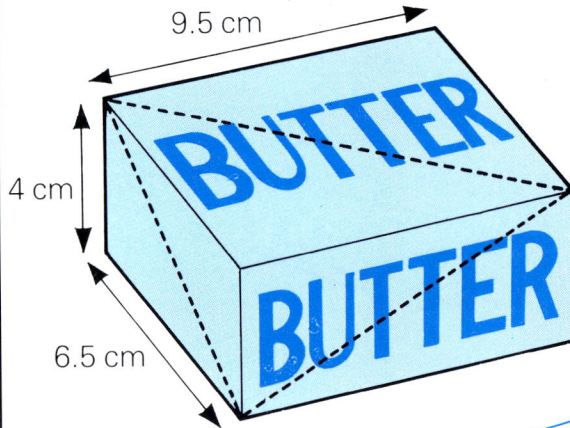

9.5 cm
4 cm
6.5 cm

a) What kind of solid is this?

b) What shape is this cut surface?

c) Darren now makes a vertical cut like this. Make an accurate copy of the new face made by his cut.

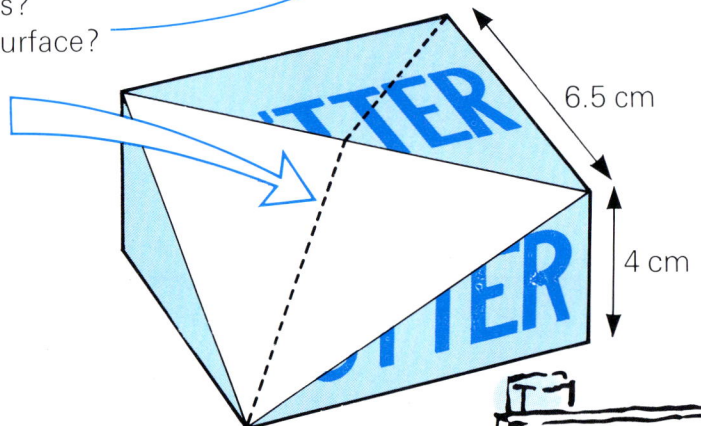

6.5 cm
4 cm

Next chapter

10

11 Using scale

A 1 This is a map of the British Isles.
Its **scale** is 1 cm to 50 km.

Each centimetre represents 50 km.

a) Use a ruler.
   Measure the map distance
   in centimetres from
   London to Glasgow.
   Write the distance
   to the nearest centimetre.

b) About how many
   kilometres is it from
   London to Glasgow?

Use your measurement from part (a).

SCOTLAND

● Aberdeen

N

● Glasgow

IRELAND

● Belfast

c) Find the greatest east–west
   distance across the mainland on the map.
   Write your result to the
   nearest centimetre.

Irish Sea

d) Approximately, what is the
   greatest east–west distance
   in kilometres across
   the mainland?

● Birmingham

WALES

ENGLAND

11

Challenge

e) You need a piece of string
   about 1 m long.
   Every year there is a
   'Round Britain' yacht race.
   The yachts travel through
   the Irish Sea.
   The approximate distance
   the yachts travel is one
   of these:

   Use your string to
   decide which one is correct.

● Cardiff

● London

● Plymouth

250 km, 2500 km, 5000 km,
7500 km, 25 000 km, 50 000 km.

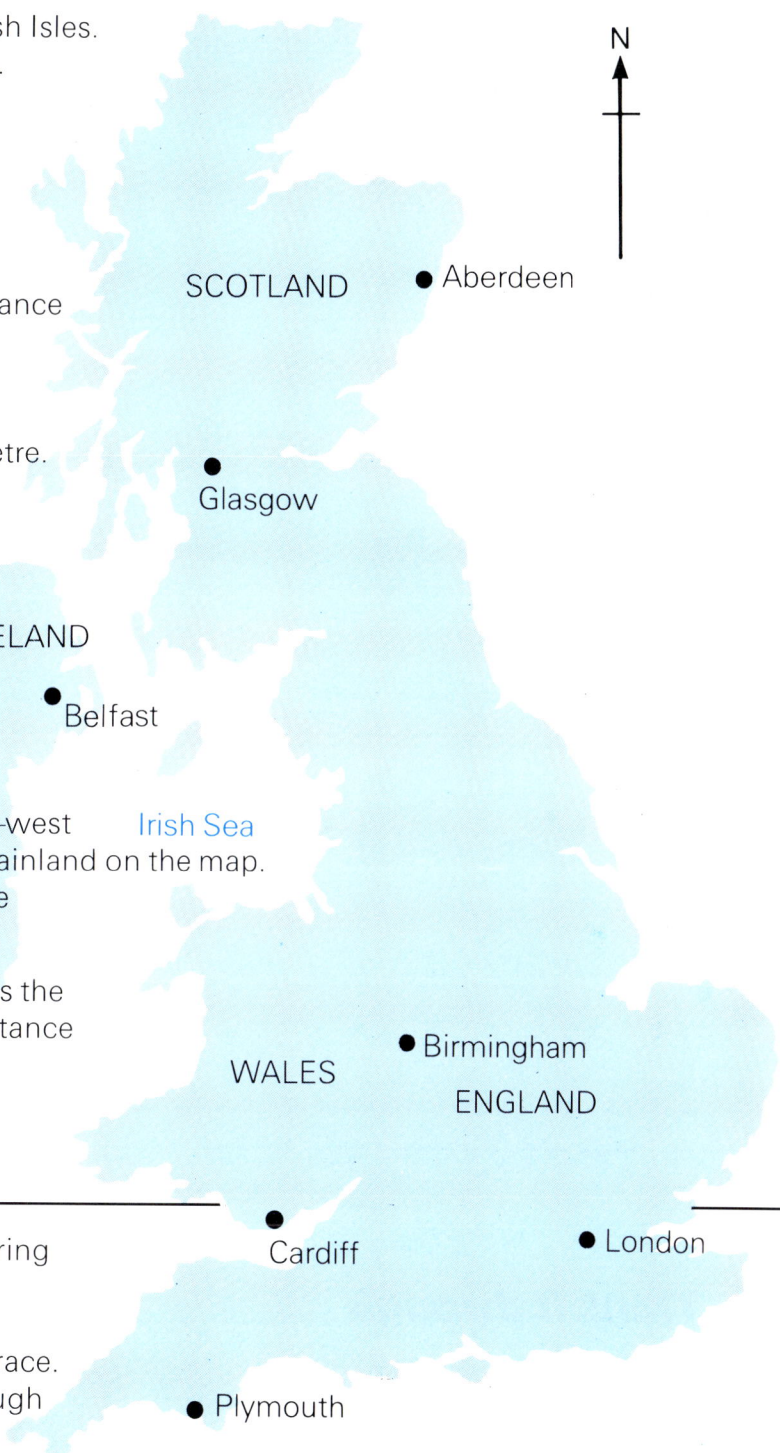

● 95

2 This is a map of Guernsey.
Its scale is 1 cm to 2 km.

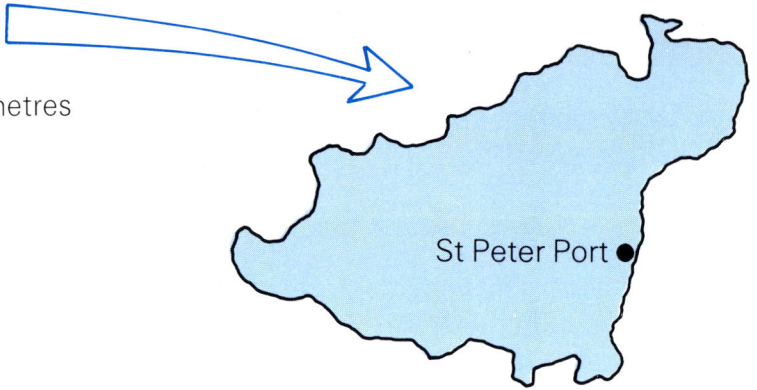

a) About how many kilometres
is it around the island?

*Use a piece
of string.*

b) Here are three more
maps of the island.
Write down the approximate scale of each one like this: 1 cm to ☐ km.

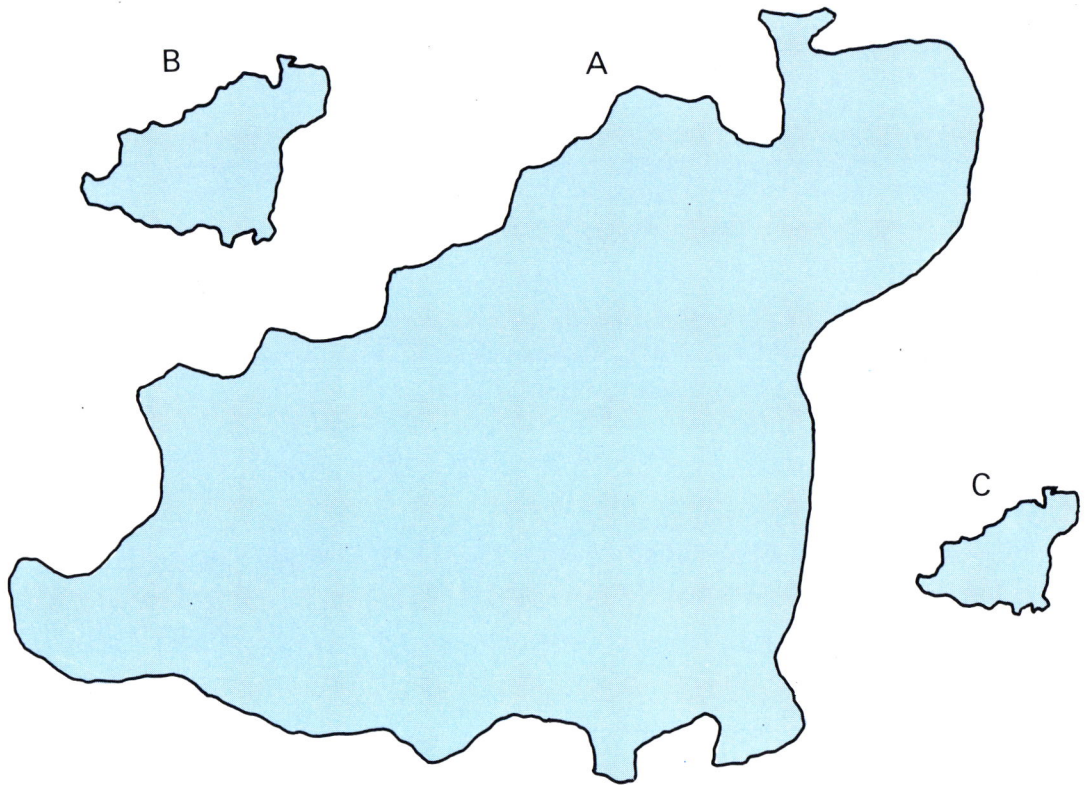

St Peter Port

B

A

C

Think it through

3 a) This map of Jersey is drawn
on 2 mm squared paper.
It is about 5 km from La Trinité
to St Helier.

Write the approximate scale of the map
like this: 1 cm to ☐ km.

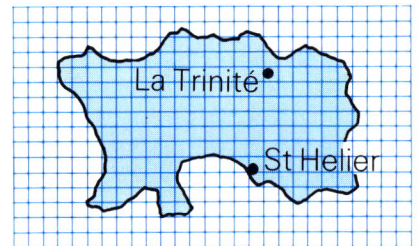

La Trinité

St Helier

b) Sketch your own map of Jersey on 2 mm squared paper.
Use a scale of 1 cm to 5 km.

11

Scale drawings

B 1 Lenny made this scale drawing
of his bedroom.
He used a scale of 2 cm to 1 m.

a) Write these in metres:
   (i) the length of the wall
       with the window
   (ii) the length of the wall
        with the door
   (iii) the length of the wall
         where Lenny has his bed
   (iv) the width and depth of
        the area where Lenny keeps
        his fish tank
   (v) the width of the window
   (vi) the width of the door
   (vii) the width and length of Lenny's bed

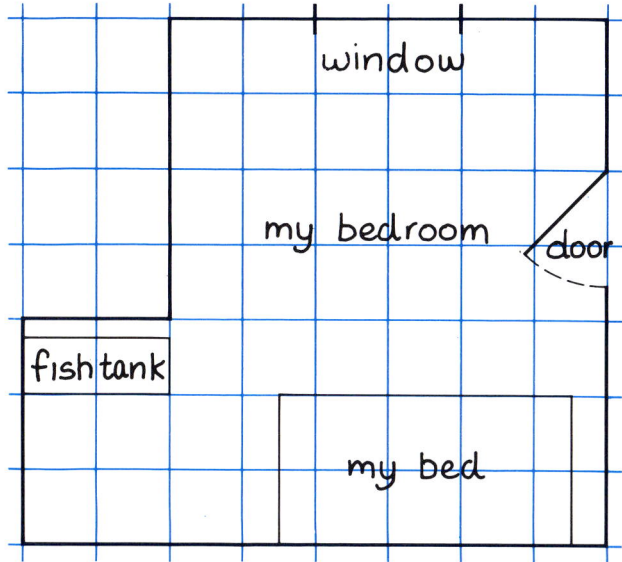

Think it through

b) Lenny's mum says his room needs a new carpet.
She decides carpet tiles would be best.
How much will it cost her to tile the room?

SALE
CARPET
TILES
25 cm × 25 cm
ONLY 30 p EACH

With a friend

2 Make a scale drawing of your classroom.
Use a scale of 2 cm to 1 m.
Work out how much it would cost to carpet your classroom . . .
in lovely, comfortable, expensive Wilton.

DEEP PILE
WILTON
3m WIDE
£18/m²

11

3 The legs of the water tower are 16 m long.
The tank is 6 m square; it is 4 m tall.

a) Two of these are correct
scale drawings of it.
Which two?

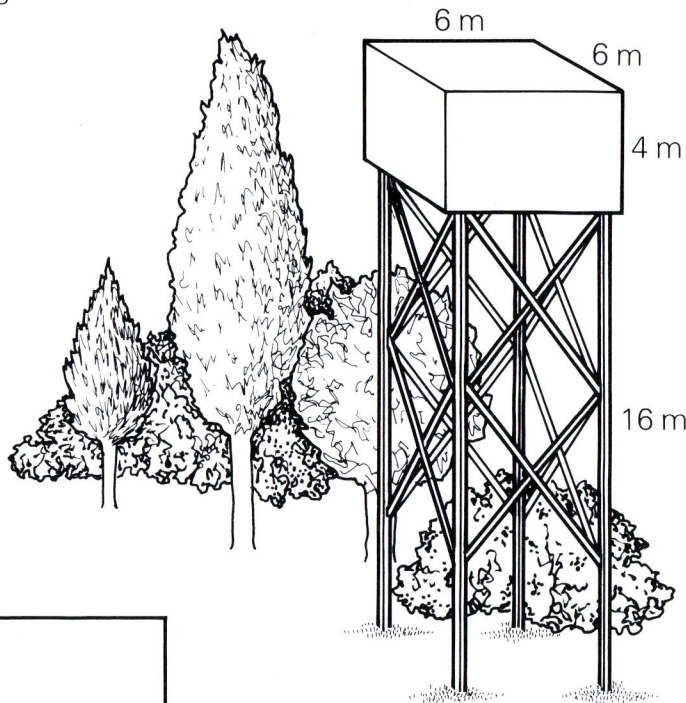

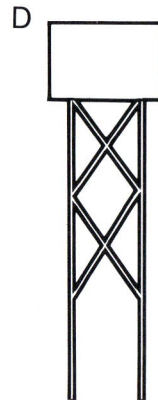

6 m

6 m

4 m

16 m

A

B

C

D

■ 99
▲ 101
● Next chapter

Write ☐ *cm to* ☐ *m.*

b) In the two scale drawings, what scales are used?

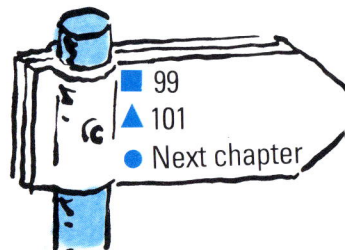

11

Drawing to scale

A 1 The drawings are on 1 cm squared paper.
The scale of each drawing is different.
Copy and complete the table.

Item	Height in drawing (cm)	Scale	Real height (cm)
Bottle	7	1 cm to 5 cm	
Coffee table			
Beagle			
Flagpole			
Bushes			
House			

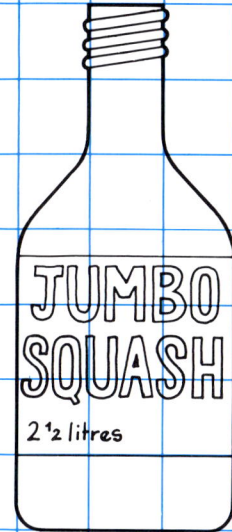

1 cm here represents 5 cm in real life.

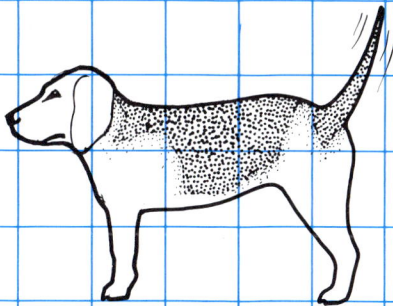

JUMBO SQUASH 2½ litres

Scale: 1 cm to 5 cm

Scale: 1 cm to 10 cm

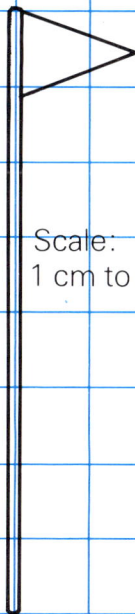

Scale: 1 cm to 15 cm

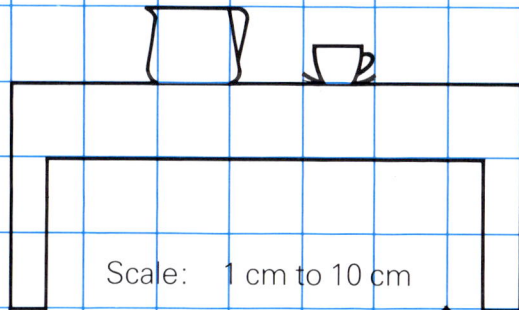

Scale: 1 cm to 12 cm

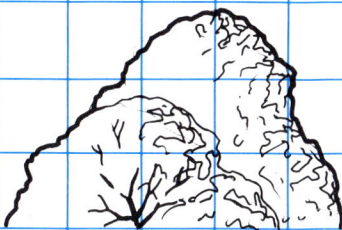

Scale: 1 cm to 40 cm

Scale: 1 cm to 2 m

B 1 This is a scale drawing of the fishpond.

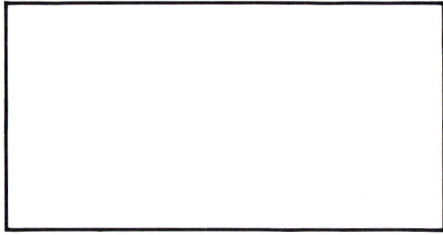

Which of these are also correct scale drawings?
(Three of them are.)

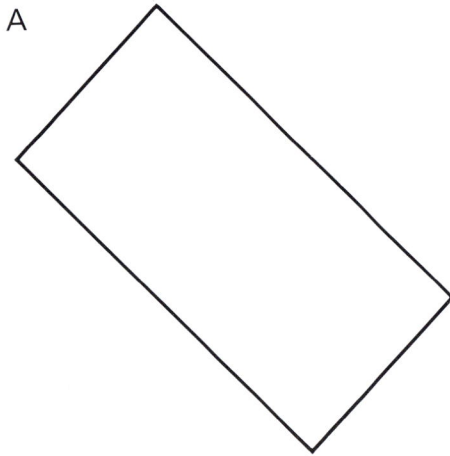

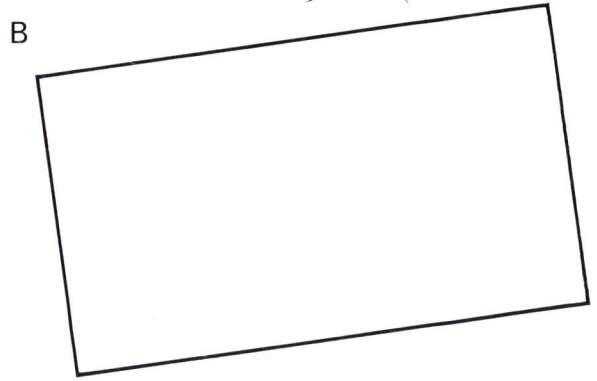

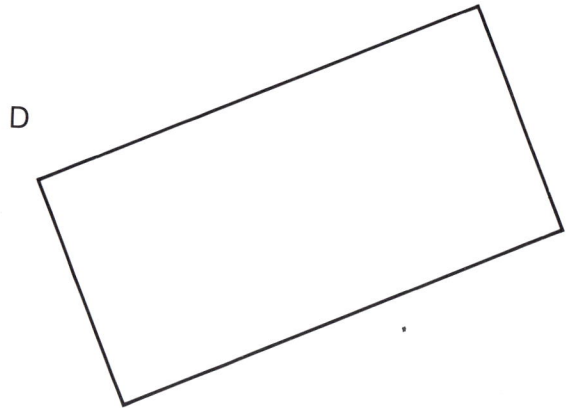

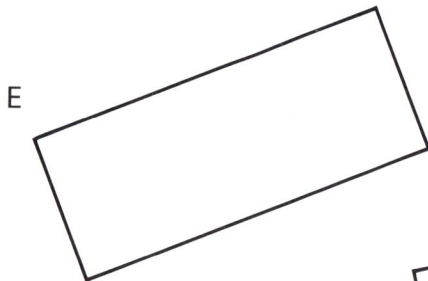

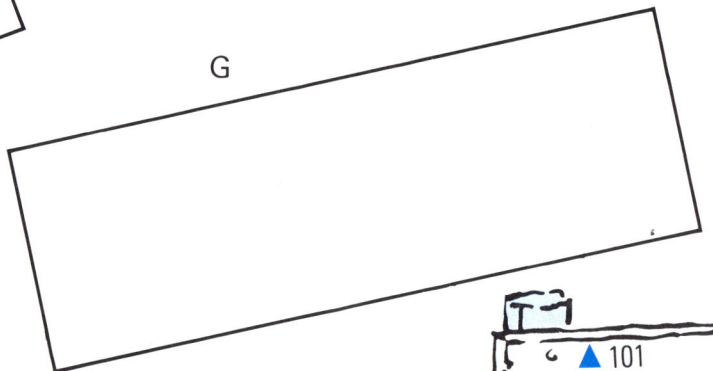

A

B

C

D

E

G

F

▲ 101
● Next chapter

Scaling up and scaling down

C 1 Lenny made this scale drawing of a water tower.
He used a scale of 1 cm to 3 m.

   a) How tall is the water tower
      in real life?

   b) How wide is it at the top?

   c) Midge uses Lenny's drawing to
      make her own, larger version.
      She enlarges Lenny's drawing
      using a scale factor of × 3.
      What is the scale of Midge's drawing?

   d) Use 1 mm squared paper.
      Make your own scale drawing
      of the tower.
      Reduce Lenny's drawing by
      a scale factor of ÷ 2.
      Underneath your drawing,
      write its scale.

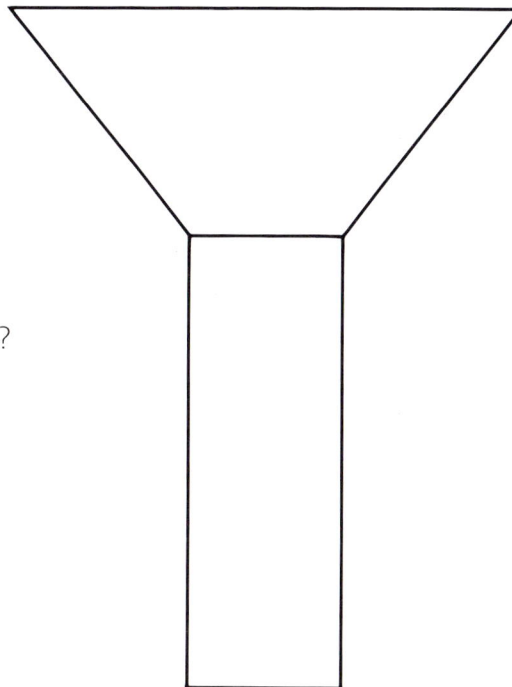

Challenge

2 These are scaled-up drawings.
   Estimate the scale used for each one.
   Write your estimates like this:
   '1 cm represents about ☐ mm.'

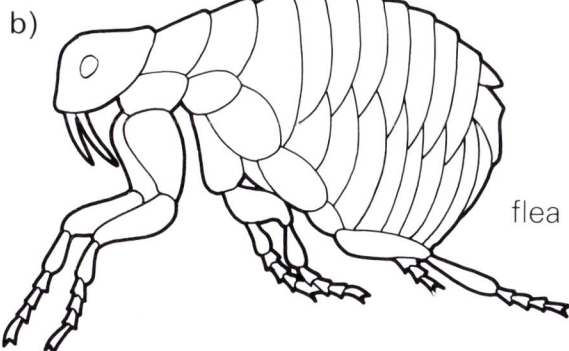

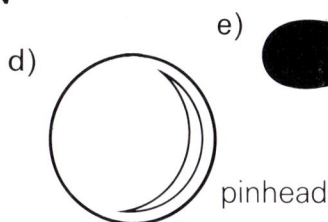

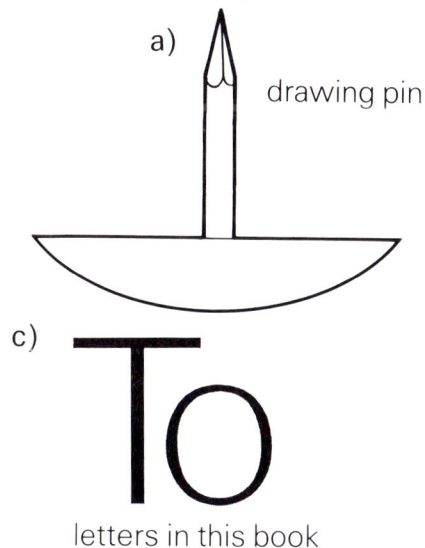

   a) drawing pin

   b) flea

   c) To
      letters in this book

   d) pinhead

   e) matchstick

● Next chapter

12 Solving problems

A In these challenges you will have to work carefully and slowly.
Make sure you do not miss any possibilities.
Try to work out a strategy so that you do not miss any.

Challenge: *Making money*

1 We can make up 5p in these ways:

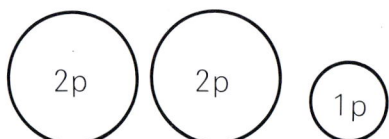

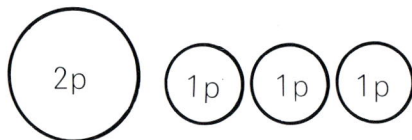

(1p) (1p) (1p) (1p) (1p)

(2p) (2p) (1p)

(2p) (1p) (1p) (1p)

(5p)

There are more than 10, but fewer than 15.

List all the ways in which we can make up 10p.

Challenge: *Making words*

2 You can use one, two, three or all four letters.

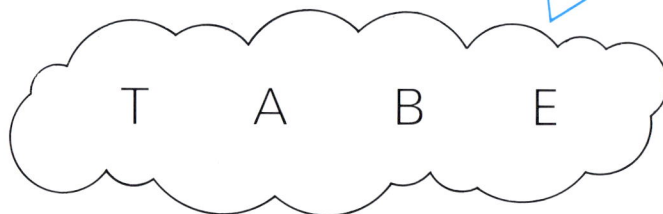

T A B E

Don't use the same letter twice in the same word.

There are more than 10, but fewer than 18.

How many dictionary words can you make?
Write them all down.

3 a) Ron is cycling from Swanston to Tingley.

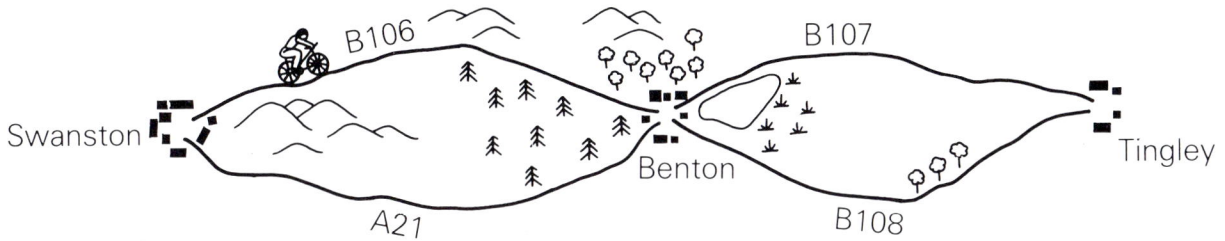

One route he can take is B106, then B108.
List all the routes he can take.

b) Meg is cycling from Wedgewood to Plover.
One route she can take is B37, then B32.
List all the routes she can take.

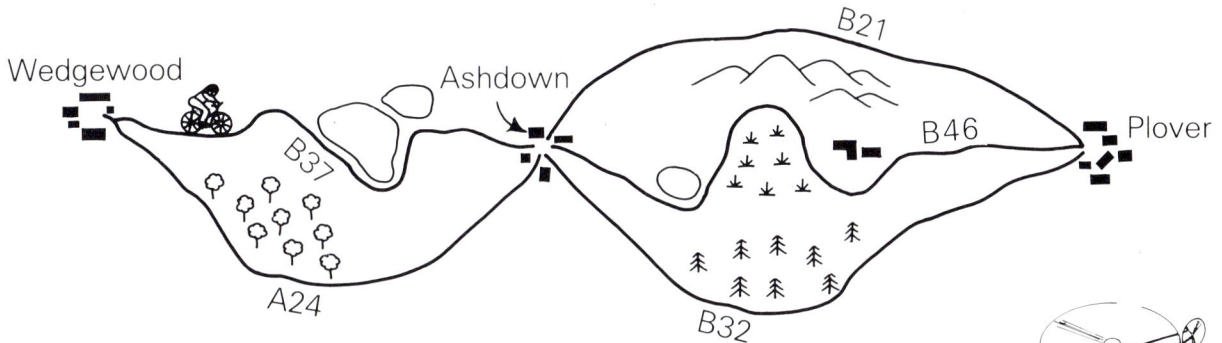

4 The helicopter has to visit the three oil rigs.
One possible order is

STORM → WIND → HAIL.

Another is

WIND → STORM → HAIL.

a) Make a list of all the possible orders.

b) How many are there?

Challenge

c) How many orders are
there for visiting
four oil rigs?

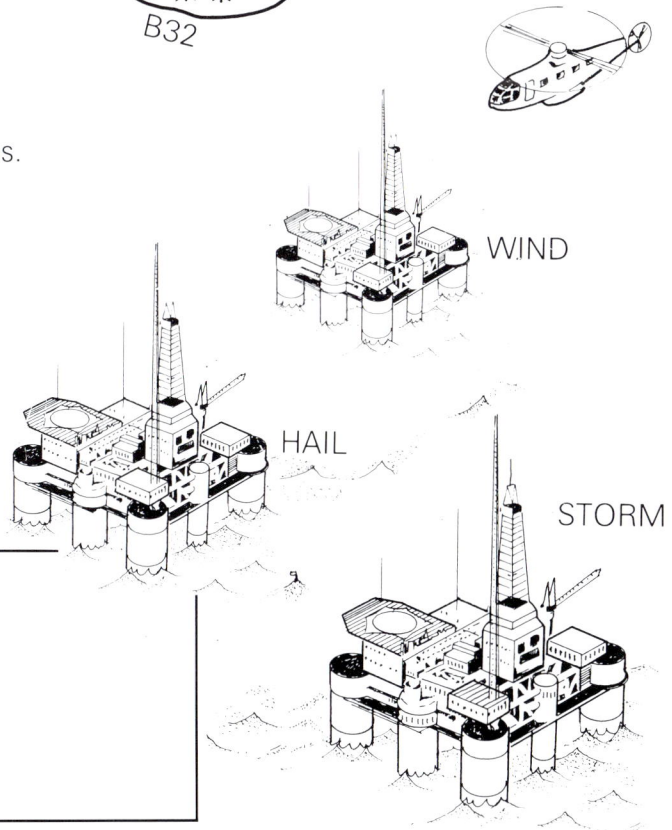

12

5 This is a wire framework. Wallace the spider wants to crawl from A to G. One possible route is
A → B → C → G.
List all the routes which Wallace can take. He is not allowed to touch A, B, C, D, E, F, G and H more than once.

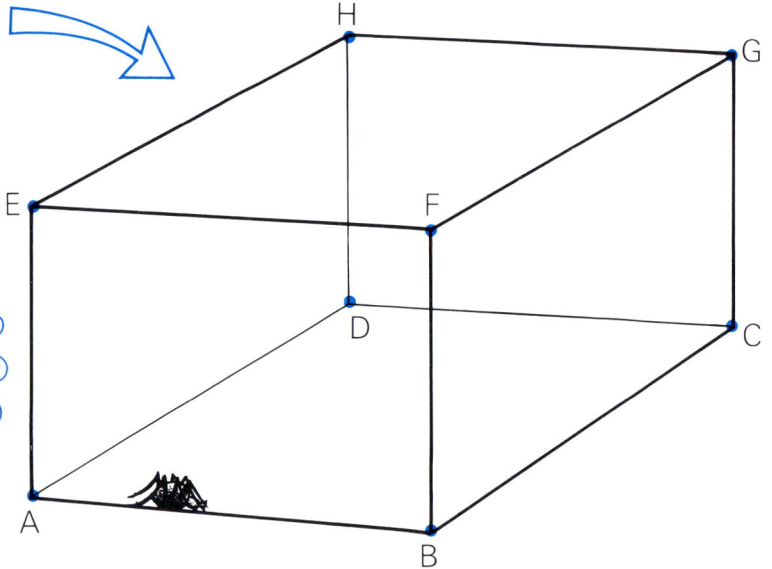

There are more than 15 ways, but fewer than 20.

Activity: *Making tetrahedrons*

6 You need some 1 cm isometric dotted paper and a pair of scissors.

a) Check that this net will make a tetrahedron.

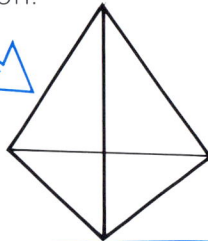

b) Check that this net will **not** make a tetrahedron.

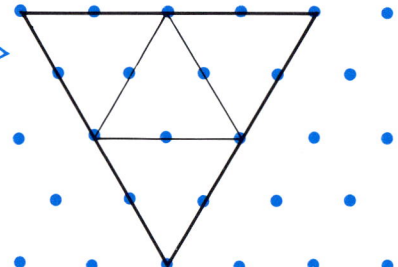

c) Make all the nets which will fold to make a regular tetrahedron.

Stick them in your book.

All edges 2 cm long.

■ 105
▲ 107
● Next chapter

12

Finding all the possibilities

A   **Challenge:** *Making flags*

1   You need squared paper.

Meg has designed this flag.
She wants to use red, white
and yellow stripes . . .
but she can't decide
how to arrange them.

Here is another possibility:

Draw all the possibilities.

*There are more
than 3, but
fewer than 10.*

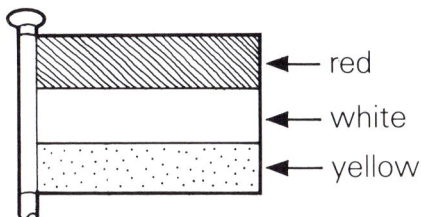

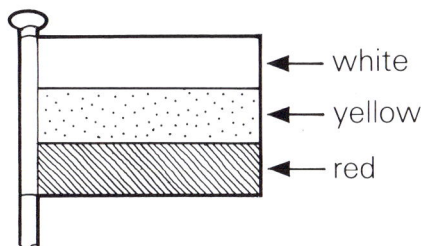

red
white
yellow

white
yellow
red

Challenge: *Making numbers*

2   Using **9** and **7** we can make two numbers:

97 and **79**

a) Using **9**, **4** and **6** we can make six numbers:

946, 964, 469, 496, ?, ?

Write down the two missing numbers.

b) Write down all the numbers which can be made from

(i)   **2**, **2**

(ii) **4**, **1**, **4**

(iii) **3**, **3**, **5**, **5**

(iv) **1**, **2**, **3**, **4**

12

3 Here are three possible scores with two dice:

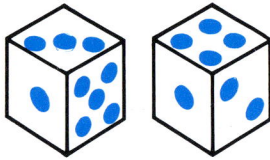
7

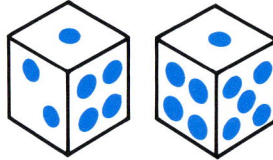
2

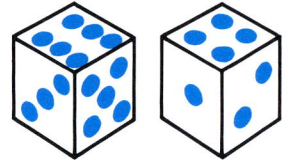
10

a) Make a list of **all** the possible scores.

b) How many are there?

c) What is the smallest possible score with **three** dice?

d) What is the greatest possible score with three dice?

e) How many different scores are possible with three dice?

4 Midge's dog has three puppies.
Midge wants to name them

Bonzo   Happy   and   Clegg

but she can't decide
which should be which.

Here are two possibilities:

Bonzo   Clegg   Happy

Clegg   Happy   Bonzo

▲ 107
● Next chapter

How many different ways are there of naming them?

Being systematic

B 1 a) In how many ways can 30 be made by adding consecutive even numbers? Write down all the possibilities.

Example:
8 + 10 + 12

  b) In how many ways can 40 be made by adding consecutive even numbers? Write down all the possibilities.

2 In the hockey match the final score was 2–2. The half-time score might have been

Wasps Dragons
  0   0

or

Wasps Dragons
  0   1

or

Wasps Dragons
  1   2

or ...

  a) Write down all the possibilities.

  b) List the possible half-time scores for a 3–3 draw.

  c) How many possible half-time scores are there for a 4–4 draw?

3 Glenda is dressing a dummy in a shop window. Glenda has
    3 umbrellas: white, black and blue
    3 hats: white, black and blue
    3 coats: white, black and blue.

In how many different ways can she dress the dummy?

Example:
black umbrella,
white hat,
blue coat.

12

4 There are five people in Winston's family. Here is their breakfast table.

Dad always sits here.

a) Winston arrives first for breakfast.
In how many different places can he sit?

b) Molly arrives next.
In how many places can she sit?

c) Winston's dad says there are 4 × 4 ways in which the family can arrange themselves at the table.
Do you agree?
Explain your answer.

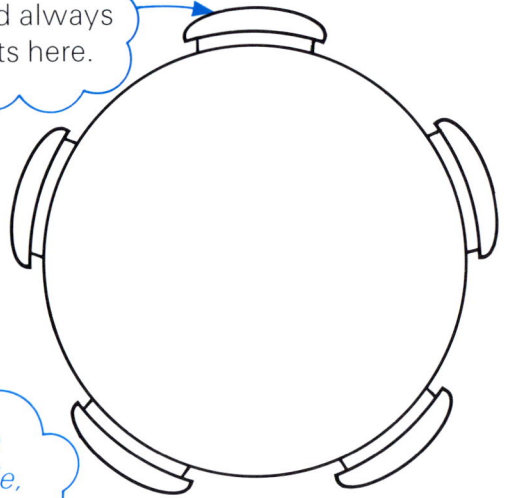

Because there are 4 seats and 4 people, not counting Dad.

Activity: *Nets for cubes*

5 You need 1 cm squared paper, scissors and glue.

a) Check that this net will make a cube.

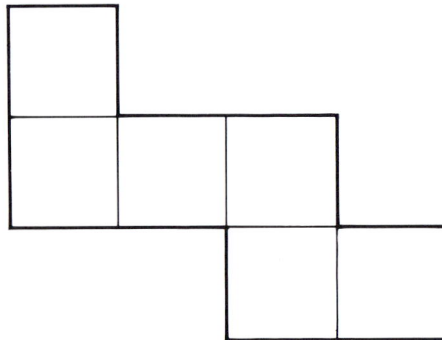

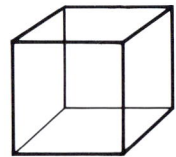

b) Check that this net will **not** make a cube.

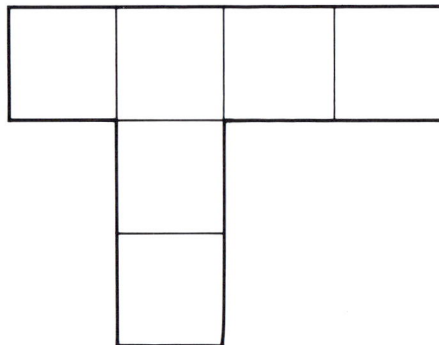

c) Draw all the nets which make a cube.
Stick them in your book.

There are 11 of them.

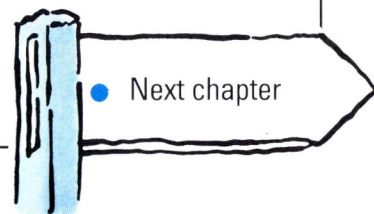

Next chapter

12

13 Working with numbers

A *With a friend:* *The Cutting Game*

1 This is a wooden rod, 1 m long.
In the Cutting Game, you take
turns to cut a piece off the rod.
You can cut off 0.1 m, 0.2 m or 0.3 m.
You **win** if you can force your friend
to leave you with 0.1 m, 0.2 m or 0.3 m.

Here is a game played by Vince and Pepper.

	Vince's cuts	Pepper's cuts
1st	0.2 m	
2nd		0.3 m
3rd	0.3 m	
		0.2 m left. *I win!* *I score 0.2 points.*

Pepper wins.
She is left with 0.2 m.
She scores 0.2 points.

a) Vince could have won by making
a different second cut.
What cut should he have made?

b) But Pepper could have made sure
he scored only 0.1 points.
How?

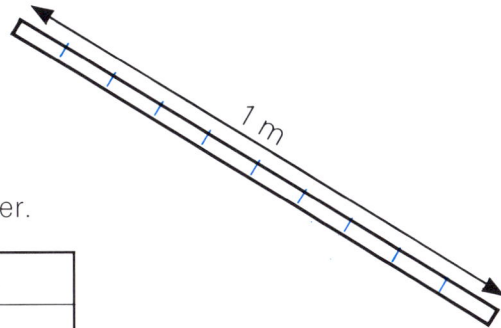

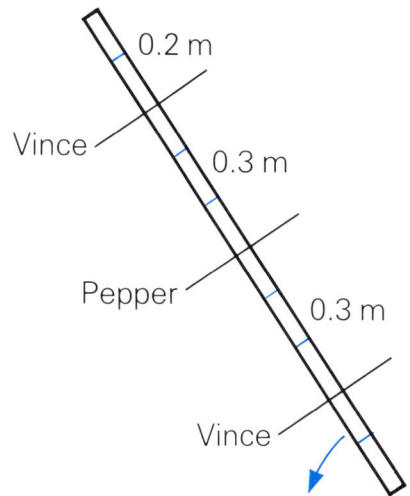

2 Play the Cutting Game.

> RULES
> Take turns to start.
> Play six times altogether.
> The one with most points altogether wins the game.

Draw a rod to help you each time.

1 m

0.2 m
Vince
0.3 m
Pepper
0.3 m
Vince

0.2 m left
Pepper claims this.
She scores 0.2 points.

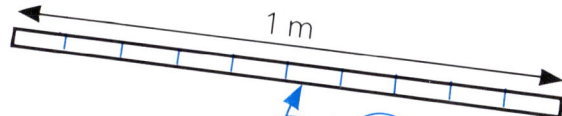

1 m

3 Copy and complete these sentences.

a) $1\,m - 0.3\,m = \square\,m$

b) $5 \times \square\,m = 1\,m$

c) $0.7\,m - \square\,m = 0.3\,m$

4 Do not use a calculator.
This 1 kg bag of sugar can be packed into

ten 0.1 kg bags

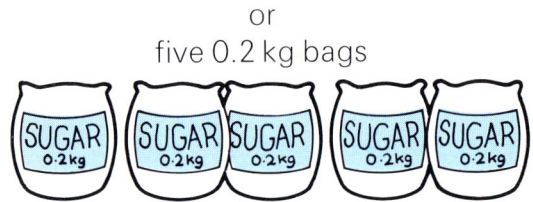

or
five 0.2 kg bags

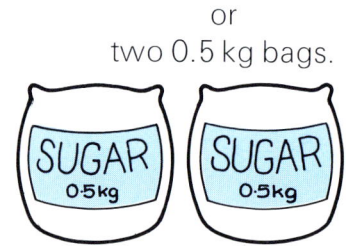

or
two 0.5 kg bags.

. . using bags which hold an exact number of tenths of a kilogram.

These are all the ways a 2 kg bag can be packed . . .
using bags which hold an exact number of tenths of a kilogram.

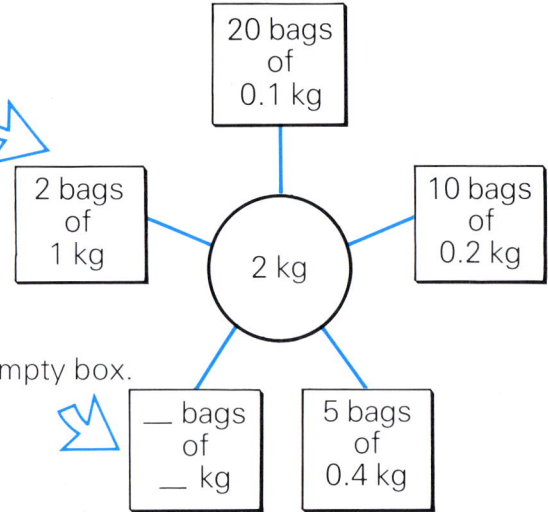

| 20 bags of 0.1 kg |
| 2 bags of 1 kg | 2 kg | 10 bags of 0.2 kg |
| __ bags of __ kg | 5 bags of 0.4 kg |

a) Write down what is missing from the empty box.

b) Investigate how these amounts can be packed . . .
using bags which hold an exact number of tenths of a kilogram
Draw a diagram for each amount.

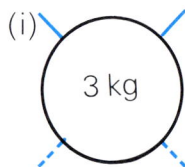

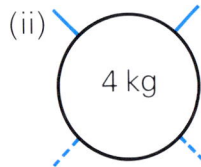

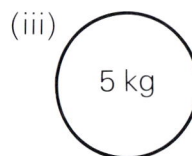

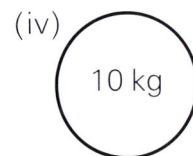

(i) 3 kg

(ii) 4 kg

(iii) 5 kg

(iv) 10 kg

13

5 These three sentences go with box **A** in the diagram.

$$10 \times 0.1 \, l = 1 \, l$$

$$1 \, l \div 10 = 0.1 \, l$$

$$0.1 \, l + 0.1 \, l + 0.1 \, l + 0.1 \, l + 0.1 \, l +$$
$$0.1 \, l + 0.1 \, l + 0.1 \, l + 0.1 \, l = 1 \, l$$

Write three sentences which go with
a) box **B**   b) box **C**.

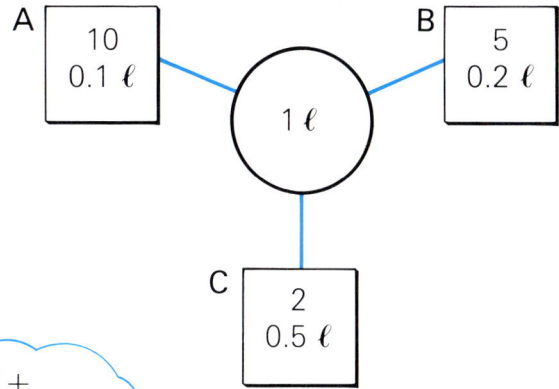

6 a) Copy and complete this diagram.

b) Write three sentences to go with
(i) box **C**   (ii) box **D**.

A
| 10 |
| 0.1 ℓ |

B
| 5 |
| 0.2 ℓ |

(1 ℓ)

C
| 2 |
| 0.5 ℓ |

A
| 20 |
| 0.1 m |

E
| 2 |
| ? m |

B
| 10 |
| ? m |

(2 m)

D
| ? |
| 0.5 m |

C
| 5 |
| ? m |

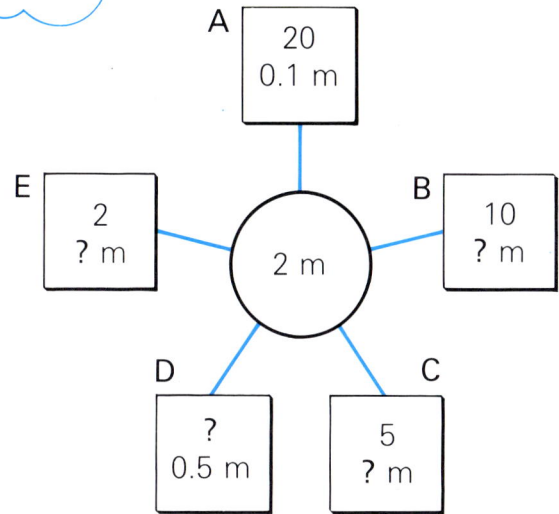

Think it through

7 Do not use a calculator.
a) Copy and complete the diagrams.

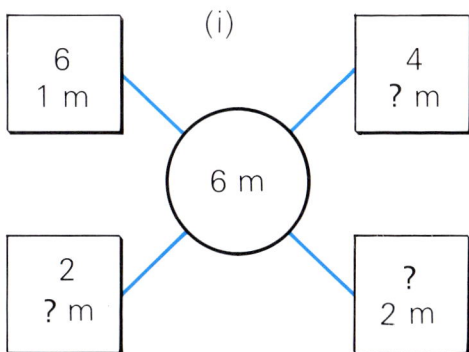

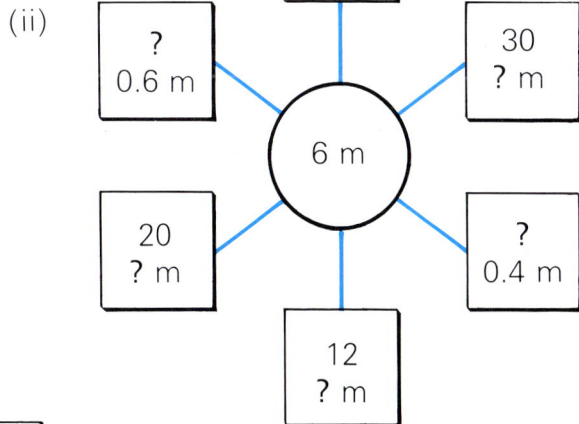

(i)

| 6 |
| 1 m |

| 4 |
| ? m |

(6 m)

| 2 |
| ? m |

| ? |
| 2 m |

(ii)

| 60 |
| 0.1 m |

| ? |
| 0.6 m |

| 30 |
| ? m |

(6 m)

| 20 |
| ? m |

| ? |
| 0.4 m |

| 12 |
| ? m |

b) Write down three sentences for the [4 / ? m] box in diagram (i).

Use ×, ÷ and +.

● 111

B Do each question *without* a calculator
Then check each result with your calculator

1 How many kilograms of
Breakmeal make each breakfast?

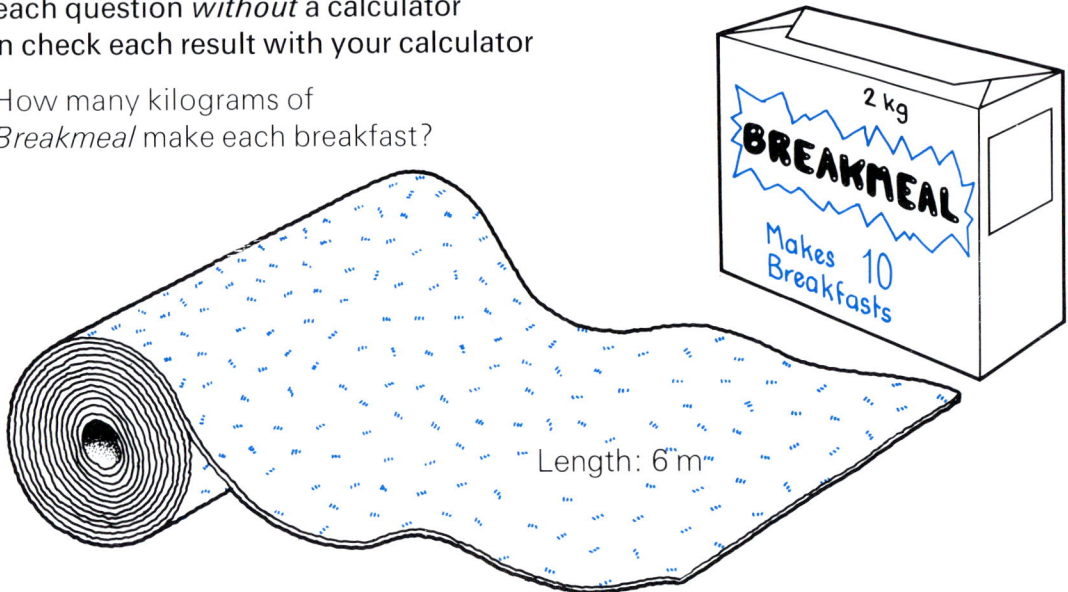

2 kg

BREAKMEAL

Makes **10**
Breakfasts

Length: 6 m

2 a) How many 0.8 m lengths can be cut from the carpet?
   b) How many metres will be left?

0.6 tonne

NMP REMOVALS
maximum load
14 tonnes

3 Each roll of steel wire
weighs 0.6 tonne.
The lorry can carry 14 tonnes.
How many rolls of wire can
it carry?

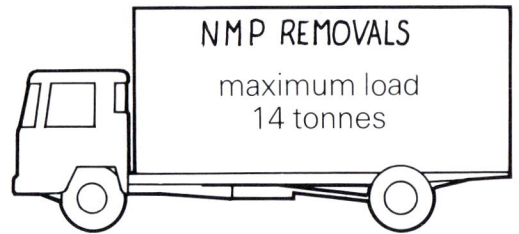

4 Meg is bottling ginger beer.
She has made 12 litres.
Her bottles hold 0.7 litres.
How many bottles does she need?

Challenge

5 These are metal weights.

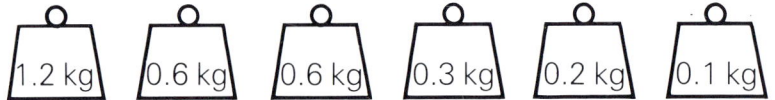

| 1.2 kg | 0.6 kg | 0.6 kg | 0.3 kg | 0.2 kg | 0.1 kg |

They can be
balanced like
this.

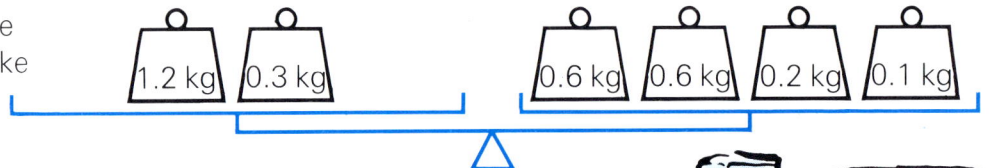

| 1.2 kg | 0.3 kg | | 0.6 kg | 0.6 kg | 0.2 kg | 0.1 kg |

a) Find another way in which they can be balanced.
   Draw a picture to show your way.

Use all the weights.

b) Try to find other ways.
   Draw a picture of each one you find.

■ 113
▲ 114
● Next chapter

13

Decimals and decimal amounts

A
B

1 Do not use a calculator.

This 0.8 m plastic strip
is cut in half . . .
and each part
in half again.

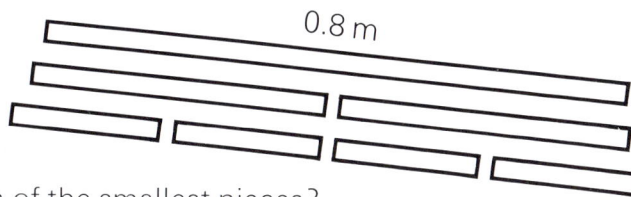

0.8 m

a) How many metres long is each of the smallest pieces?

b) Copy and complete:

$0.8\,m \div 2 =$
$0.2\,m \times 4 =$

$0.8\,m \div 4 =$
$0.4\,m \times 2 =$

$0.8\,m \div 8 =$
$0.1\,m \times 8 =$

2 Check your results in question **1b)** with your calculator.

3 Do not use a calculator.

a) Meg needs 4.5 kg of rice.
She can only find 0.5 kg bags.
How many should she buy?

b) Vince weighs out a bag of flour into 0.3 kg piles.
He has seven piles.
How many kilograms of flour were in the bag?

c) Horace cuts 1 m of string into five equal parts.
How many metres long is each part?

d) Each cup holds 0.2 l of lemon.
How many cups will the bottle fill?

LEMON DRINK 2.4 l

4 Check your results in question **3** with your calculator.

Challenge

5 Do not use a calculator.

↗ means 'subtract 0.4'; ↗ means 'multiply by 2'; ↗ means 'subtract 0.2'.

a) Copy and complete the pattern.

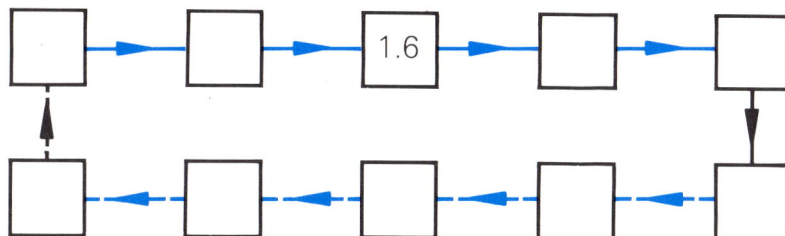

1.6

b) Copy and complete: ↗ means 'multiply by ☐'.

▲ 114
● Next chapter

13

113

Decimal calculations

C 1 Do not use a calculator.

a) Work out how much milk there is in the two bottles.

1.4ℓ

2.7ℓ

NMP DAIRIES

NMP DAIRIES

b) Meg calculates how much milk is in the two bottles.
   She writes the addition like this.

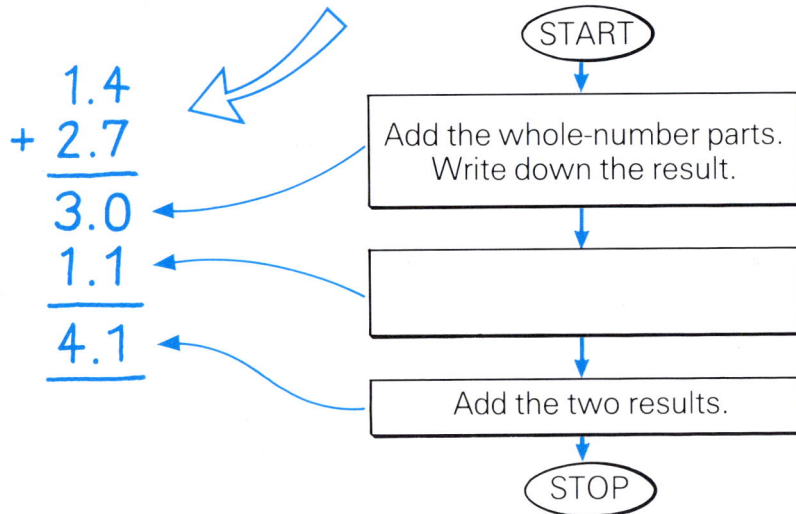

START

```
   1.4
 + 2.7
 ──────
   3.0
   1.1
 ──────
   4.1
```

Add the whole-number parts.
Write down the result.

Add the two results.

STOP

Write down what is missing from the flow chart.

c) Use Meg's method to do these additions:

(i) 4.6 + 3.7   (ii) 7.7 + 8.4   (iii) 9.2 + 2.9

d) What is the total amount of sugar in these bags?

Sugar 0.8 kg   Sugar 1.4 kg   Sugar 2.8 kg

e) Check your additions in (c) and (d) with your calculator.

13

2 Do these calculations in your head.
Write down only the answers.

    a) $1\,kg + 0.6\,kg$ **b)** $0.8\,kg + 0.3\,kg$ **c)** $1.9\,l + 0.6\,l$

    d) $1\,l - 0.4\,l$ **e)** $2.0\,km - 0.5\,km$ **f)** $3.1\,km - 0.4\,km$

3 **Do not use a calculator.**

Meg pours 2.5 l of milk from the bottle.
a) Calculate how much milk is left
in the bottle.

b) Meg calculates how much is left, like this.

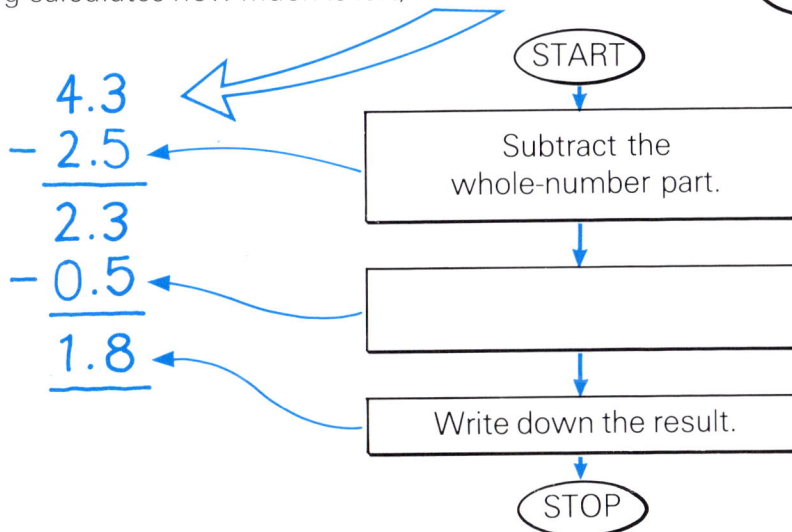

$$
\begin{array}{r}
4.3 \\
-\;2.5 \\
\hline
2.3 \\
-\;0.5 \\
\hline
1.8 \\
\hline
\end{array}
$$

START
Subtract the whole-number part.

Write down the result.

STOP

    Write down what is missing from the flow chart.

c) Use Meg's method to do these subtractions:

    (i) $2.6 - 1.4$ (ii) $7.1 - 4.2$ (iii) $8.2 - 5.7$

d) How much more honey is there
   (i) in a large pot of honey than in
      a medium sized pot?
   (ii) in a large pot than in a small pot?

SMALL HONEY 1.5 kg   MEDIUM HONEY 2.8 kg   LARGE HONEY 4.2 kg

e) Meg bought a medium pot and a small pot.
Did she buy *more* or *less* honey than there is in a **large** pot?

f) Check your results in **(c)**, **(d)** and **(e)** with your calculator.

4 **Do not use a calculator.**

   a) Jake marks out this flowerbed.
   What is the total distance around it?

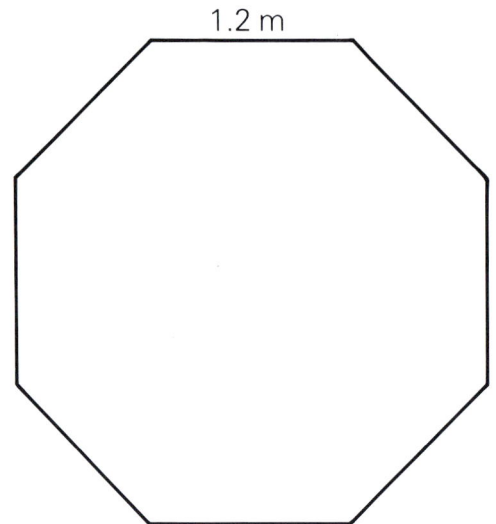

1.2 m

   b) Jake works out its perimeter like this.

   $$\begin{array}{r} 1.2 \\ \times\ 8 \\ \hline 8.0 \\ 1.6 \\ \hline 9.6 \end{array}$$

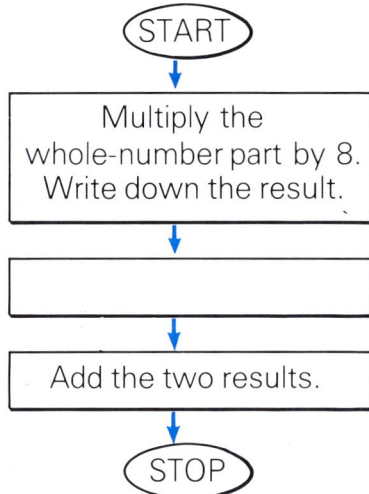

START

Multiply the
whole-number part by 8.
Write down the result.

Add the two results.

STOP

   Write down what is missing from the flow chart.

   c) Use Jake's method to do these:

   (i) 2.4×3 (ii) 3.8×5 (iii) 4.9×9

   d) The multiplication 2.4×3 could have come from this situation.

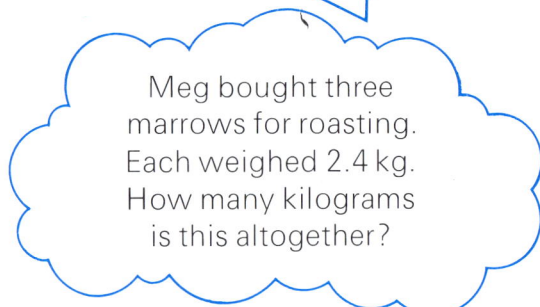

   Write down your own situation
   for these multiplications:
   (i) 16.6×5 (use centimetres)
   (ii) 9.4×6 (use tonnes)

   Meg bought three
   marrows for roasting.
   Each weighed 2.4 kg.
   How many kilograms
   is this altogether?

   e) Write down the result for each
   multiplication in **(d)**.

   f) Jake buys six packs of roof tiles.
   There are eight tiles in each pack.
   Each tile weighs 0.3 kg.
   What do the tiles weigh altogether?

   g) Check your results in **(c)**, **(e)**
   and **(f)** with your calculator.

13

With a friend

5   Each of you make an estimate, in your head,
    for each of these calculations.
    Write your estimates down.
    Compare each estimate with your friend's.

a) The cost of nine
   crystal glasses.

*Your estimate should
be within £5 of the
true cost.*

Crystal Glasses
£5·57 each

*Your estimate should
be within 2 km of the
true distance.*

Dover to London: 114·2 km

b) The distance travelled in
    six journeys between Dover
    and London.

c) The cost of one cake.

£134

6 Cakes

*Your estimate should
be within 5p of the
true cost.*

58.7 m

d) The length of each
    railway truck.

*Your estimate should
be within 0.5 m of the
true length.*

6   Work together on this calculation.
    The tree trunk has to be sawn
    into six equal lengths.
    Decide between you how long
    each part will be.
    Each of you write down how
    you calculated the lengths.

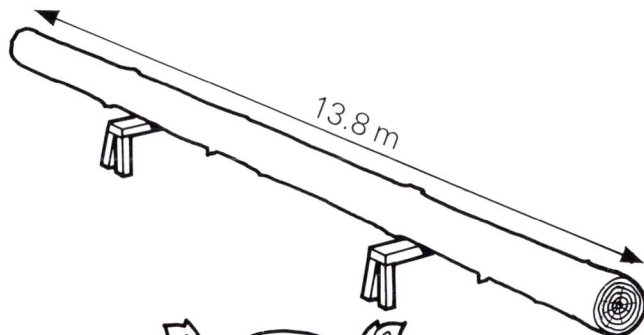

13.8 m

7   Work together on this explanation.
    Meg needs to divide the sand in
    this bag into three equal piles.
    She does the calculation like this.

$$14 \cdot 1 \div 3$$

$$3\overline{)14}\;^{4} \qquad 3\overline{)21}\;^{7}$$

$$4 \cdot 7$$

Copy her calculation.
Explain how she arrived at her result.

SAND
14.1 kg

13

8 Here are two more of Meg's divisions.
Copy and complete each one.

a)

$$7.5 \div 5$$

$$\begin{array}{r} 1 \\ 5\overline{)7} \end{array} \qquad \begin{array}{r} ? \\ 5\overline{)25} \end{array}$$

$$1.\,?$$

b)

$$8.4 \div 6$$

$$\begin{array}{r} ? \\ 6\overline{)8} \end{array} \qquad \begin{array}{r} 4 \\ 6\overline{)?} \end{array}$$

$$?.4$$

Think it through

9 Do not use a calculator.

a) Copy and complete each calculation.

WEIGHT CHART
Before 98.76 kg
After 92.89 kg

A
$$\begin{array}{r} 2.64\,\text{kg} \\ +\,4.97\,\text{kg} \\ \hline 6.00\,\text{kg} \\ .\quad\text{kg} \\ 0.11\,\text{kg} \\ \hline \quad\text{kg} \end{array}$$

B

$$\begin{array}{r} 98.76\,\text{kg} \\ -\,92.89\,\text{kg} \\ \hline \quad\text{kg} \\ -\,0.80\,\text{kg} \\ \hline \quad\text{kg} \\ -\,0.09\,\text{kg} \\ \hline \quad\text{kg} \end{array}$$

Weight lost

C
$$\begin{array}{r} 2.8\,\text{m} \\ \times\,12 \\ \hline 20.0\,\text{m} \\ 8.0\,\text{m} \\ 4.0\,\text{m} \\ \hline .\quad\text{m} \\ \hline \quad\text{m} \end{array}$$

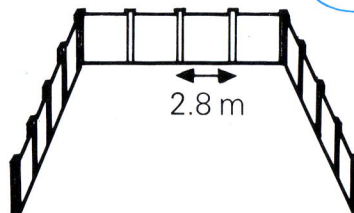

2.8 m

Total length of fencing

D To be shared equally into six jugs.

$$4.14 \div 6$$

$$\begin{array}{r} 0 \\ 6\overline{)4} \end{array} \qquad \begin{array}{r} ? \\ 6\overline{)41} \end{array} \qquad \begin{array}{r} 9 \\ 6\overline{)?} \end{array}$$

$$0.\,??\,l$$

MILK
4·14 l

b) Check your results in (a) with your calculator.

● Next chapter

13

14 Moving rods

A 1 George Painter is in trouble.
His ladder is sliding away from under him!

a) Think about the top end of the ladder.
It travels vertically down the wall.
What happens to the bottom end?

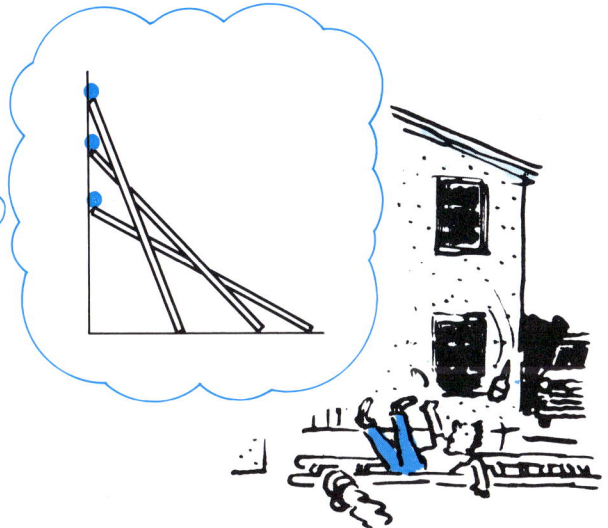

b) George's feet are on the middle rung of the ladder.
Guess what path his feet trace out
as the ladder slides.

Make a sketch to explain
your guess.

*George
hangs on.
He's too scared
to move his feet!*

Activity

2 a) Cut out an 8 cm strip of paper like this.

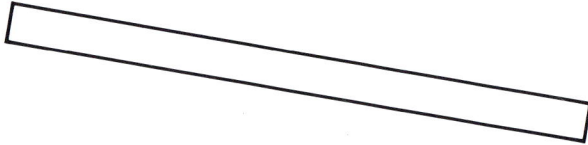

This is your 'ladder'.
Copy this drawing of the wall and ground.

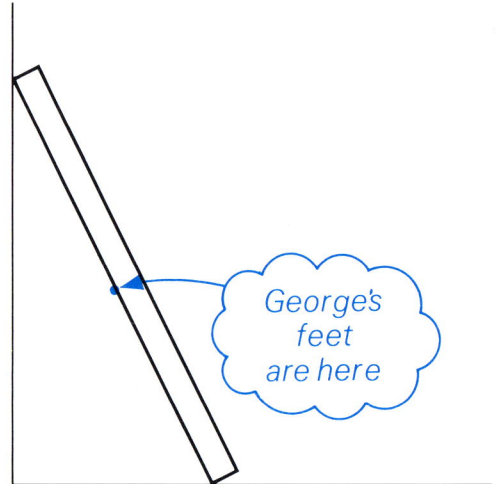

b) Lean your ladder against the wall.
Now slide your ladder down the wall.
Mark **ten** different positions
for George's feet.
Draw a curve to show the path
of George's feet.

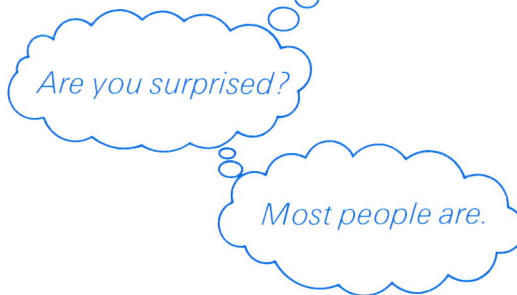

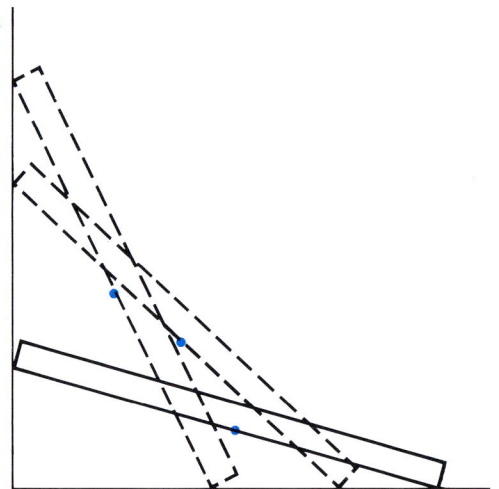

Are you surprised?

Most people are.

c) Choose another point on the ladder.
For example, the point one-quarter
of the way up from the bottom.
Find the path which this traces out.

d) Choose at least **two** more
points on the ladder.
Draw their paths.

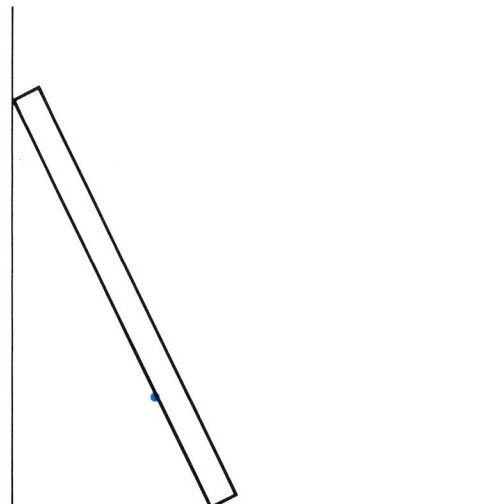

George's
feet
are here

14

3   This is a full-size drawing of the face of a small alarm clock.
    A is the centre of the clock face.
    B is the end of the minute hand.
    C is the end of the hour hand.

    This diagram shows how A and B move.

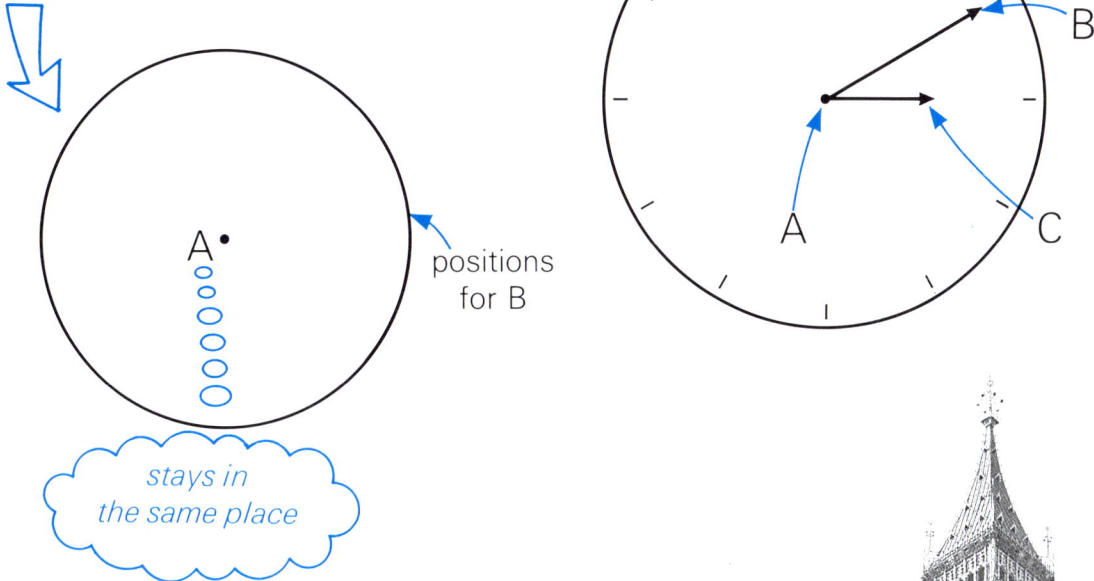

A •

positions
for B

*stays in
the same place*

Copy the diagram.
Mark the positions for C.

Take note

Diagrams which show how points move
are called **motion diagrams.**

4   Draw a motion diagram
    for the points marked
    on this watch face.

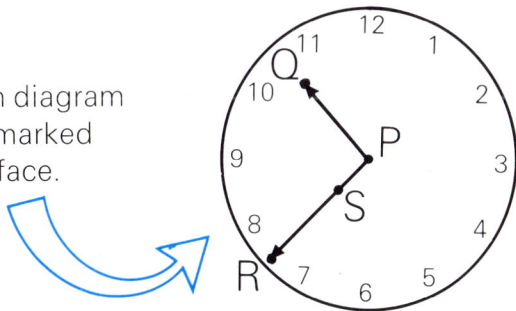

5   This is a motion diagram for a clock face.
    It is drawn full size.
    The circles are the paths of the
    middle points of the hands.
    How much longer is the minute hand
    than the hour hand?

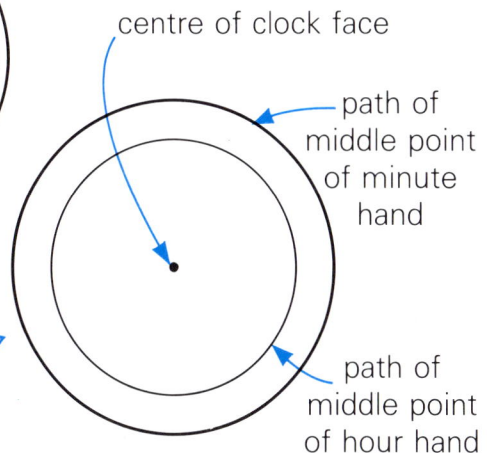

centre of clock face

path of
middle point
of minute
hand

path of
middle point
of hour hand

14

6 This is a 5 cm strip of card.
   Point P is pinned to a piece of card.
   The strip can turn about P.

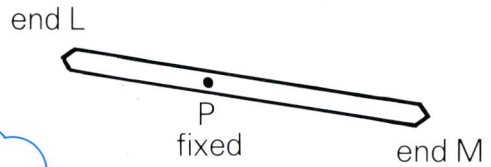

end L

•
P
fixed

end M

*They are not
drawn full size.*

Which of these is a motion diagram
for P, L and M?

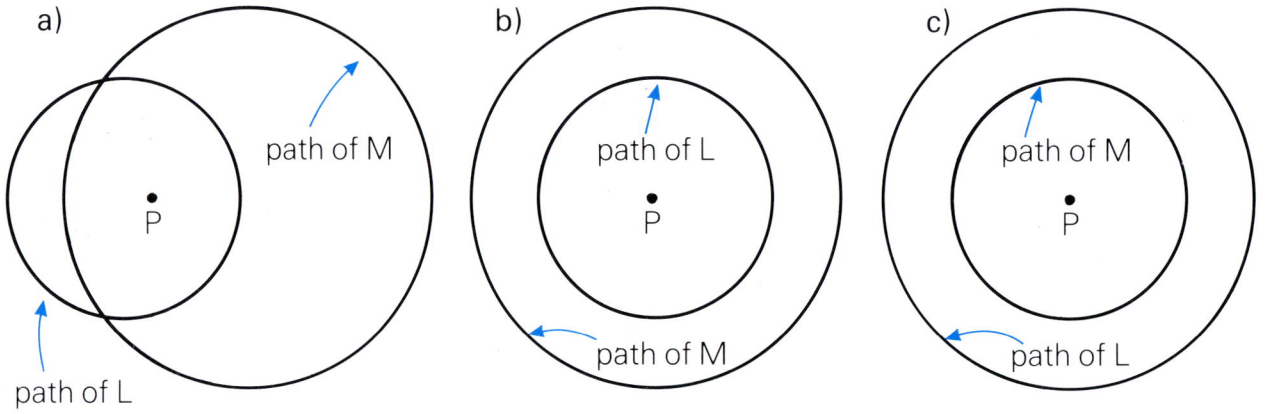

a)

path of M

•
P

path of L

b)

path of L

•
P

path of M

c)

path of M

•
P

path of L

7 Here are four motion diagrams and four strips.
   Match each motion diagram with a strip.

*These are
scale drawings.*

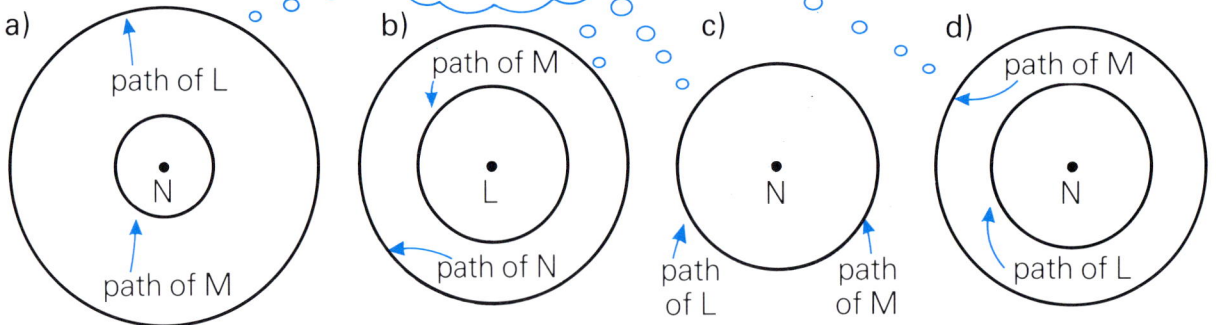

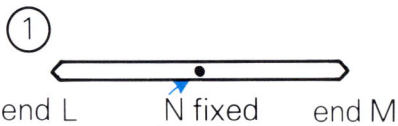

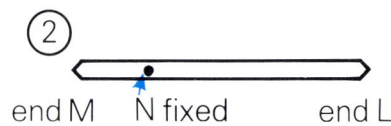

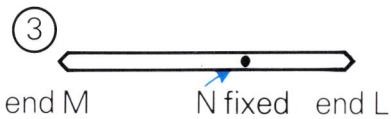

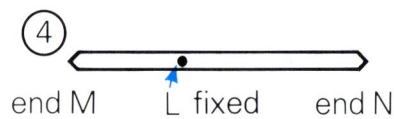

a)

path of L

•
N

path of M

b)

path of M

•
L

path of N

c)

•
N

path
of L

path
of M

d)

path of M

•
N

path of L

①

end L   N fixed   end M

②

end M   N fixed   end L

③

end M   N fixed   end L

④

end M   L fixed   end N

14

122

8 a) Ben uses these cardboard strips.
      He makes this framework.

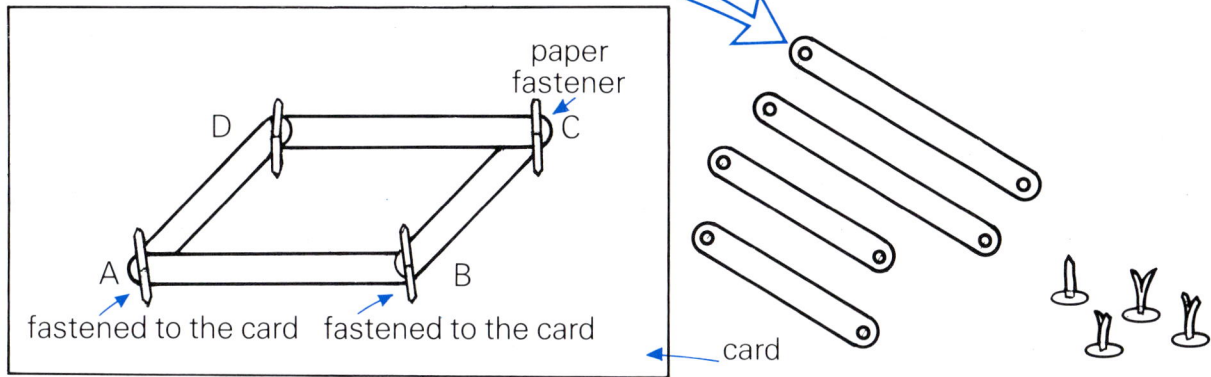

paper
fastener

D   C

A   B

fastened to the card   fastened to the card

card

Ends A and B are fastened to the sheet of card.
Ends C and D can move.
Which of these is the motion diagram
for A, B, C and D?

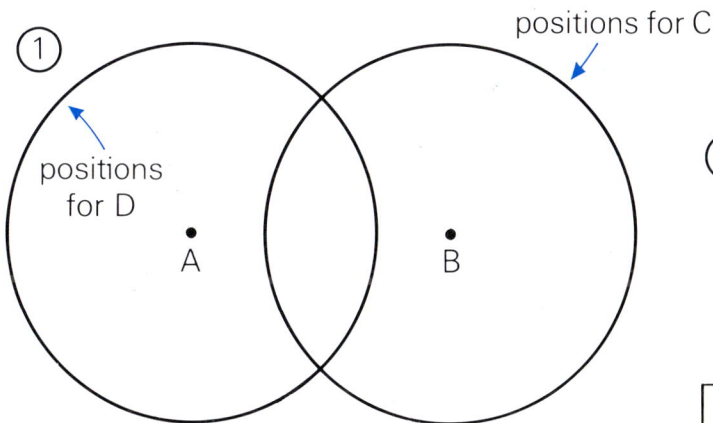

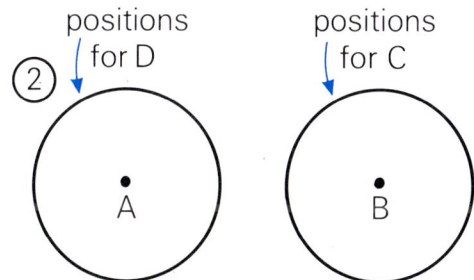

positions for C

① positions for D   A   B

② positions for D   positions for C

A   B

b) Ben fixes a cross-strip to his framework.
   Which of these is the motion diagram
   for his new framework?

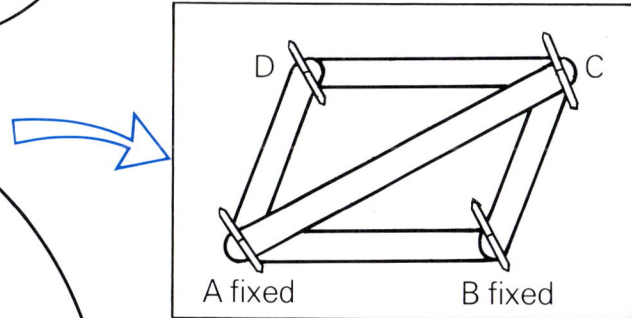

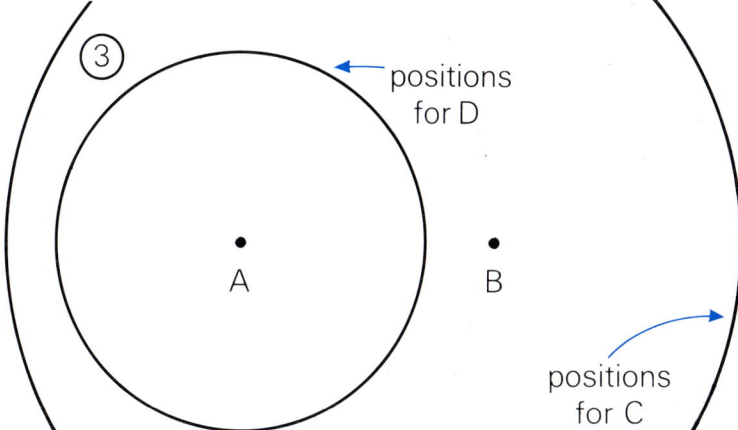

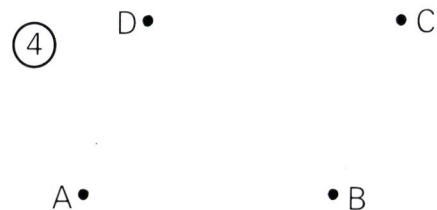

D   C

A fixed   B fixed

③ positions for D

A   B

positions for C

④ D•   • C

A•   • B

14

9 a) Ben made this triangle framework.
Draw an accurate motion diagram
for A, B and C.

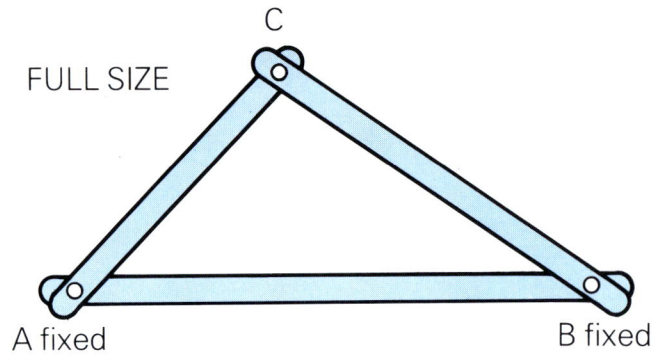

C

FULL SIZE

A fixed

B fixed

b) Five-bar gates have a crossbar.
Write down why you think this is.

EXPLORATION

10 You need two strips of card like this,

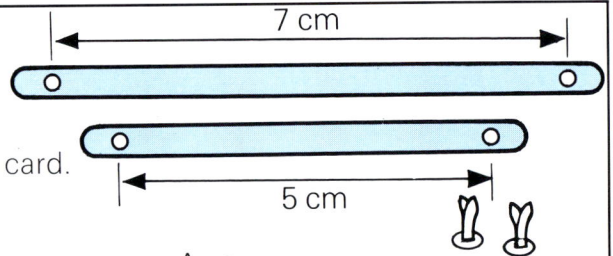

7 cm

two paper fasteners, and a small piece of card.

5 cm

A

Fix the strips together.
Fasten this end to the small piece of card.

B

Investigate how A, B and C can move.

The card cannot move.

Sketch a motion diagram for A, B and C.

C

Take note

When this dot
can move
anywhere
inside this
square . . .

. . . then this is its
motion diagram.

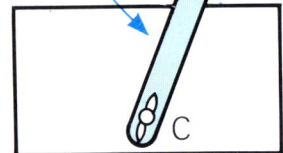

■125
▲127
●Next chapter

14

Motion diagrams for frameworks

A Activity

1 You need four paper fasteners, four strips of card like this, and another piece of card.

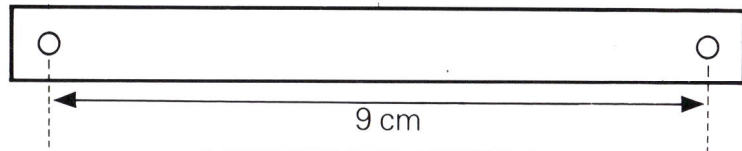

9 cm

Make this framework.
Fix A and B to the piece of card.

Move your framework around.
Keep A and B still.

a) Here are some possible positions.

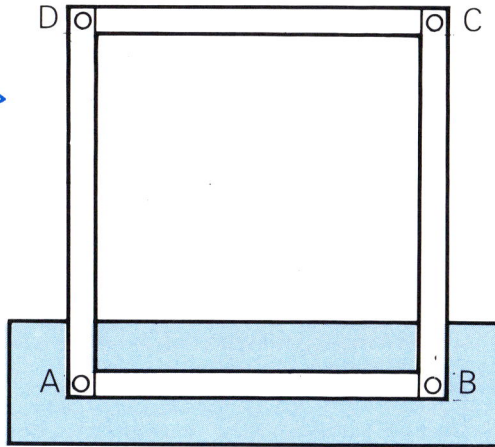

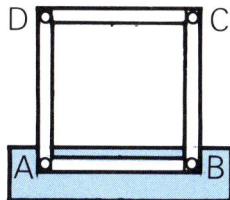

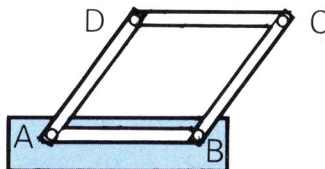

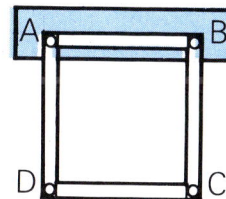

Sketch three more.

b) Mark two points 9 cm apart.
Draw a diagram to show all the positions A, B, C and D can have.

*Remember:
This is called a
motion diagram.*

Challenge

2 a) Make a hole at the midpoint of DC.
Call it point X.
Guess what path X follows as the framework moves.

b) Put your pencil through the hole.
Now **draw** the path of X.
Was your guess correct?

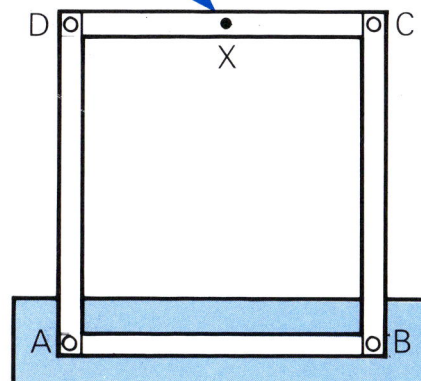

14

3 This is the framework from question 1.
   A new strip has been fixed between B and D.

   a) Guess the motion diagram for A, B, C and D.
   Make a sketch to show what you think.

   b) Fix a strip across your own framework.
   Did you guess correctly?

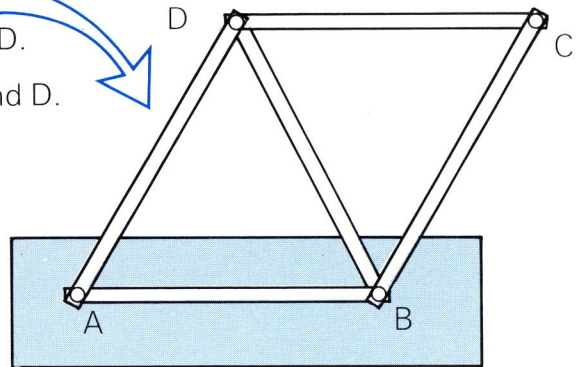

4 Draw a motion diagram for this strip.
   Use compasses.
   Measure accurately.

end B

C fixed

end A

5 Ben made this framework.
   He fixed P and Q to a piece of card.
   Use compasses.
   Draw the motion diagram for
   P, Q, R and S.

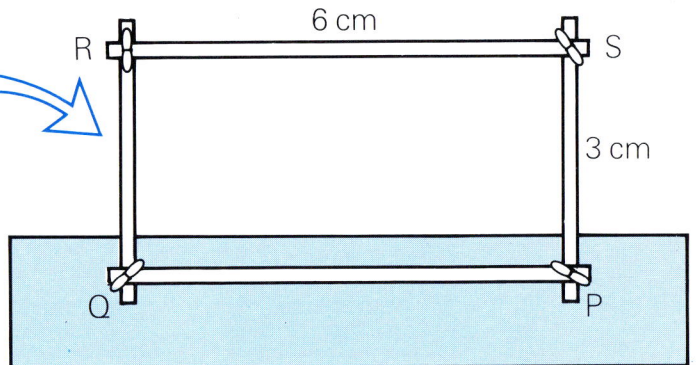

R   6 cm   S

                        3 cm

Q   P

Think it through

6 This is a motion diagram
   for a four-strip framework.
   Sketch the framework.
   Mark the distances between
   the fasteners.

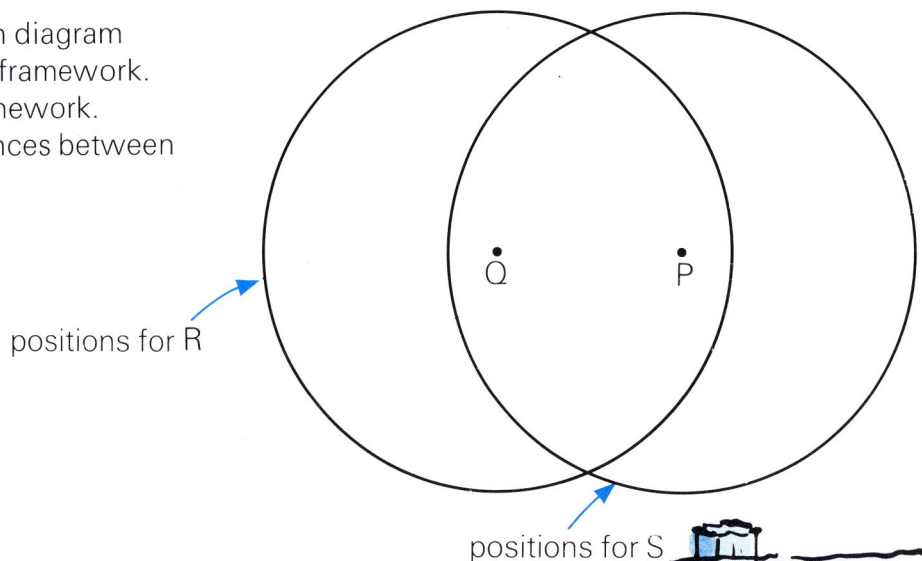

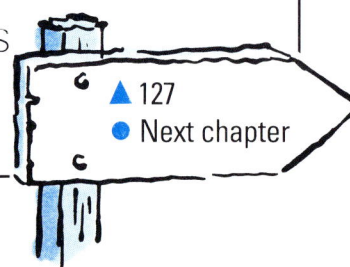

Q   P

positions for R

positions for S

▲ 127
● Next chapter

Garage doors

B 1 Look at these four pictures.
They show a garage door opening.

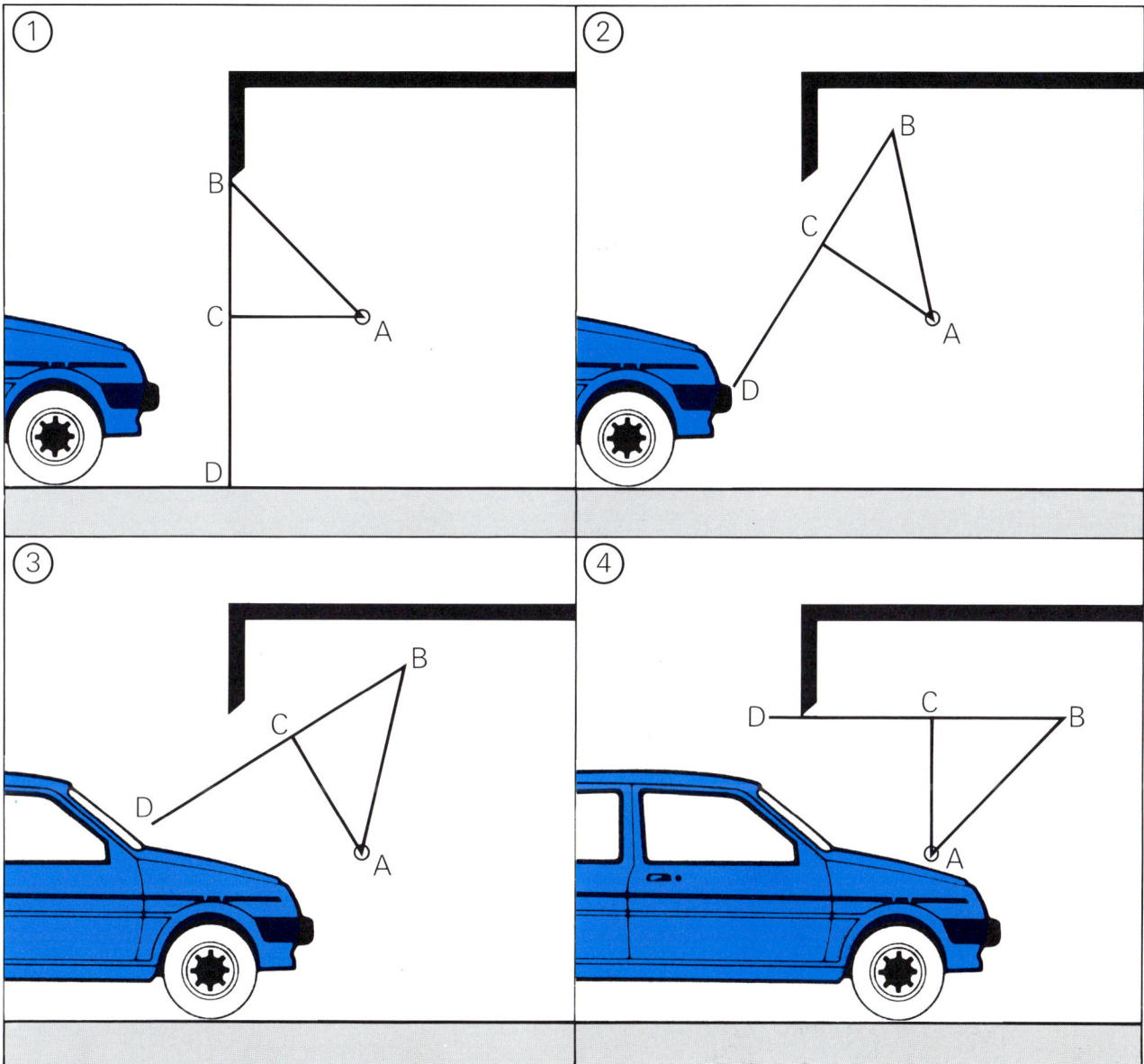

a) Mark point A in your book.

b) Measure distances in millimetres.
Measure angles with a protractor.
Draw the motion diagram for A, B, C and D.
(You should **not** get complete circles.)

c) Measure the angles these turn through:
(i) AC   (ii) AB   (iii) BD

d) Write down what you notice.

2   Work with a friend.

a) These quadrilaterals are made from strips of card.
    Each one is fixed at points A and B.
    Sketch motion diagrams for A, B, C and D in each one.

Make the framework if it helps.

Square

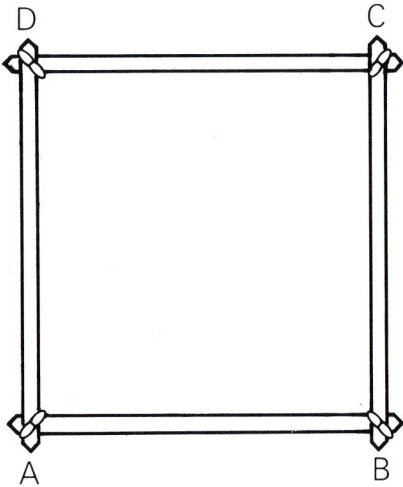

Rectangle

Kite

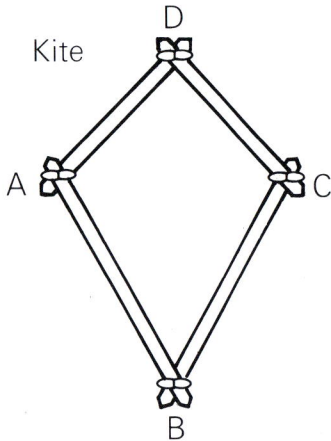

Isosceles trapezium

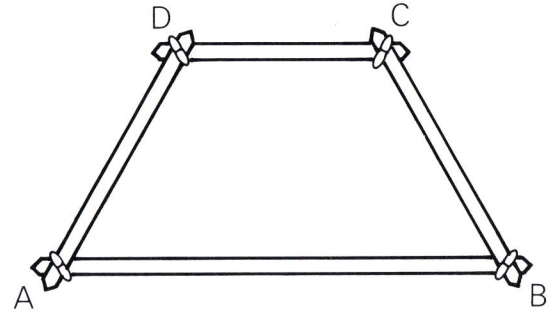

A 'general' quadrilateral (all sides of a different length)

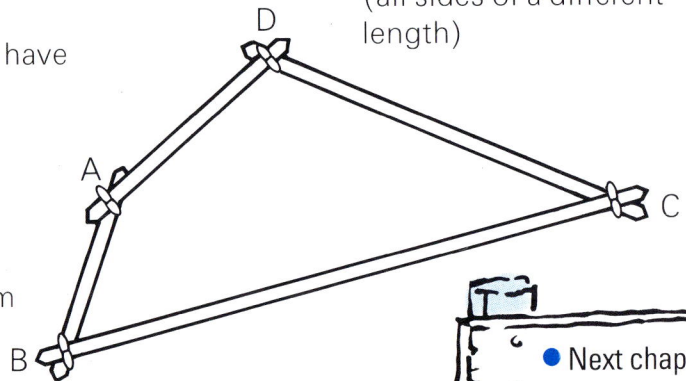

b) Which of the diagrams you have already drawn could be for
    (i) a rhombus?
    (ii) a parallelogram?

c) Suppose that B and C are fixed instead of A and B.
    Draw a new motion diagram for the corners of each quadrilateral.

● Next chapter

14

15 On average

A 1 Meg and Midge share the orange in these jugs between them.

fairly!

How many centilitres is this each?

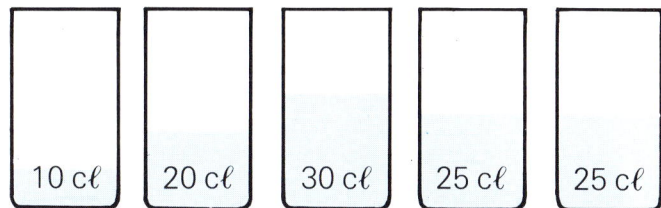

20 cℓ   50 cℓ

2 Horace, Winston and Ben share the orange in these jugs.

How many centilitres is this each?

60 cℓ   50 cℓ   10 cℓ

3 All five friends share the lemon in these glasses.

How many centilitres is this each?

10 cℓ   20 cℓ   30 cℓ   25 cℓ   25 cℓ

Take note

The **mean** is the amount we get when we 'share out equally'.

The **average** or **mean** amount of orange per jug is 40 cl.

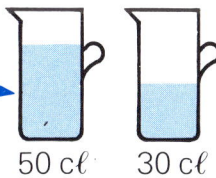

50 cℓ   30 cℓ

Here the **mean** amount is 30 cl.

10 cℓ 60 cℓ 20 cℓ

4 The mean amount of chocolate per bar is 70 g. What does a Pluto bar weigh?

PLUTO
CHOCOLATE FUDGE

CHOCOLATE
50g

Challenge

5 The mean amount of orange per glass is 20 cl.

How much orange is there in the third glass?

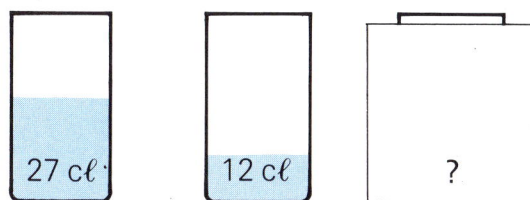

27 cℓ   12 cℓ   ?

15

● 129

6   Do these in your head.

a) The mean amount of orange per glass is ____.

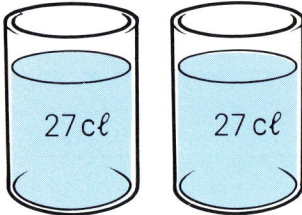

27 cℓ   27 cℓ

b) The mean amount of cheese per mouse is ____.

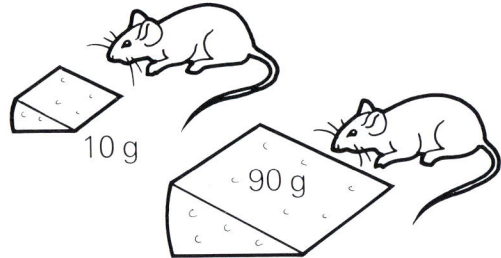

10 g   90 g

c) The mean length of the raccoons' tails is ____.

70 cm

50 cm

d) The mean length of the ladders is ____.

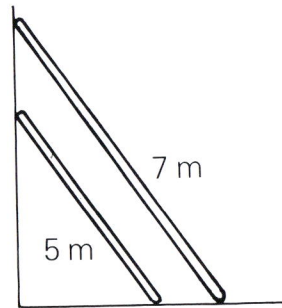

7 m

5 m

e) The mean mass of the cannonballs is ____.

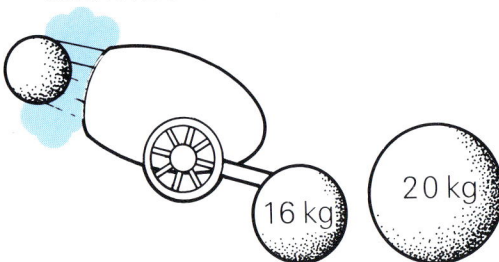

16 kg   20 kg

f) The mean amount of lemonade per glass is ____.

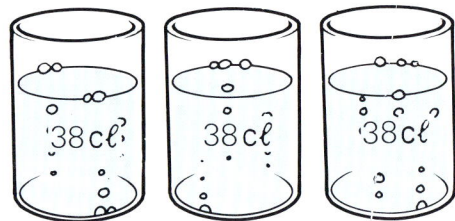

38 cℓ   38 cℓ   38 cℓ

g) The mean length of the pencils is ____.

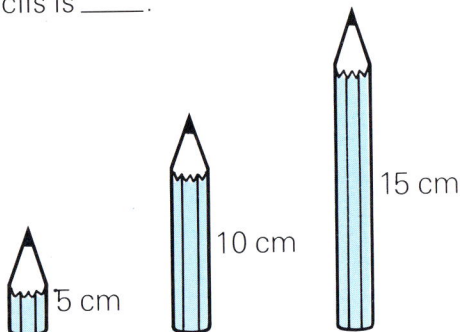

15 cm

10 cm

5 cm

h) The mean amount of potatoes per sack is ____.

14 kg

10 kg   6 kg

7   Use your calculator to check your answers to question 6.

15

Using the mean

B 1   Horace and his class are making clay models.
The teacher asks them to make two giraffes.
'Make them about 12 cm tall, on average,' she says.

a) These are Midge's giraffes.
What is their mean height,
to the nearest centimetre?

b) These are Horace's.
What is their mean height,
to the nearest centimetre?

14 cm   10 cm   1 cm   23 cm

c) Lenny's giraffes are 12 cm tall on average.
One of them is 8 cm tall.
How tall is the other?

2   The mean mass of two cats is 9 kg.
One cat weighs 2 kg.
What does the other weigh?

Think it through

3   The mean height of two trees is 21 m.
Each tree is a whole number of metres tall.
a) How much taller might one
be than the other?
Write down the greatest possible
difference.
b) What is the nearest they
might be to each other's height?

15

4 Helen Manion goes for a long run every Monday,
Tuesday and Wednesday.
Every Monday she runs 10 km.
Every Tuesday she runs 12 km.
After Wednesday's run her average is always 15 km per day.

How many kilometres does she run each Wednesday?

5 Wilson Wanderers tug-of-war team needs ten members.
The first nine team members weigh 100 kg on average.

'Get me somebody to bring the average up to 105 kg,' says Coach.

What must the tenth team member weigh?

Think it through

6 Pierre is collecting grapes from his vines.
He has 1000 vines.
These are the amounts he collects
from the first ten vines:

9 kg	16 kg	20 kg
15 kg	25 kg	22 kg
18 kg	5 kg	
12 kg	20 kg	

About how many kilograms of grapes do
you think he will collect altogether?

With a friend

7 A pine forest covers 6000 hectares
of land.
A forester wants to know
approximately how many
trees there are.

Discuss his problem with your friend.
Decide what he should do to estimate
the number of trees.
Write down what you decide.

Do you remember...?

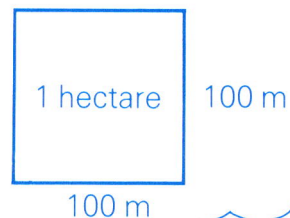

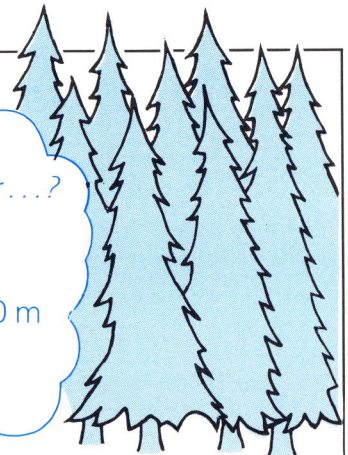

1 hectare — 100 m
100 m

15

Activity

8   You need a small piece of tracing paper with a 1 cm square drawn on it.

This is a picture of a colony of bugs.
It is full size.

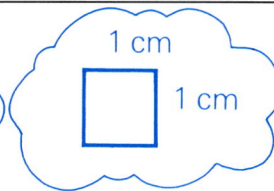

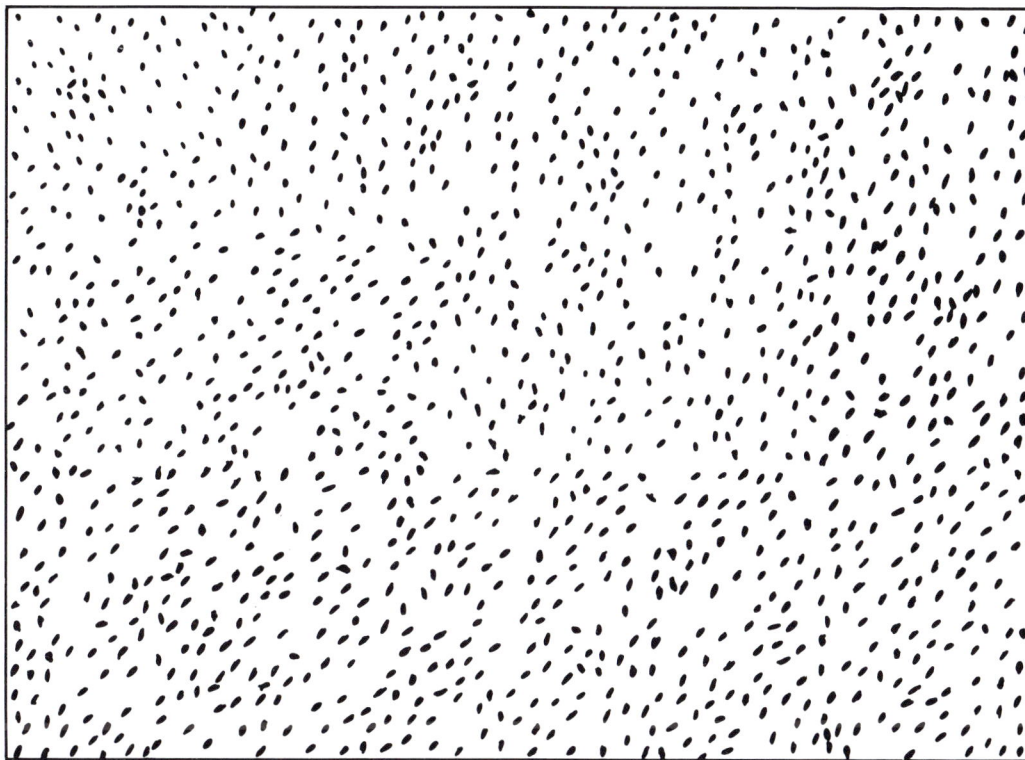

1 cm
1 cm

a) Drop your 1 cm square onto the colony ten times.
Count how many bugs are inside the square each time.
Record your results.

Number of bugs	Mean
8 10 8 ...	

b) Use your calculator to work out the mean.
Write your result to the nearest whole number.

Use your result in (b).

c) Work out the area of the picture in cm².

d) About how many bugs are there altogether?

e) Explain how you can use your 1 cm square to get a better estimate.

15

■ 134
▲ 135
● Next chapter

Working with averages

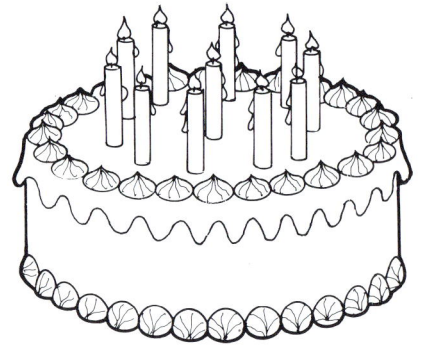

A
B
1   What is the mean age of yourself and your best friend, to the nearest year?

2   The average age of two friends is 12.
One of them is 11.
How old is the other?

3   The mean of each pair of numbers should be 3.
Write down the missing number each time.

a) 2 ; ☐   b) 3 ; ☐   c) 0 ; ☐   d) $5\frac{1}{2}$; ☐

4   The mean height of the basketball players is 200 cm.
How tall is Stilts Johnson?

200 cm   204 cm   196 cm   190 cm   Stilts Johnson
? cm

5   Lynn is taking three tests.
Her mean mark after the first two is 7.
Each test is marked out of 10.

a) How many marks has she scored altogether so far?

b) She wants her mean mark for the three tests to be at least 8.
How many marks must she score in the third test?

15

Think it through

6   The mean age of three friends is 12 years.
Another friend joins them.
The mean age is now 11.

How old is the fourth friend?

▲ 135
● Next chapter

Averages and totals

C   _With a friend_

1   **a)** Midge checked 100 packets of nails.

     40 packets had 20 nails.
     50 packets had 21 nails.
     10 packets had 19 nails.

     Discuss whether the 'Average contents' label is fair.

     Write down what you decide.

   b) Ben, Meg and Midge estimated how many nails there would be in 200 packets.

     Ben did this.

> 80 packets with 20 nails = _____
> 100 packets with 21 nails = _____
> 20 packets with 19 nails = _____
>   Total = _____

     Meg did this.

> Mean number per pack = 20
> Number in 200 packs = 200 × 20
>   = _____

     Midge did this.

> Mean number per pack for 100 packs = 20·4
>   Number in 200 packs = 200 × 20·4
>   = _____

     Do they all get the same result?
     If not, which method do you think gives the best estimate?
     Decide between you.
     Write down what you decide, and why.

15

Mean problems

Challenge

1 There is 1 mint in jar 1 and 3 in jar 2.
So the mean number of mints per jar is 2.

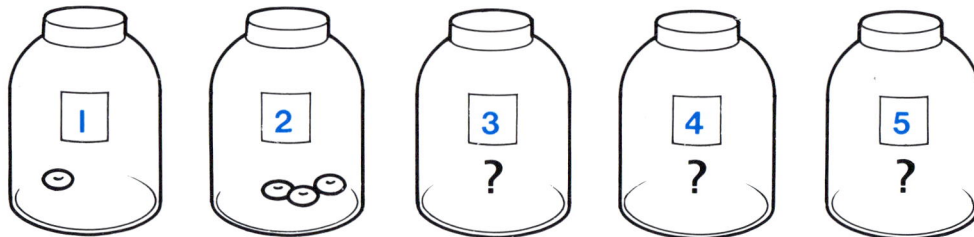

a) The mean number for the first 3 jars is 3.
How many mints must there be in jar 3?

b) The mean number for the first 4 jars is 4.
How many must there be in jar 4?

c) The pattern continues in this way.
How many mints must there be in jar 100?

d) Write down a rule which tells you how many mints there are in
a jar when you know the jar number.

Activity

2 Ask your teacher for the bag of beans.
Altogether there are 500 beans
in the bag.
Without emptying out more than ten
beans at a time, estimate how many
there are of each colour.
Write down your result.
Explain how you arrived
at your estimate.

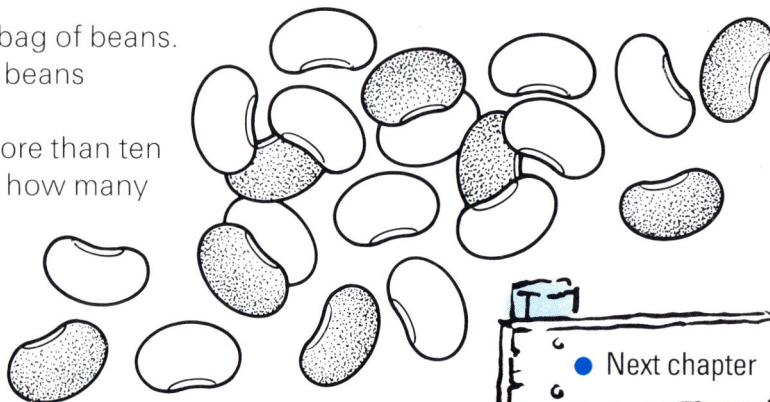

● Next chapter

15

16 Lines and angles

A

DO YOU REMEMBER ...?

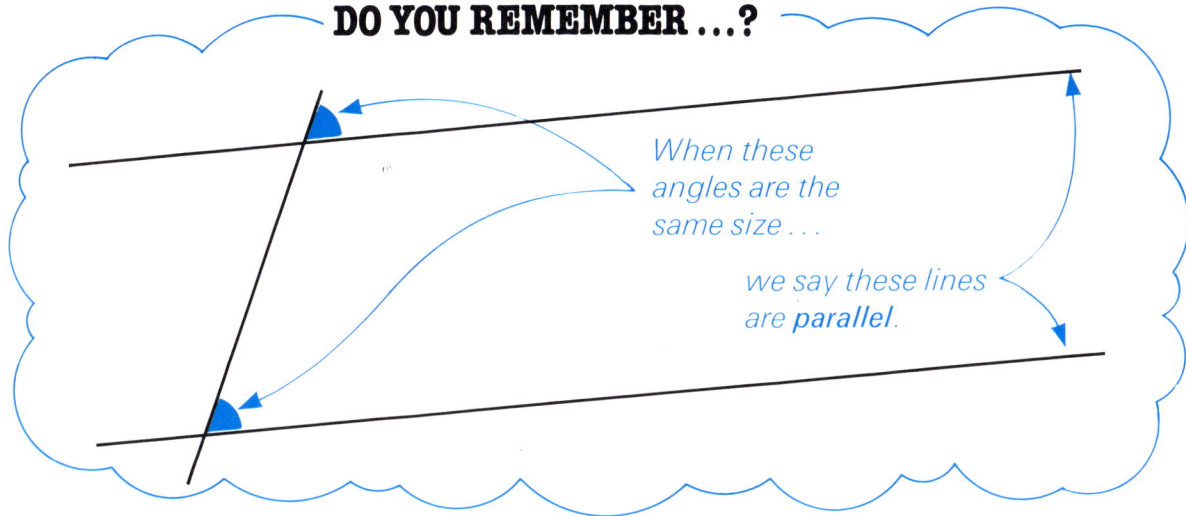

When these angles are the same size ...

we say these lines are parallel.

1 You need a protractor.

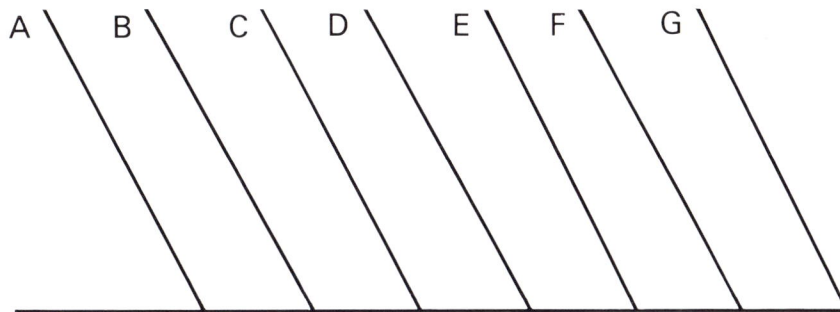

A B C D E F G

Three of these lines are parallel.
Which ones?

2 The telegraph poles have been blown over in a gale.
   a) Two of them are still parallel to each other.
      Which two?

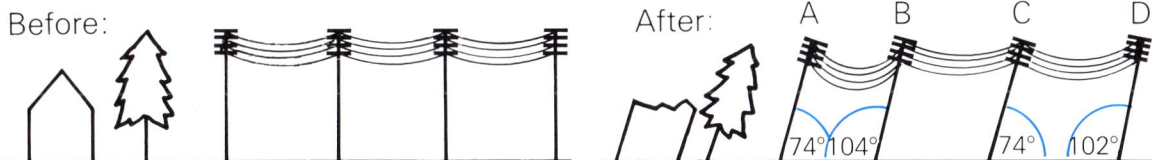

Before:

After: A B C D

74° 104° 74° 102°

   b) Through how many degrees must each pole be turned to make it vertical again?

Think it through

3 Through how many degrees should pole **A** be turned so that it is parallel to pole **B**?

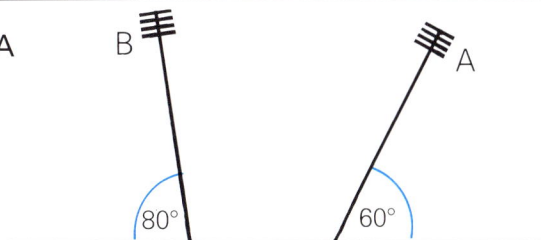

B A

80° 60°

16

● 137

4 The picture shows
two children on
a seesaw.
What is the size
of the angle
marked ● ?

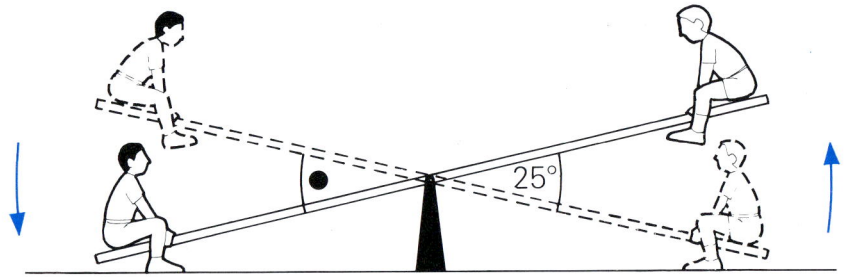

5 This is Horace's kite.
What is the size of the angle marked

a) ✳ ?

b) ● ?

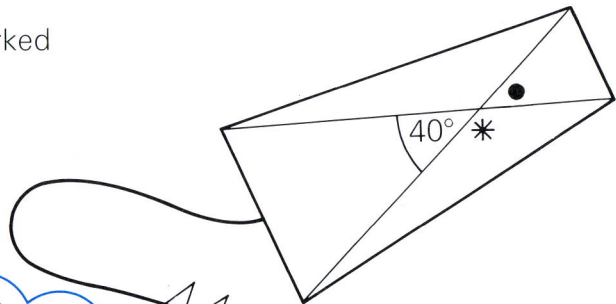

6

DO YOU REMEMBER...?

The angles of a triangle add up to 180°.

These are three straight roads.
What is the size of the angle marked

a) ● ?   b) ■ ?   c) ▲ ?

d) ✳ ?   e) ◿ ?

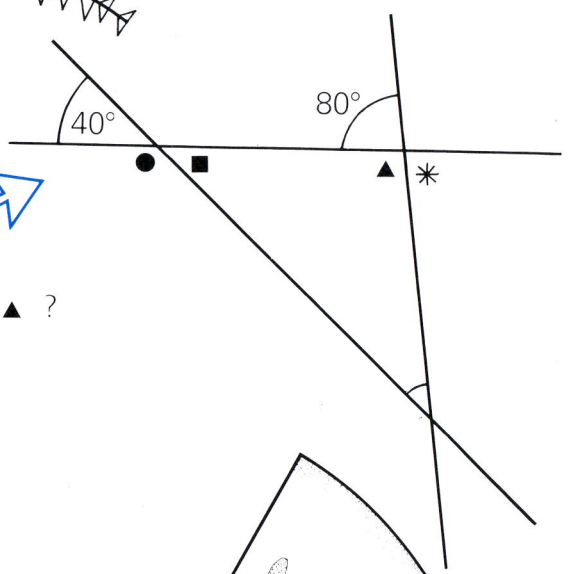

7

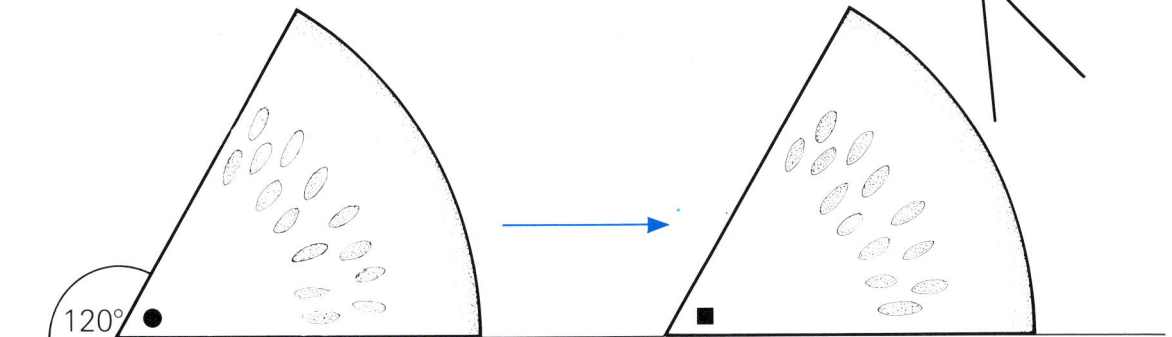

The melon slice is pushed across the table.
What is the size of the angle marked

a) ● ?

b) ■ ?

Activity

8 a) Draw a line like this.
Mark two points
A and B on it.

A ●   B ●

b) Lay your ruler across your line.
Draw a line which passes through A.

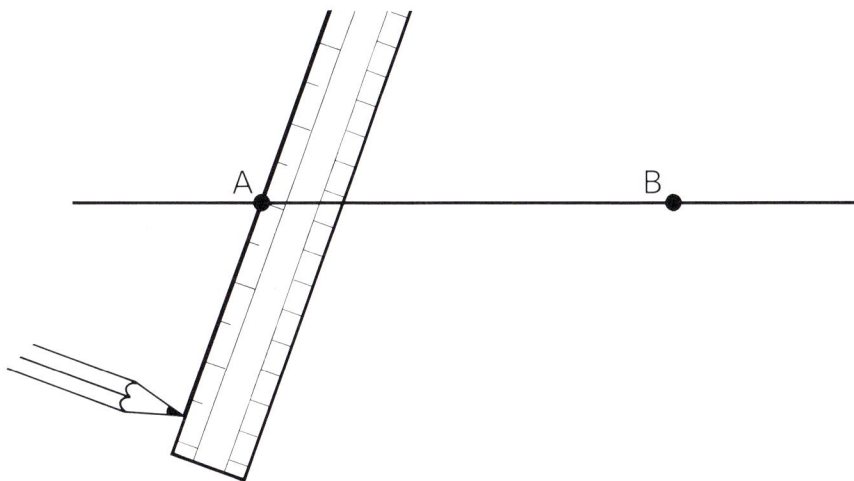

c) Slide your ruler over to B.
Try to draw a line parallel to the line you drew at A.

d) Think about the angle you made here.
Two of the angles at B should
be the same size as this angle.
Mark one of them
on your drawing.

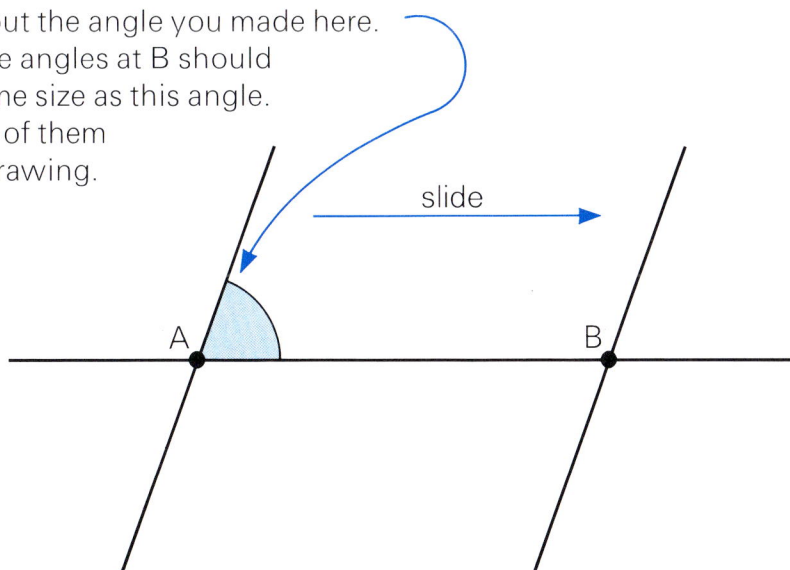

slide

A   B

e) Mark two more angles on your drawing which should be the same size as ◁

f) There are some other angles which should be the same size as each other.
Mark them on your drawing with a dot: ◁●

16

● 139

9

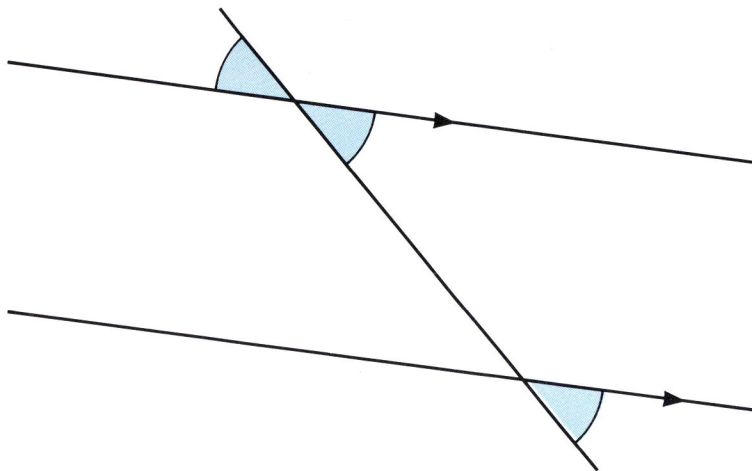

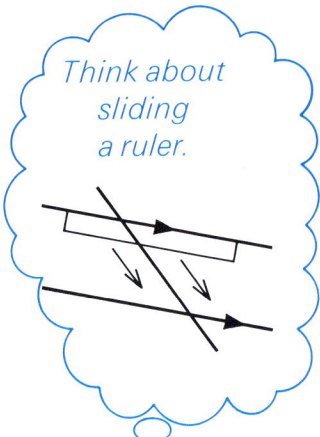
Think about sliding a ruler.

The three angles marked 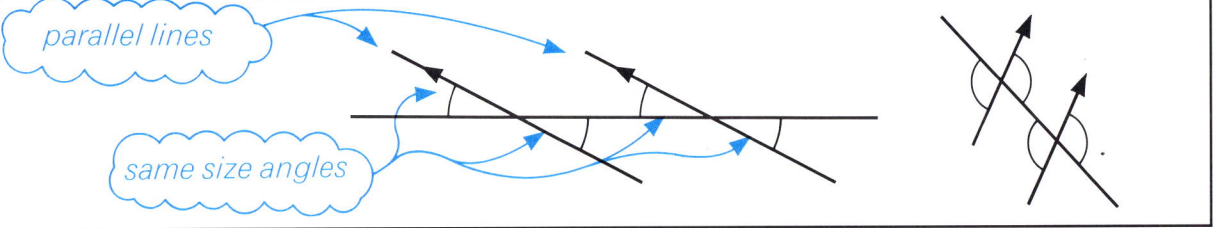 are the same size.

a) Sketch the drawing.

b) Mark another angle the same size as the three marked ∠.

c) Mark any other angles that are the same size as each other.
   Mark them like this ∠.

10 This is a cast-iron stairway.
    It is made of parallel iron bars.
    Without measuring, write down the size
    of each angle marked

a) ● .

b) ╪ .

c) ✳ .

d) ⸦ .

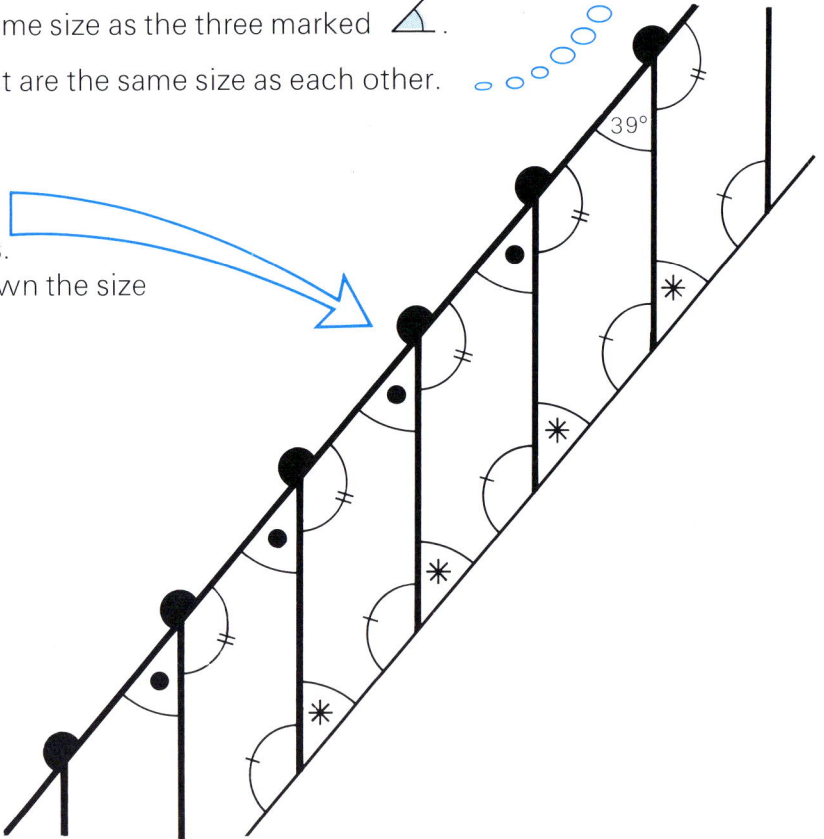

39°

Take note

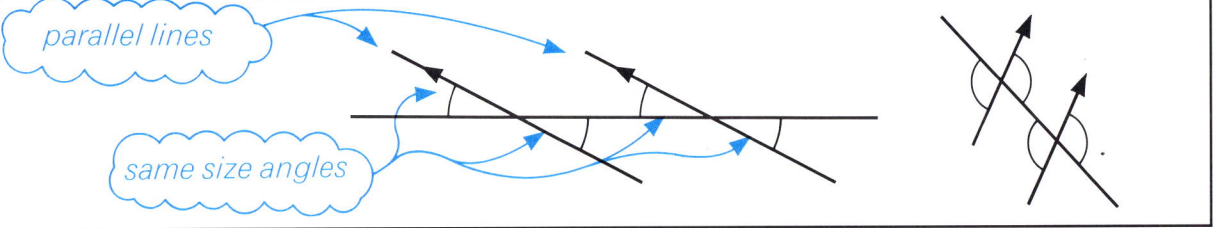

parallel lines

same size angles

Challenge

11 Use **only** a **ruler** and a **set square**.

  a) Is line a′ parallel to line a″?

  b) Is line b′ parallel to line b″?

  c) Explain how you decided.
  You might like to draw a sketch
  to help you explain.

  d) What special kind of quadrilateral
  is ABCD?

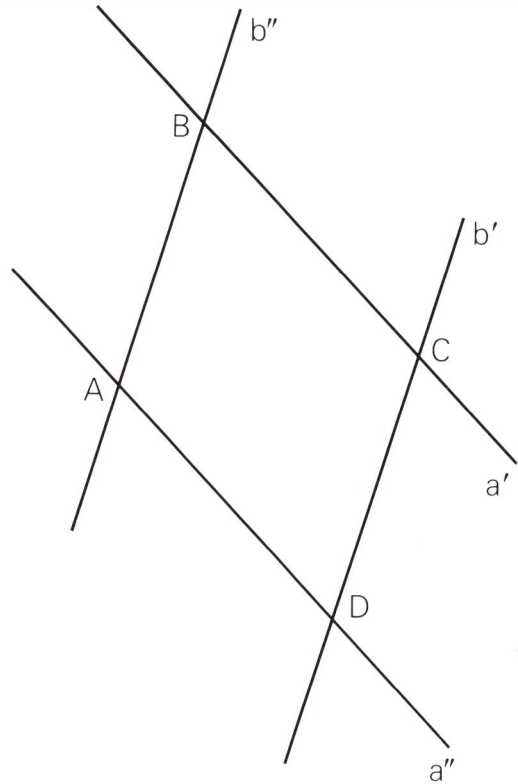

12 Use **only** a **protractor**.

  a) Is k′ parallel to k″?

  b) Is m′ parallel to m″?

  c) Explain how you decided.
  You might like to draw
  a sketch to help
  you explain.

  d) What special kind of
  quadrilateral is PQRS?

  e) Horace wants to know if these
  lines are parallel.

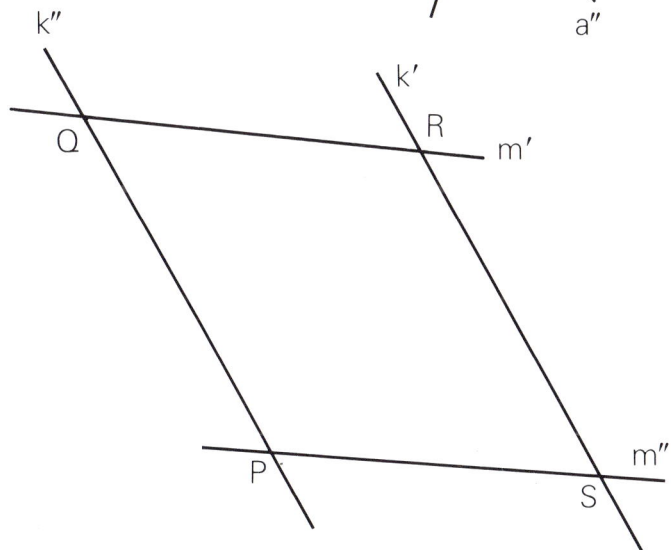

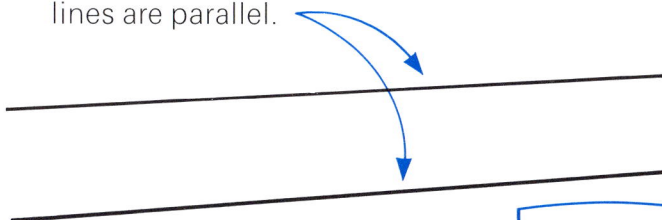

  He only has a protractor like this
  and a pencil.
  Explain how he can decide if the lines are parallel.
  Draw a sketch to help you explain.

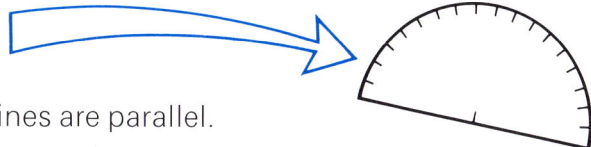

16

13 Which of these pairs of lines are parallel?
For those which are not, show on which side of
the 'crossing' line they meet.

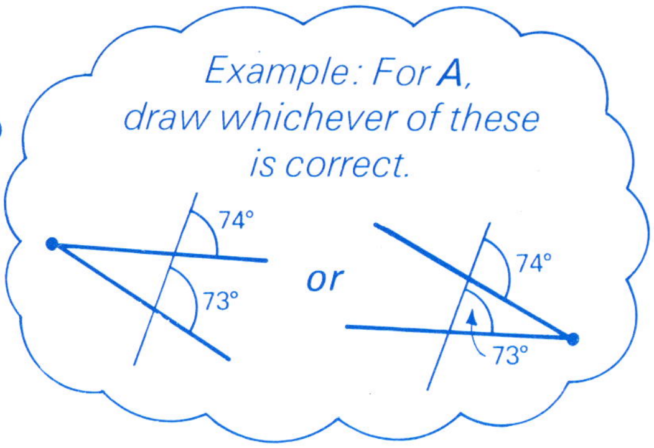

Example: For **A**,
draw whichever of these
is correct.

74° 73° *or* 74° 73°

A
74°
73°

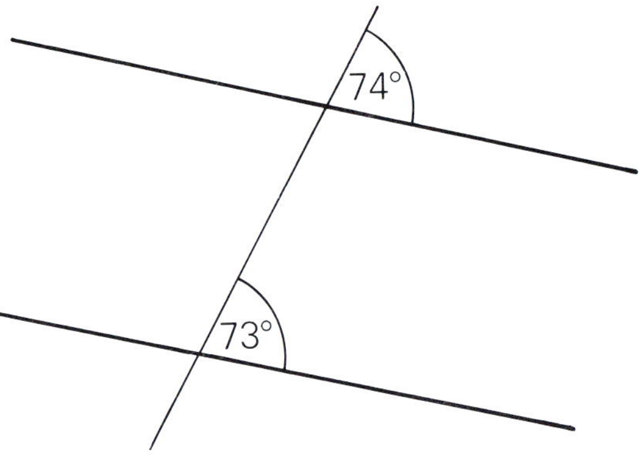

B
81° 82°

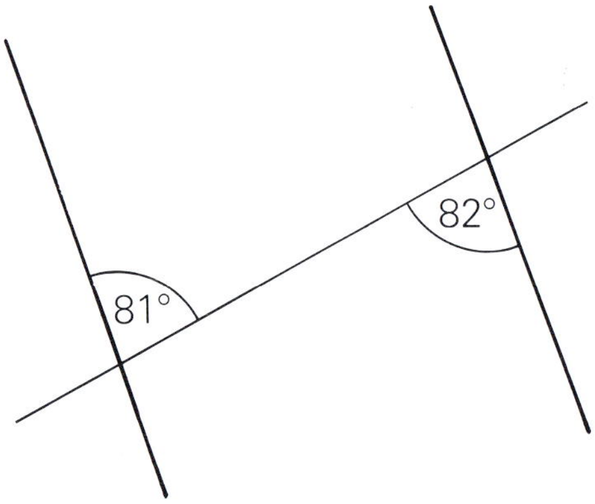

C
72°
72°

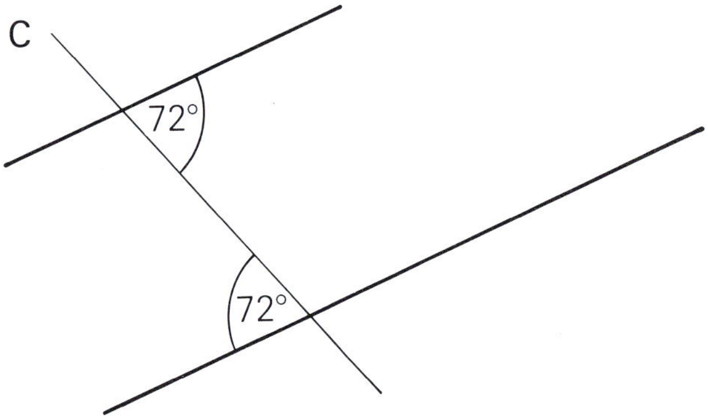

D
88°
88°

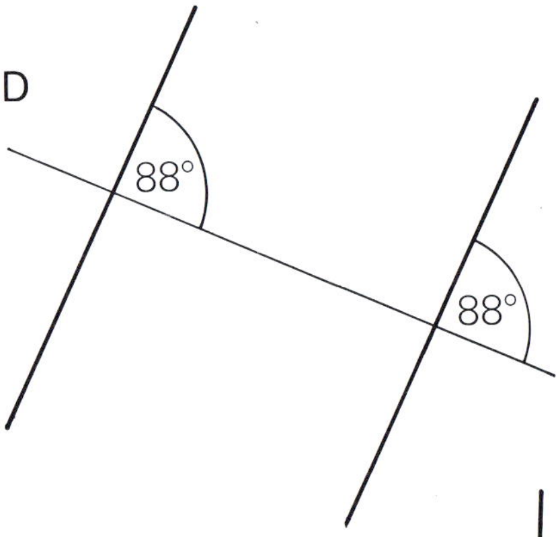

E
120°
60°

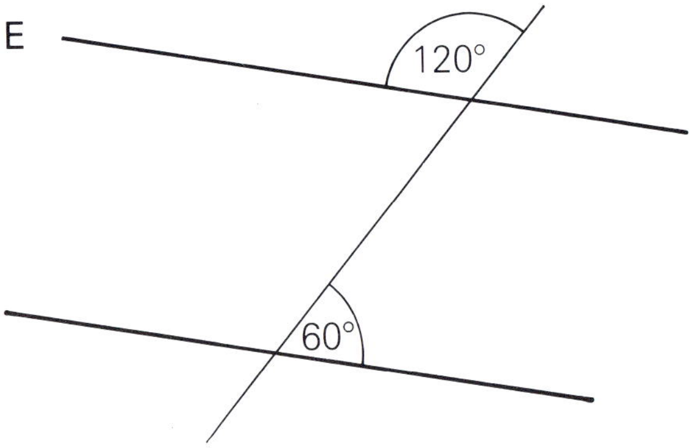

F
114°
66°

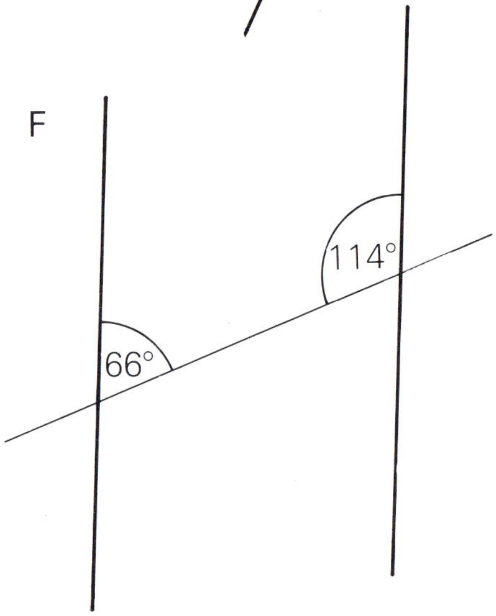

G
61°
109°

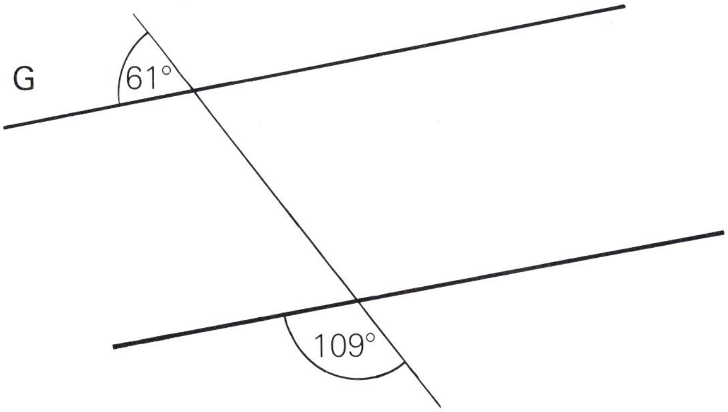

16

142

Angles and directions

B 1 The map shows five villages.

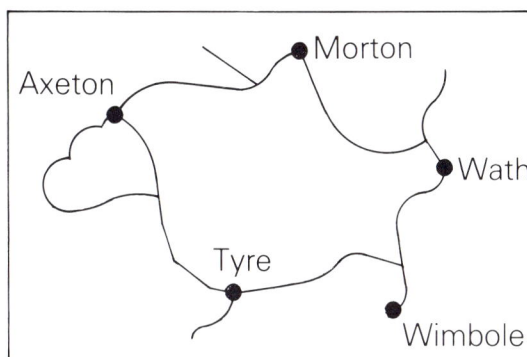

a) Direction compasses are drawn at Tyre and Wath.

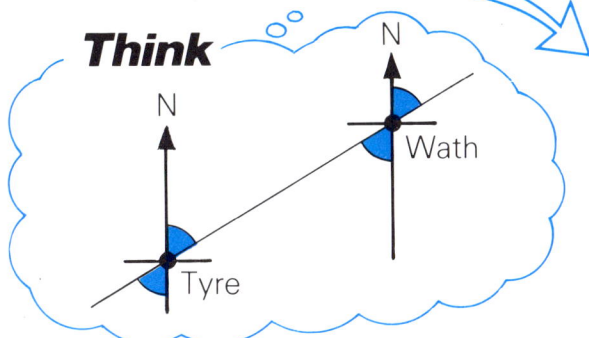

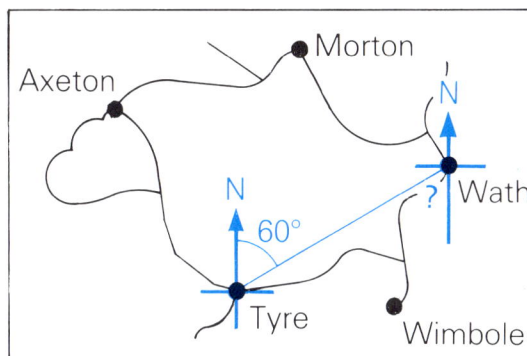

Think

What is the size of the angle marked '?' at Wath?

b) Copy and complete:
   (i) Direction of Wath from Tyre: N ☐° E.
   (ii) Direction of Tyre from Wath: S ☐° W.

c) This map shows direction compasses at Wath and Morton.
   Copy and complete:
   (i) Direction of Morton from
        Wath: N ☐° W.

   (ii) Direction of Wath from
        Morton: S ☐° E.

Think it through

d) Copy and complete:
   (i) Direction of Wimbole from
        Axeton: S ☐° E.
   (ii) Direction of Axeton from
        Wimbole: N ☐° W.

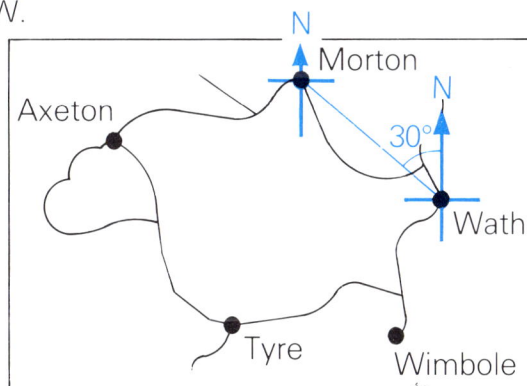

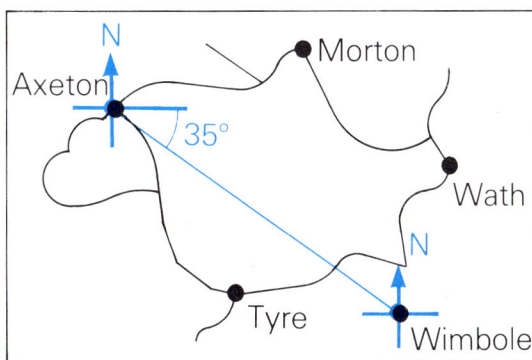

16

■ 144
▲ 146
● Next chapter

Angles and parallel lines

A 1

Draw a line.	Try to draw a line parallel to it.	Check by ● drawing any line across your pair of lines, ● measuring this pair of angles.

*Use **only** your **ruler** and **pencil**.*

Write 'Yes, my lines are parallel'
    or 'No, they are not parallel'.

2

Draw a line.	Try to draw a line parallel to it.	Check by ● drawing a 'crossing' line, ● measuring this pair of angles.

Write 'Yes, parallel'
    or 'No, not parallel'.

3

A

B

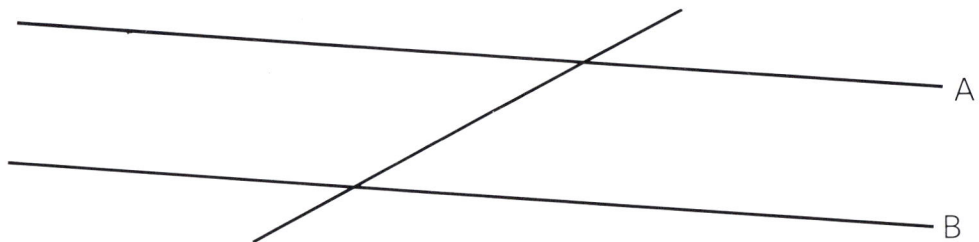

   a) Measure two angles to decide if lines A and B are parallel.

   b) On a sketch, mark with a ◀● the angles you measured.

   c) Are lines A and B parallel?

4 Use a protractor.
   Decide if these pairs of lines are parallel.

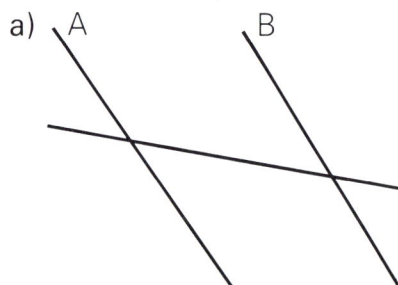

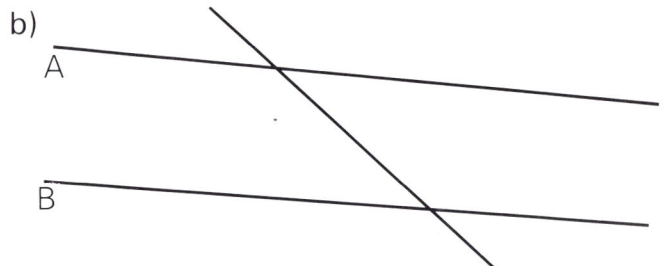

   a) A   B   **b)**

                                A

                                B

16

Write 'Yes' or 'No' for each pair.

144 ■

B 1

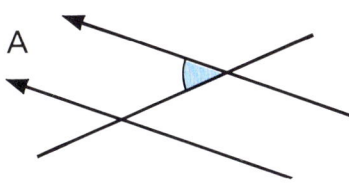

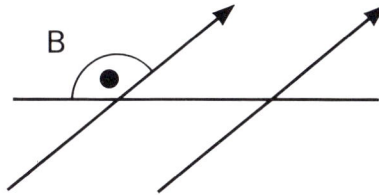

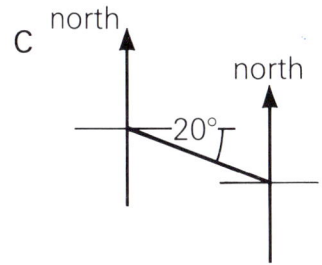

a) **Sketch** each diagram.
On diagram **A** mark three angles the same size as the one marked ▷.

b) On diagram **B**, mark four equal size angles, using 👁.

c) On diagram **C**, mark two 70° angles.

2

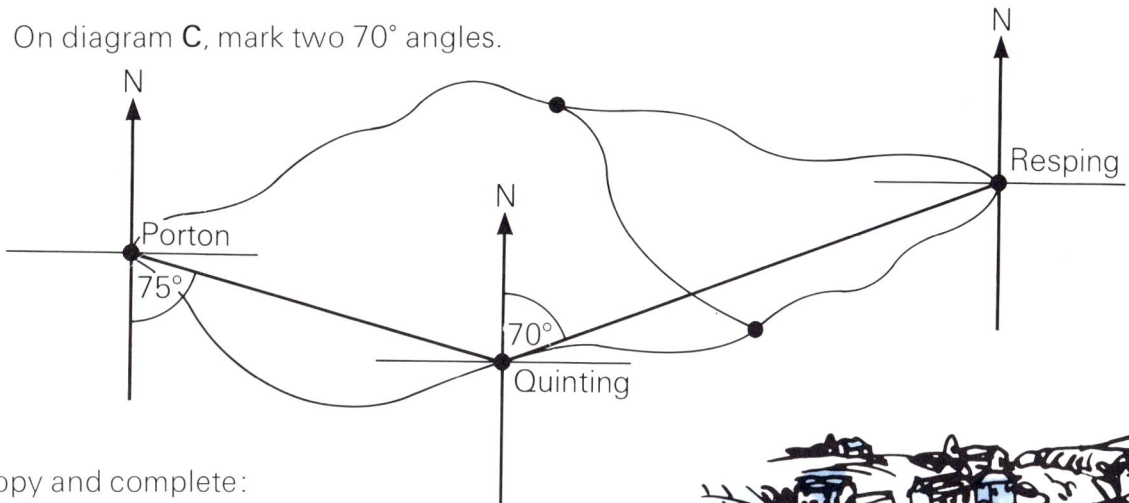

Copy and complete:

Direction of	from	
Quinting	Porton	S ☐° E
Porton	Quinting	N ☐° W
Resping	Quinting	N ☐° E
Quinting	Resping	S ☐° W

Think it through

3 a) Horace is in a rowing boat.
He is S 40° E of 'Wainsote' lighthouse.
In what direction is the lighthouse from Horace's boat?

b) Meg is in another rowing boat.
She is due north of the lighthouse.
Which of these **might** be her direction relative to Horace?

A N 40° W   B N 20° W   C N 60° W

▲ 146
● Next chapter

16

■ 145

Parallel lines

Challenge

1 Caldwell, Hemsworth and Roxton are three villages joined by straight roads.

Roxton is S 20° E of Caldwell.

Hemsworth is N 70° W of Roxton.

Caldwell is N 35° E of Hemsworth.

N

Caldwell

Hemsworth

What size is the angle between the roads at

a) Caldwell   b) Roxton   c) Hemsworth?

Roxton

2

T   R   P

B ——————————————————————— A
   62°   63°   62°

               62°   63°   61°

D —————————————————————————— C

U   S   Q

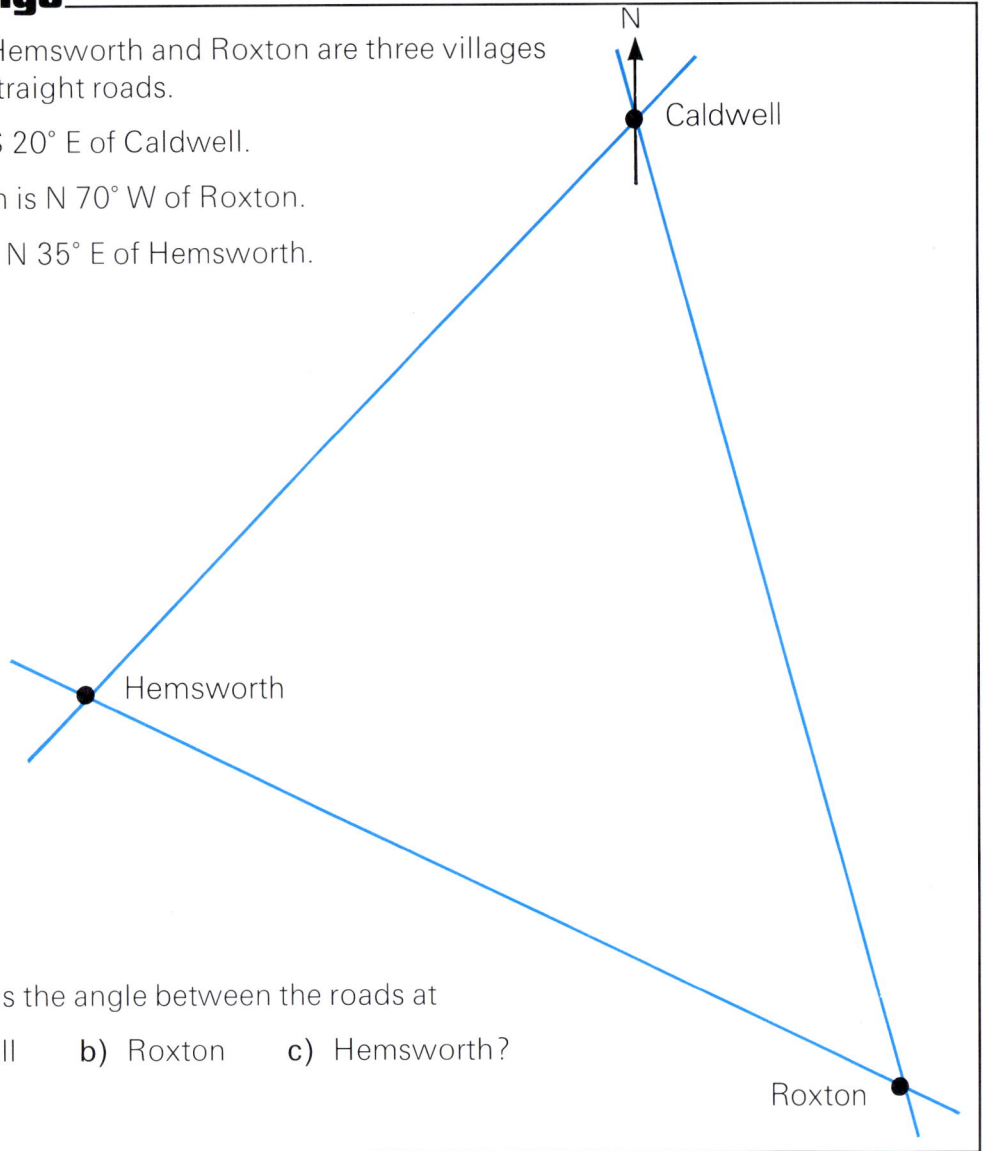

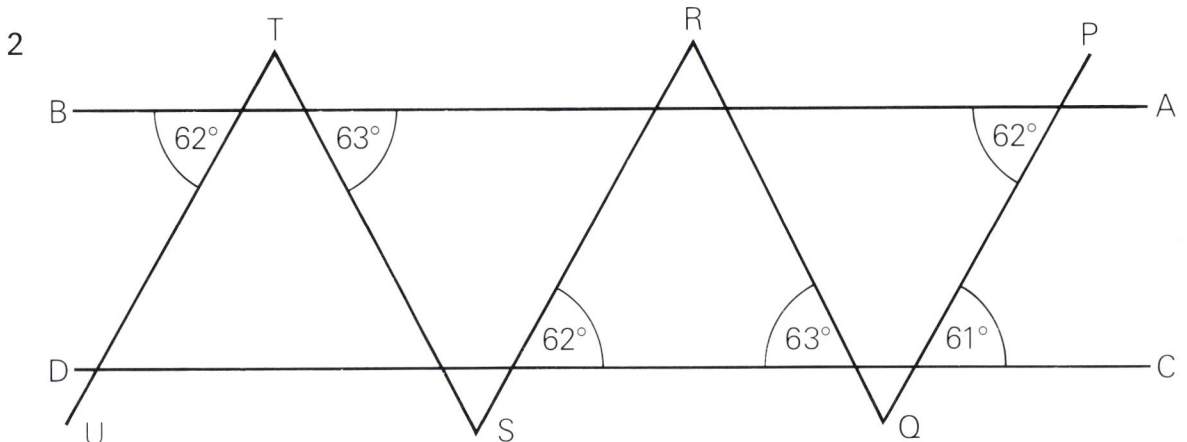

Name as many pairs of parallel lines as you can.

16

Think it through

3   Do not use a set square or a protractor.
Which of lines B to H are parallel to line A?

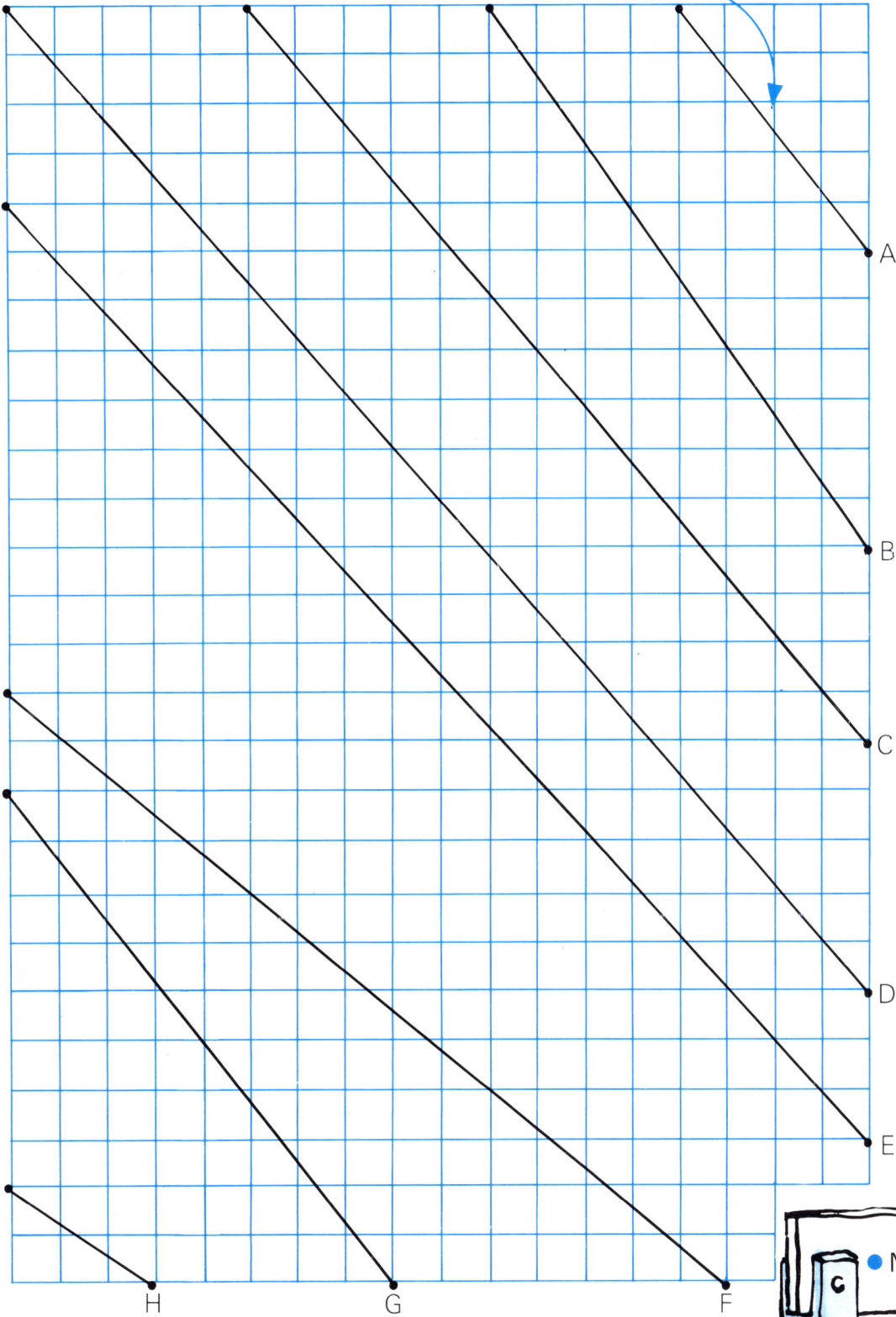

17 Rounding and estimating

A Challenge

1   The policeman is about 2 m tall.
    Estimate these distances, in metres:

    including his helmet

    a) the height of the bus

    b) the height of the lamp post

    c) the height of the tree on this side of the road

    d) the height of the tree on the far side of the road

    e) the height of the Bank

    f) the height of the bus stops

NMP BANK

2   Ben has given an **overestimate**
    of the giraffe's height.
    Meg has given an **underestimate**.

HE'S ABOUT 6m TALL!

HE'S ABOUT 5m TALL!

6 m

5 m

4 m

3 m

2 m

1 m

Write down
    (i) an overestimate
    (ii) an underestimate
         for each of these:

a) the rhino's height, in metres

2 m

1 m

b) Ben's mass, in
   kilograms

c) the mass of potatoes in
   the sack, in kilograms

d) the height of the
   tree, in metres

POTATOES
13.6kg

e) the age of the car,
   in years

A149 XYP

f) the area of the
   ink blot, in cm²

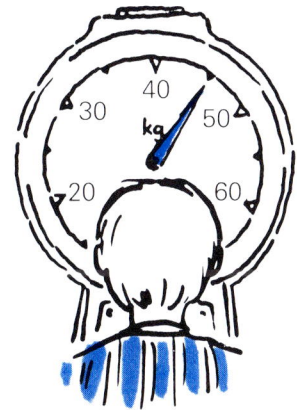

17

With a friend

3 Sometimes it is better to **overestimate** an amount.
Sometimes it is better to **underestimate**.
Decide between you which is more sensible for each of these.
Each of you write down why you made each choice.

b) the amount
of wallpaper needed to
paper a room

a) the height of a road bridge

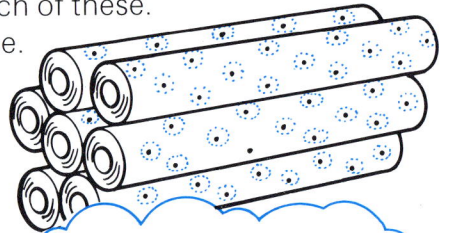

c) the time it
will take
to get to the
bus stop

e) the distance to record
for a javelin throw

d) the depth of water in a harbour

71·6 m

f) the time it will take
you to cross the road

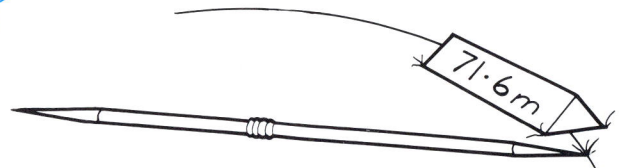

g) the time to record for a 100 m race

Challenge

4   Do **two** of A, B and C.

A   The distance from London to Miami is 4425 miles.
Here are four estimates of the flight time
for Concorde.

ABOUT 1 HOUR

ABOUT 12 HOURS

ABOUT 4 HOURS

ABOUT 8 HOURS

Which do you think is the best estimate?
Explain how you made your choice.

B   You need weighing scales and dried peas.
Here are four estimates for the number of dried peas in 1 kg.

ABOUT 400

ABOUT 2000

ABOUT 6000

ABOUT 800

Which do you think is the best estimate?
Explain how you made your choice.

C   Here are four estimates of the number of people in the picture.

ABOUT 500

ABOUT 1000

ABOUT 20000

ABOUT 6000

Which do you think is the best estimate?
Explain how you made your choice.

17

● 151

Take note

When we estimate we find rough answers.
The sign ≈ means 'is about'.

Llama's height ≈ 2 m.
Llama's mass ≈ 120 kg.

2 m

1 m

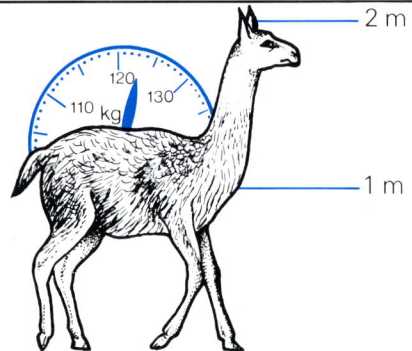

5   Copy and complete these for yourself.

My height ≈ _____ cm.
The length of my right arm ≈ _____ cm.
My weight ≈ _____ kg.
My age ≈ _____ years.
The volume of my head ≈ _____ cm³.
The number of times my heart beats
in 1 hour ≈ _____.

6   **Do these estimations in your head.**

a) Copy and complete:

Total mass of cheese ≈ _____ kg.
Total cost of cheese ≈ £_____.

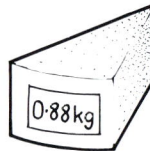

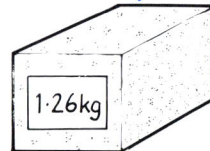

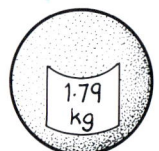

b) Write down if your estimates are
overestimates
or underestimates.

0.88 kg   1.26 kg   1.79 kg

£1.89   £3.16   £2.98

7   Phil's dad is a salesman.
He draws this map to show Phil
his weekly journey.

From Leeds to London is 193 miles.

a) About how far does Phil's
dad travel each week?
Do the estimate in your head.

b) Use an atlas.
Find out if you underestimated
or overestimated the
total distance.

START AND FINISH
LEEDS

BIRMINGHAM

LONDON

17

Rounding up and rounding down

B 1 Horace and Meg estimate the total cost
of the boots and scarf.

a) Explain how you think Horace reached his estimate.

b) Explain how you think Meg reached her estimate.

c) How would you make the estimate?
Write down what you would do and why.

d) Fiona estimates the cost of the boots and scarf:

Explain how you think she made her estimate.

e) Lenny estimates the cost:

Explain how you think he made his estimate.

f) Which of the four friends' estimates is nearest to the true cost?

Take note

There are many ways to estimate . . .

overestimate **round up**
each amount . . .

$£5.27 + £3.86 \approx £10$ *These are rounded up to the next pound.*

$£6$ $£4$

underestimate **round down**
each amount . . .

$£5.27 + £3.86 \approx £8$ *These are rounded down to the next pound.*

$£5$ $£3$

round down some amounts,
round up others . . .

$£5.27 + £3.86 \approx £9$

$£5$ $£4$

17

2 a) Estimate the total cost of the scarf and sweater
    (i) by rounding each amount up to a whole number of pounds.
    (ii) by rounding each amount down to a whole number of pounds.

b) Which estimate is nearer the true cost?

CLASSIC CLOTHING FROM BLUEBERRYS

Cashmere scarf
ONLY
£29.95

Lambswool sweater
ONLY
£23.95

3 a) This record has four tracks on each side.

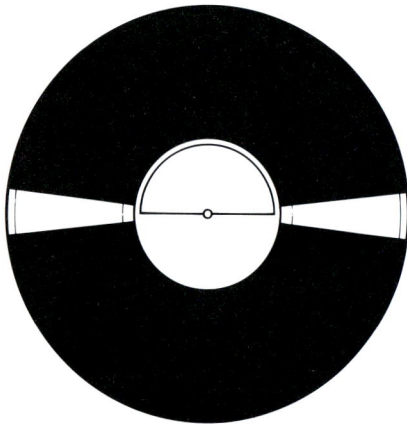

Estimate the total time for
(i) side one   (ii) side two.

b) Write down whether each estimate is
    an overestimate
or an underestimate.

SIDE ONE	
Every picture tells a story	5:58
Seems like a long time	4:00
That's all right	6:02
Tomorrow is another day	3:44

SIDE TWO	
Maggie May	5:46
Mandolin wind	5:32
I'm losing you	5:22
Reason to believe	4:07

Think it through

4 a) Estimate the total length of wood needed for this picture frame.

b) If you were planning to make the frame, would you
    round each distance up
or round each distance down
or do something else?
Why?

c) Suppose you now actually have to make the frame. What length of wood would you order? Explain how you made the calculation.

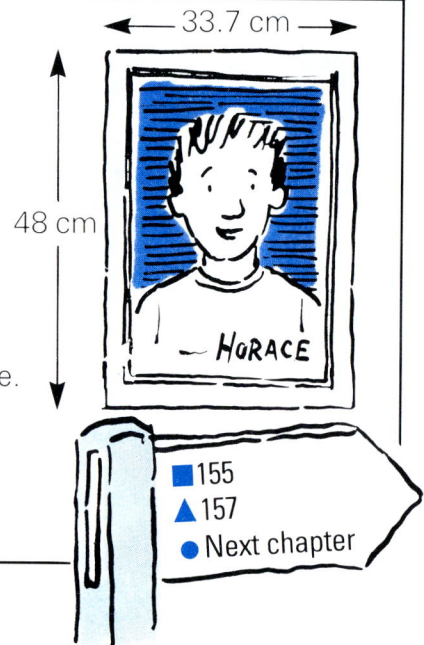

← 33.7 cm →

48 cm

HORACE

■155
▲157
● Next chapter

17

Sensible estimates

Activity: *Measuring fingers*

1 a) Measure the length of each of your fingers.
       Draw two diagrams.

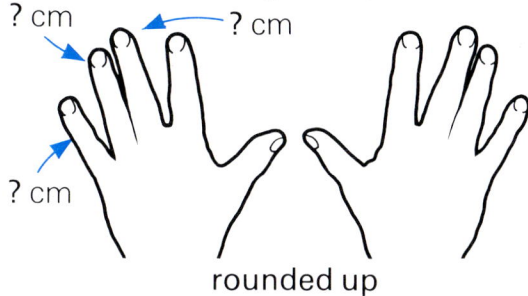

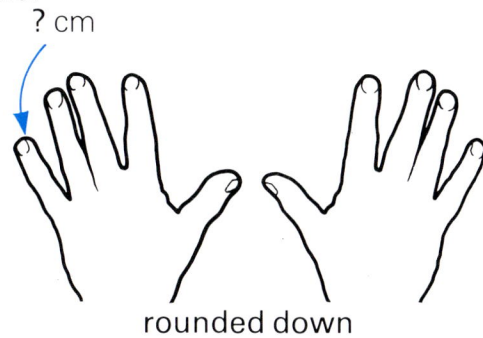

? cm

? cm

? cm

? cm

rounded up   rounded down

On one, round up the lengths to a whole number of centimetres.
On the other, round down the lengths to the next centimetre.

Measure in between your fingers like this.

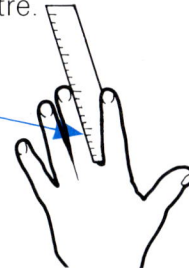

   b) Your grandma is knitting
       you a new pair of gloves.
       Which set of measurements
       would you give her?

2 a) **A** You have measured your bedroom.
          You need $6\frac{1}{2}$ rolls of wallpaper.
          Would you round up
                  or round down
          to find the number of rolls to buy?
          Why?

       B You are walking into town
          to meet your friend.
          Should you overestimate
                  or underestimate
          the time this will take?
          Why?

       C You have built a gateway
          into a car park.
          You have measured the
          clearance as 2.76 m.
          Would you write 2.8 m
                  or 2.7 m
          on the headboard?
          Why?

CLEARANCE 2·?m

   b) Write down **two** examples of your own where you need to
       overestimate an amount.

   c) Write down **two** examples of your own where you need to
       underestimate an amount.

17

155

3 **a)** Estimate the distance around the M25
  (i) by rounding up each section distance.
  (ii) by rounding down each section distance.
  (iii) by rounding some distances up and some down.

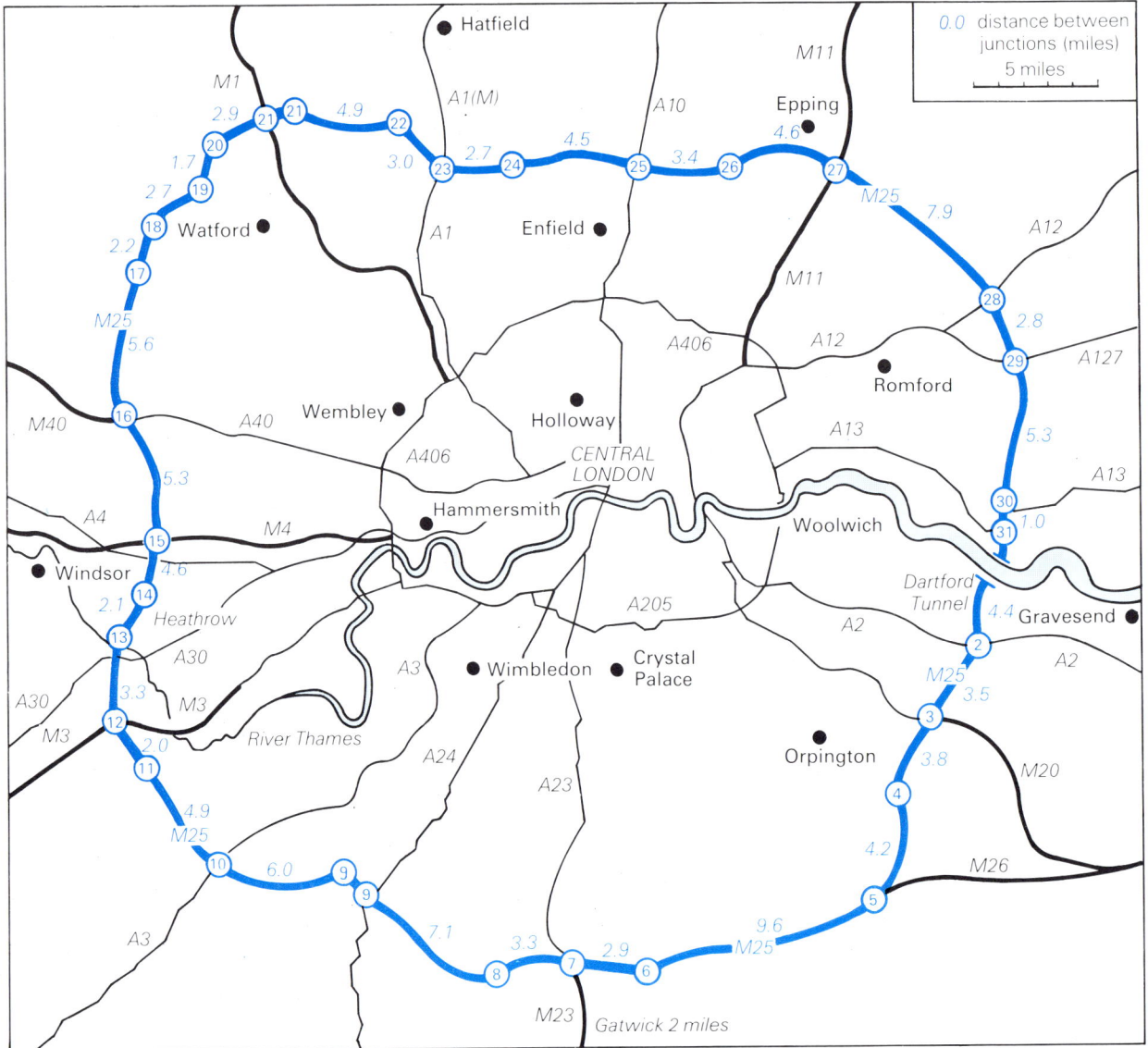

b) Use your calculator.
Calculate the distance
using the figures on the map.

c) Which of your estimates
is nearest to this total?

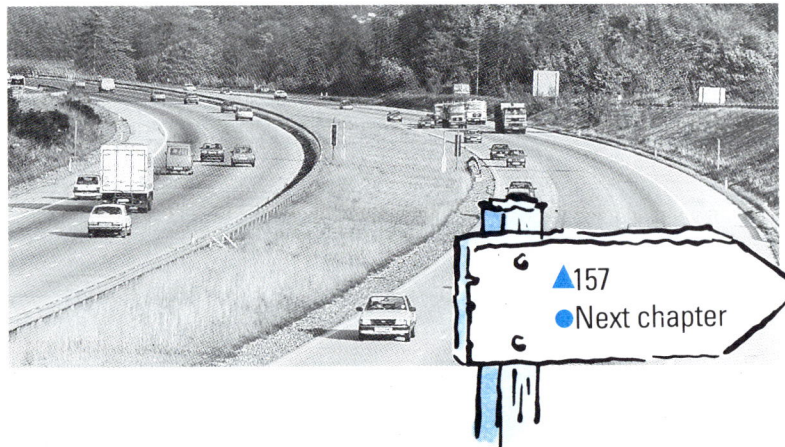

To the nearest . . .

C 1 Meg estimates the total amount of milk
in the containers, like this:

nearer 1 l than 2 l

nearer 3 l than 2 l

nearer 4 l than 3 l

MILK 1.1 ℓ

MILK NMP DAIRIES 2.7 ℓ

MILK NMP DAIRIES 3.8 ℓ

$$1.1 l + 2.7 l + 3.8 l$$

1 l   *3 l*   *4 l*

That's about 8 l

She rounds each amount to the nearest whole litre . . .
then she adds.

Use her method to estimate:

a) the total mass of steel

2.7 tonne

1.4 tonne

5.9 tonne

b) the total distance around
this circuit

Codford — 7.2 km — Bleaksop

4.8 km

5.2 km

Staunton

With a friend

2 Discuss this question with your friend.
Each of you write down what you decide.

Here are three estimates for the total
distance around this circuit.

ABOUT 44 km

ABOUT 42 km

ABOUT 45 km

Horace

Midge

Lenny

Corshall

13.6 km

18.2 km

Bathweston

11.7 km

Chippenage

a) Are they overestimates
or underestimates?

b) Explain how you think each estimate was made.

3 Discuss **two** of these situations.
Each of you write down what you decide.

A COFFEE LABELS

a) Do you think the amount of coffee
shown on the label is
    a rounded up figure
or a rounded down figure?

b) Explain why you think this.

c) What do you think is the largest amount
of coffee you might find in a jar?

COFFEE

550g

B 100 m RACE

a) Do you think the time given for the 100 m is
    a rounded up figure
or a rounded down figure?

b) Explain why you think this.

100 m FINAL
Winner: J. Jenks
9.96 s

C SHOT PUT

SHOT FINAL
Winner: K. Armstrong
20.24 m

a) Do you think the distance
given for the shot put is
    a rounded up figure
or a rounded down figure?

b) Explain why you think this.

c) How many millimetres longer do you think the winner's
throw might have been than the distance on the scoreboard?

17

● Next chapter

18 Thinking about area

A *With a friend:* Area exploration

1 You need squared paper.
Work together on this exploration.
Each of you record what you discover.

Jim has 12 fencing panels.
Each one is 1 m long.
He is building an enclosure
for his goat.

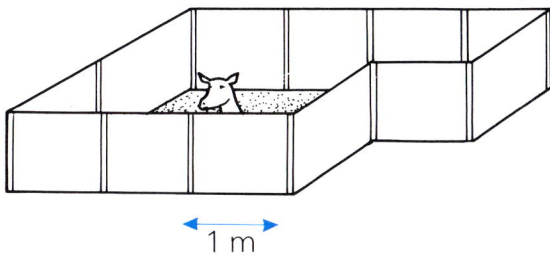

DO YOU REMEMBER ...?

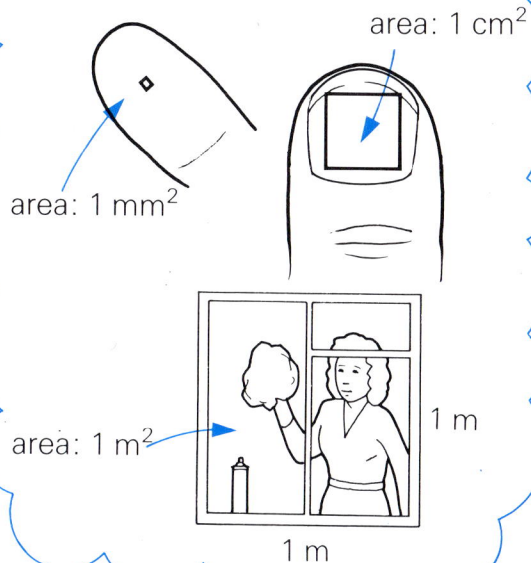

area: 1 mm^2

area: 1 cm^2

area: 1 m^2

1 m

1 m

1 m

The panels have special fixing brackets.
Jim can fix them in line with each other
... or at right angles to each other
... but **not** like this.

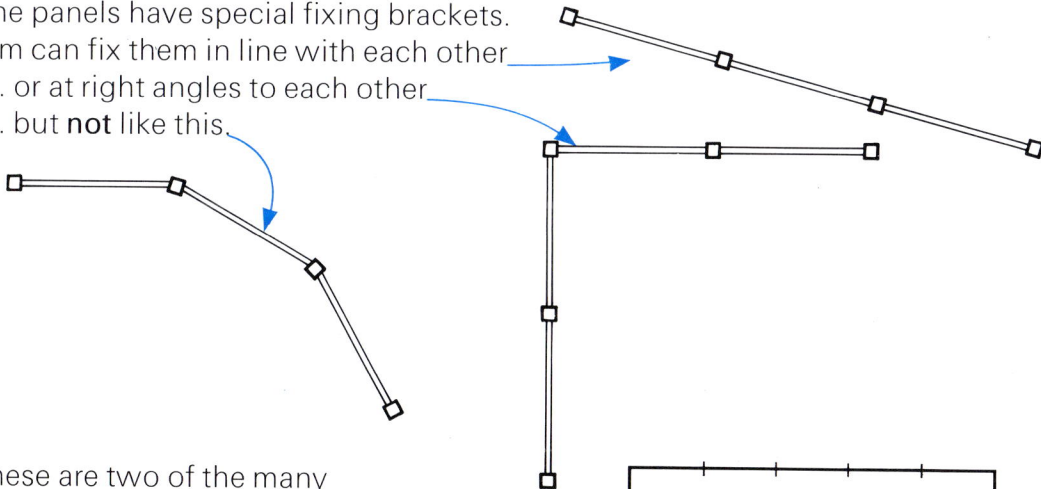

These are two of the many
shapes Jim can make.

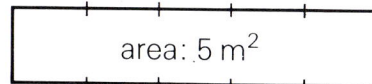

area: 5 m^2

a) Which of the shapes Jim can make
will give the largest area for his goat?
Draw it.

b) Investigate for other numbers of panels.
Try to find a rule which tells you
which shape gives the largest area.

area: 6 m^2

● 159

2 Horace has a pile of 1 cm square tiles.

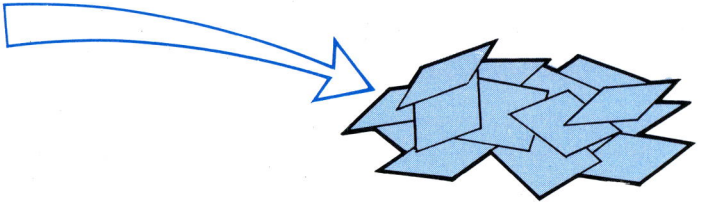

He covers this label with them.

a) Which of these tell you how many tiles he will use?
   A 4 + 5   B 4 × 5   C 5 + 4   D 5 × 4

b) Which of these tell you the area of the BUY NEW WOOL label?
   A 4 + 5 cm²   B 4 × 5 cm²   C 5 + 4 cm²   D 5 × 4 cm²

c) Horace will need 4 × 7 tiles to cover this label.
   Explain why.

d) What is the area of the label in cm²?

e) Write down the area of each of these labels.
   Write each area using ×, like this:
   2 × 7 cm², 8 × 4 cm², ...
   Measure distances with a ruler.

f) What is the perimeter of each label?

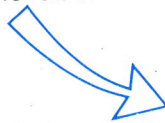

CRESTA RUN

50% cotton
50% polyes

DRY CLEAN

SPEC

GO BY TRAIN

placed below.

3 Copy and complete the **Take note**.

Take note

a) The area of
the rectangle is

$\underline{\square \times 7}$ cm²

$= \underline{7 \times \square}$ cm²

$= \square$ cm².

in cm² *in cm* *in cm*

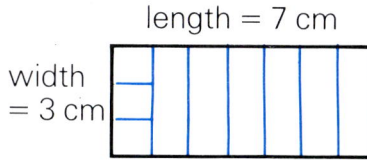
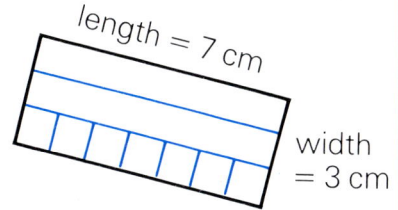

b) Area of rectangle = length × _____.

4 Without measuring, estimate in m²

a) the area of the floor of your classroom.

b) the area of the door to your classroom.

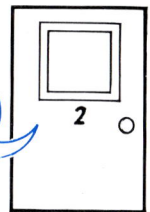

c) the area of the longest wall in your classroom.

Write your estimates like this: Area $\approx \underline{\square \times \square}$ m² $= \square$ m²

5 In this stamp you will find **nine** rectangles.

Name each one.
Write down the area of each rectangle in mm².

ABHG, ACIG, ...

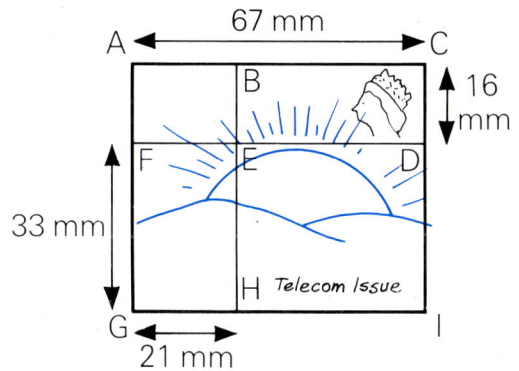

Challenge

6 These are helicopter views of two buildings.
What is the area of each roof?

a)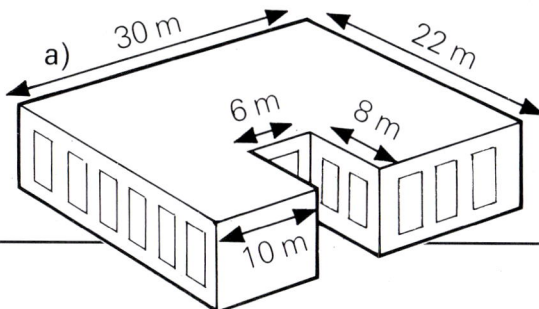

b)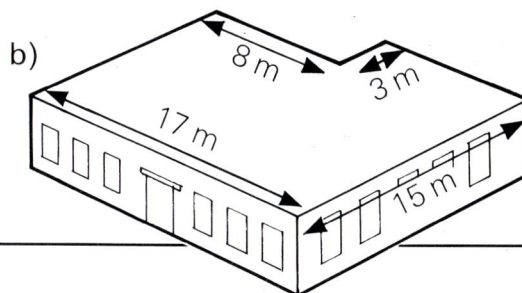

Hectares

B 1 The field measures 100 m × 100 m.
What is its area in m²?

Take note

An area of 10 000 m² is called a **hectare**.

We write: 1 ha.

One **hectare** is about the size
of 1½ football pitches.

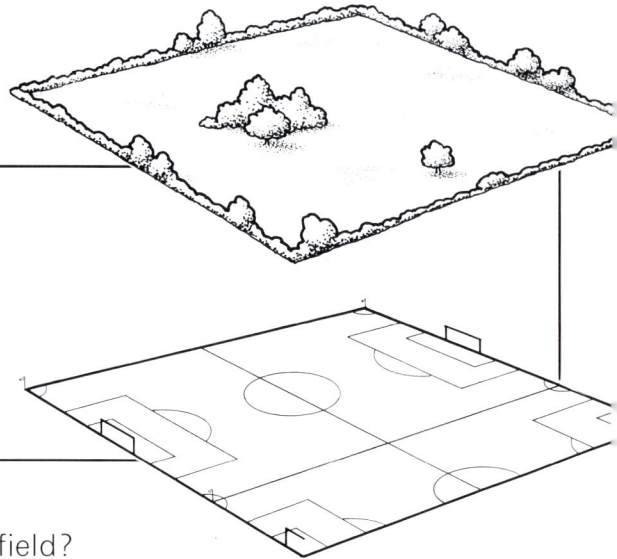

2 a) About how many hectares is this playing field?

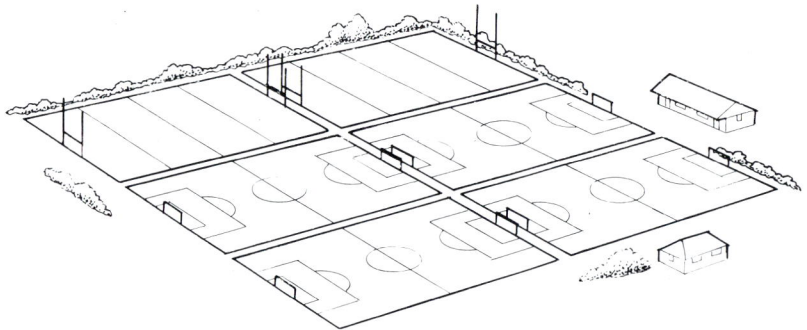

b) Make your own drawing of a playing field.
The field must have lots of football, hockey and rugby pitches.
Its area should be about 12 hectares.

With a friend: Scramble car park

3 a) Think of a playing field you both know.
Make a sketch of it.
Estimate its area in hectares.

b) The car park for a motorcycle scramble
meeting is a 1 hectare field.
About how many cars
do you think can be parked on it?
Don't forget . . .
● cars must be able to drive around the park
easily;
● there must be room to open the car doors.

You will need to estimate
● *how much space each car needs*
● *how much space is needed for driving around*
● *how much space to leave between cars.*

Problems with area

C 1 Use squared paper.
Draw **three** different rectangles, each with an area of 10 cm².

2 A square lawn has an area of 144 m².
How long is each side?

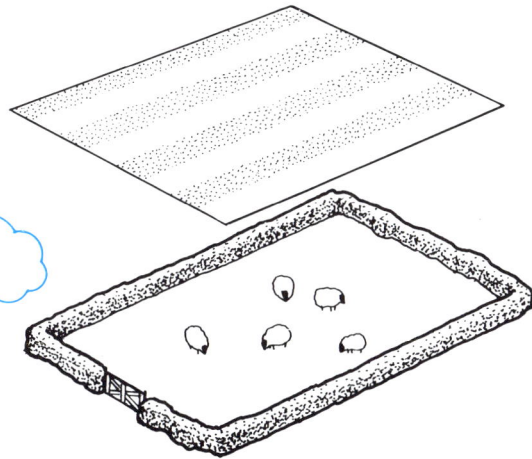

3 The area of a rectangular field is 2 ha.

hectares

a) How many square
metres is this?

b) The width of the
field is 50 m.
How long is it?

4 Advertisements in a local newspaper cost £15 per cm².
Estimate the cost of each of these.

a)

SUPAFRAMES

Britain's largest producer
of uPVC window and
door systems

Sole manufacturers:
W. Payne Limited
Unit 5, Greenhills Industrial
Estate, Glastown
Tel: 06543 11722

b)

WEST LONDON

New Self-Contained
Air-Conditioned
Office Development
10,000 sq. ft.

FREEHOLD FOR SALE

Apply Box No: T 6281

5 Use squared paper.
Draw these rectangles.

a) Perimeter: 16 cm
Area: 15 cm²

b) Length: 6.5 cm
Area: 26 cm²

6 About how many tins of paint are needed to paint the side of the hotel?

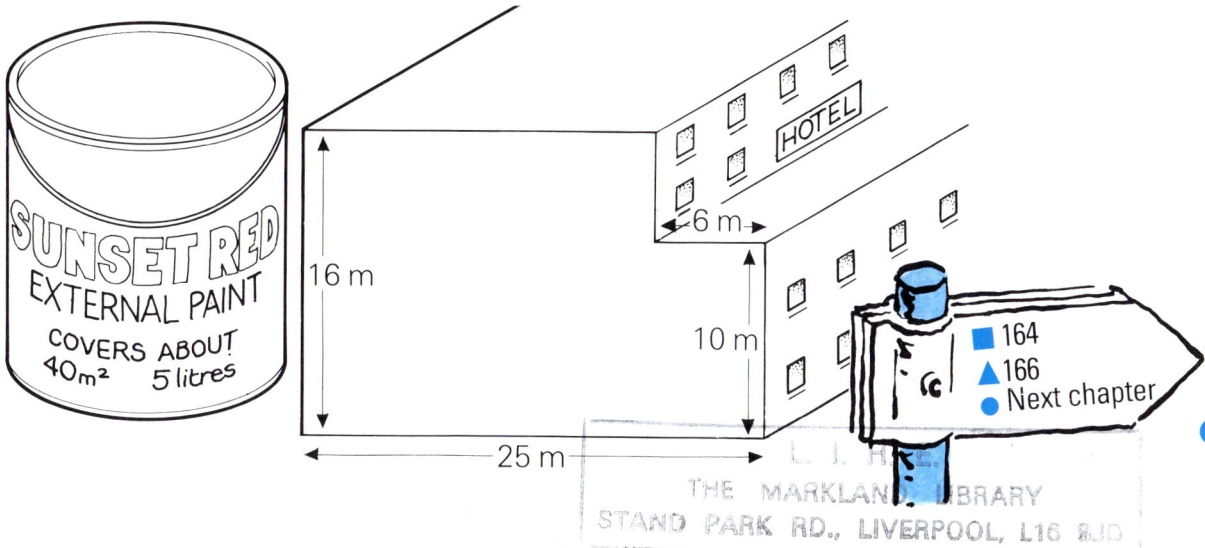

164
166
Next chapter

Rectangles and L-shapes

A 1 Parts of each rectangle have been cut away.

What is the area of each whole rectangle?

a)

b)

c)

d)

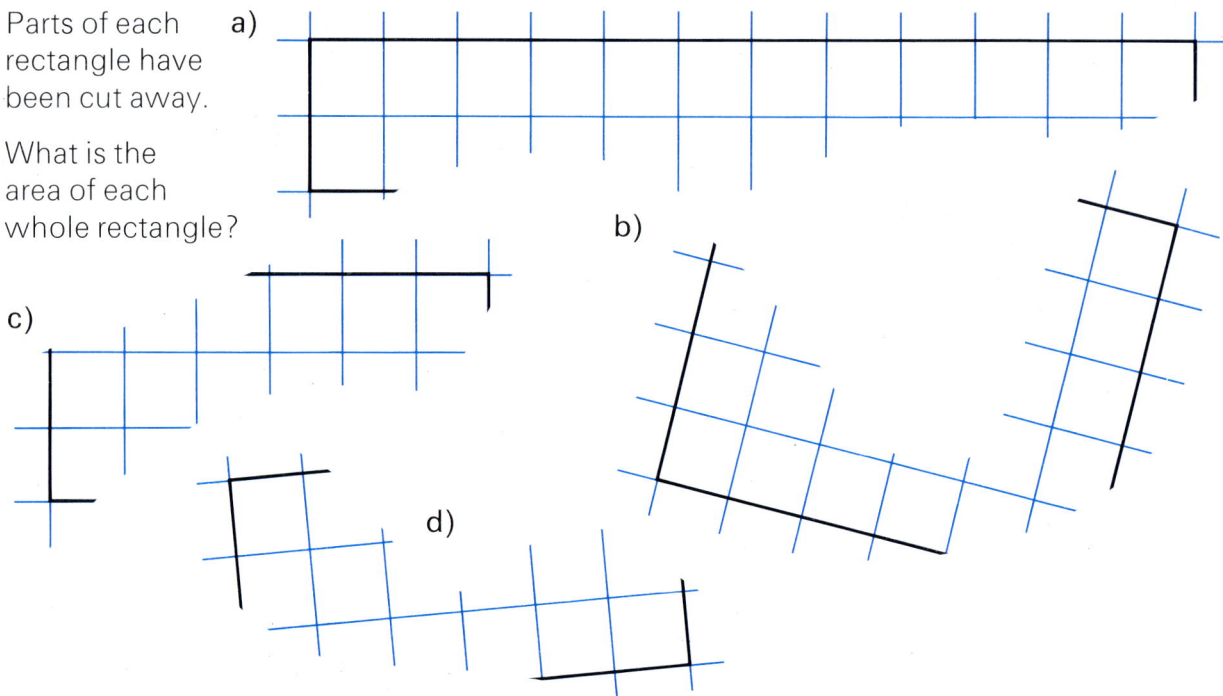

2 a) Horace's teacher said:

DESIGN A LETTER L. MAKE ITS AREA 11 cm².

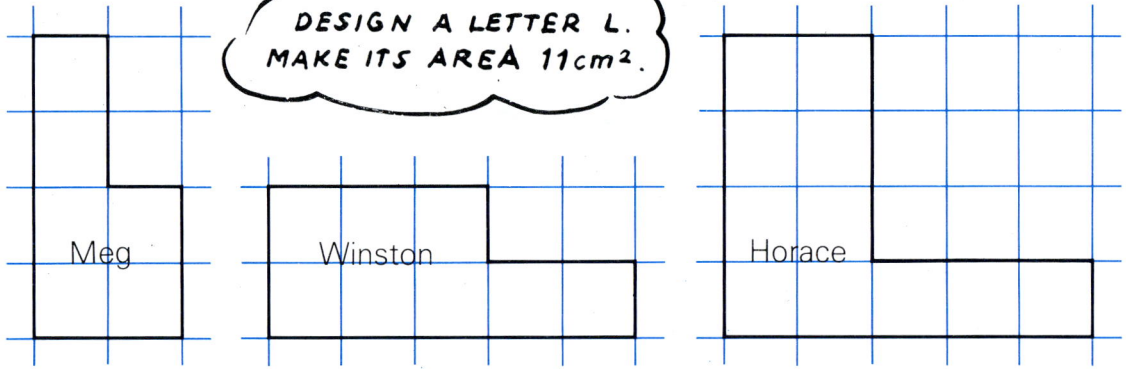

Meg

Winston

Horace

Here are three attempts.
Who is correct?

b) Each of these letter **L**s will have an area of 18 cm².
Copy and complete each sketch.
Mark the length of each part in centimetres.

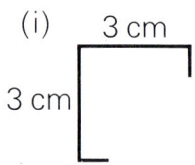

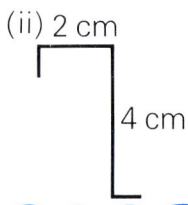

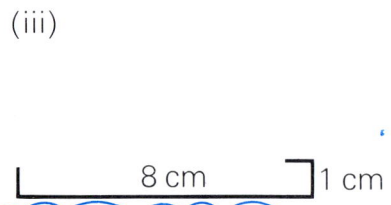

(i) 3 cm
3 cm

(ii) 2 cm
4 cm

(iii)
8 cm ⌐1 cm

3 Use squared paper.
Draw these rectangles.

a) Area: 18 cm²
Width: 3 cm

b) Area: 8 cm²
Perimeter: 12 cm

Larger areas

B 1 a) Copy each field onto squared paper.
Use a scale of 1 cm to 50 m.

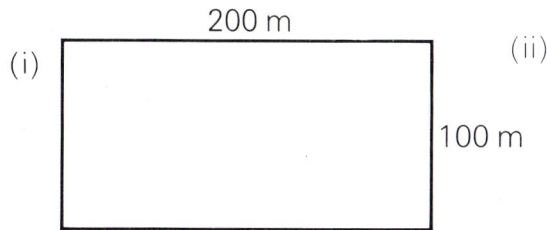

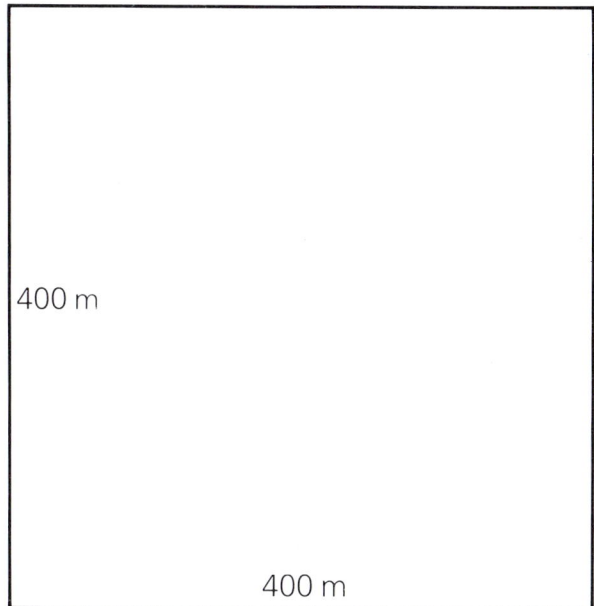

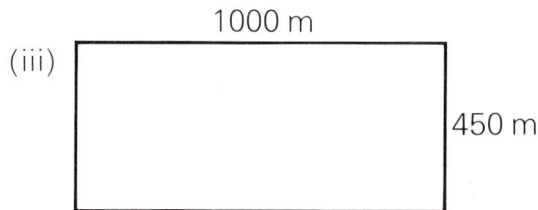

(i)

200 m

100 m

(ii)

400 m

400 m

(iii)

1000 m

450 m

b) Divide each field into hectares.
Write down the area
of each field in hectares.

Examples of a hectare:

100 m

100 m | 1 hectare
(1 ha)

200 m

1 hectare (1 ha) | 50 m

1 hectare is about 1½ football pitches.

c) On squared paper, draw these rectangular fields.

(i) Area: 7 hectares
Length: 700 m

(ii) Area: 10 hectares
Length: 400 m

Write down the scale you used for each field.

Think it through

2 About how many hectares is Lower Ditch building site?

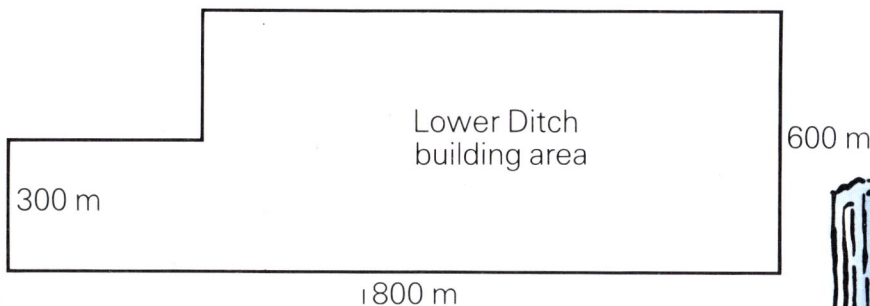

Lower Ditch
building area

600 m

300 m

800 m

▲ 166
● Next chapter

Triangles

D Challenge

1 You need 1 cm squared paper.
   This is a plan of Meg's bedroom.

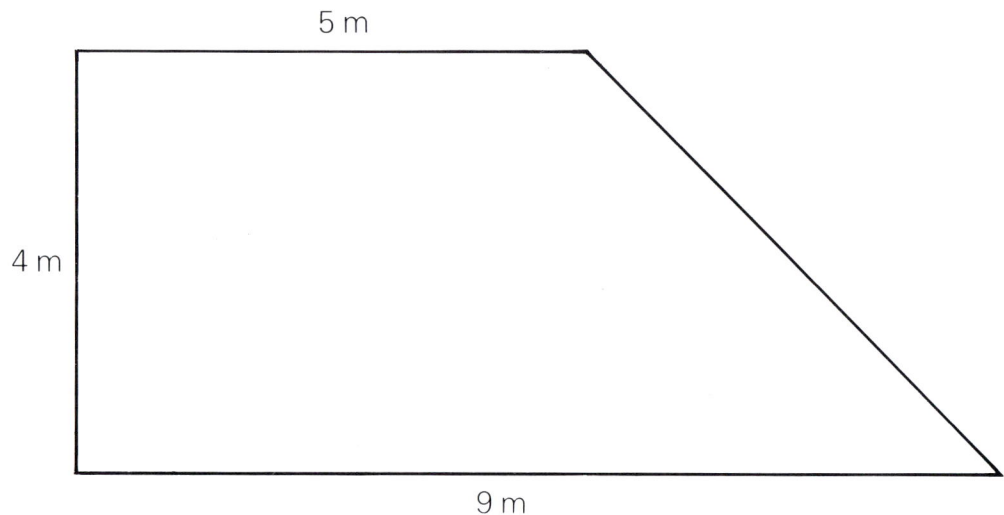

5 m

4 m

9 m

Make a scale drawing of it on squared paper.
Choose your own scale.
Use your drawing to find the area of the bedroom in m².

2 a) The teacher asks Glenda's class to draw triangles.

> THEIR AREA MUST BE 6 cm².

Here are some attempts.
Three of them are correct.
Which three?

b) This is Pepper's triangle.
   It is covered by
   her book.
   Its area is correct.
   Draw it full size.

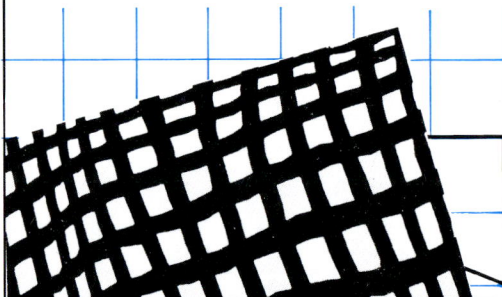

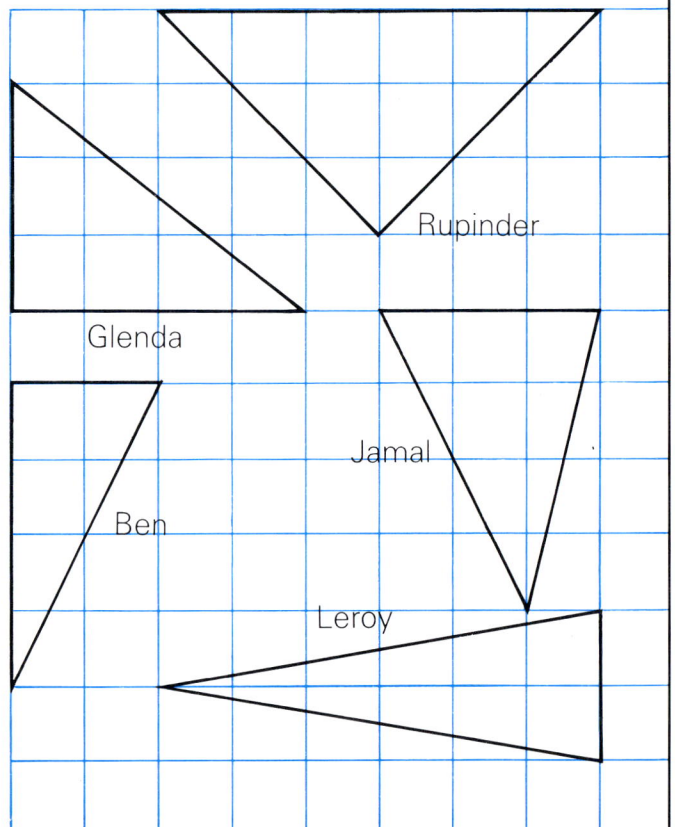

Rupinder

Glenda

Jamal

Ben

Leroy

Activity: *Triangle areas*

3 You need a pair of scissors and some squared paper.

a) On squared paper, make a drawing like Ⓐ.
   Follow the big arrows.
   Do what the drawings tell you.

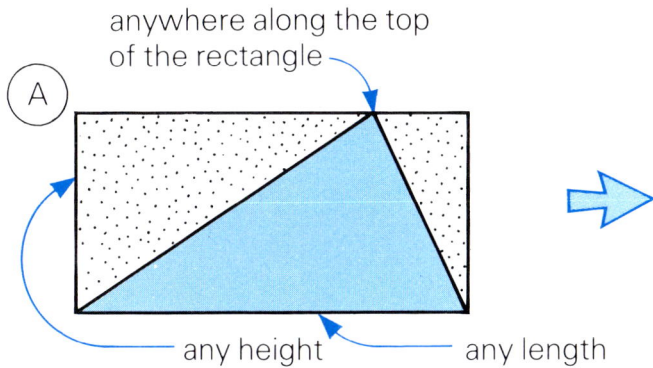

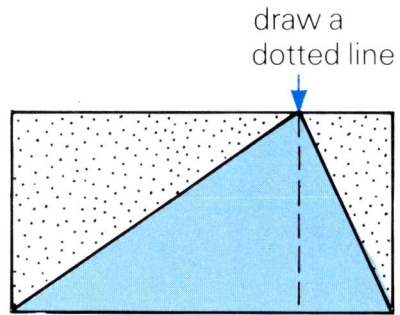

Ⓐ

anywhere along the top of the rectangle

draw a dotted line

any height — any length

b) The drawings tell you something about the area of the coloured triangle

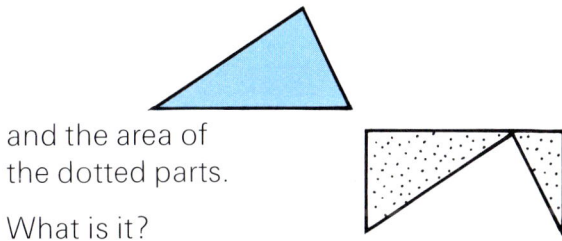

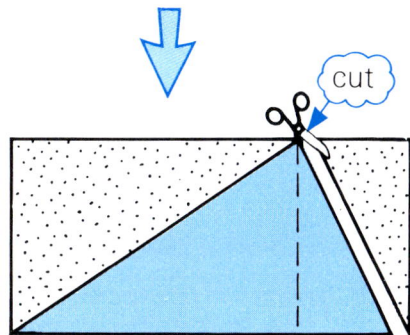

and the area of the dotted parts.

What is it?

cut

fit the dotted triangle here

c) The drawings tell you something about the area of the rectangle

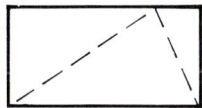

and the area of the coloured triangle.

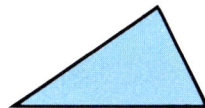

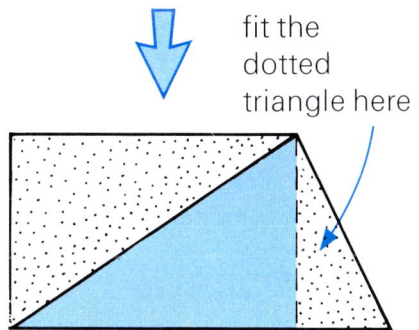

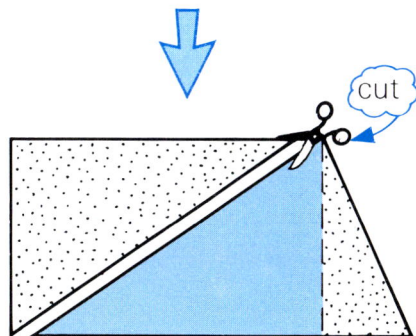

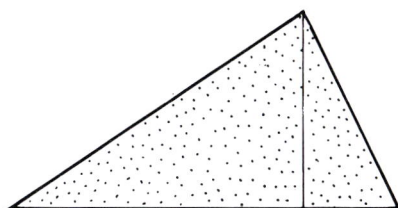

What is it?

cut

4 Copy and complete the sentence
'The area of . . .' in the *Take note*.

5 The triangles are drawn on 1 cm squared paper.
What is the area of each triangle?

a)

b)

Take note

To find the area of
a triangle . . .

. . . draw a rectangle
around it.

The area of the triangle
is _____ the area
of the rectangle.

Challenge

6 **a)** These rectangles are all exactly the same size.
List the triangles in order of area.
Start with the one with smallest area.

Measure with a ruler, if you wish.

A B C D E F

b) What can you say
about the area
of rectangles
(i) and (ii)?

Write down
the exact area
of each one.

(i)

(ii)

7 Draw a triangle
for each of these
instructions.

a)
Area: 10 cm².
One side 5 cm long
One right angle

b)
Area: 12 cm²
One side 3 cm long
No right angles

c)
Area: 16 cm²
Height: 4 cm

height

● Next chapter

19 Using proportion

A 1 Basil makes good coffee.
He serves it in mugs or cups.

Mugs hold twice as much as cups.

Midge has a mug.
She takes 2 lumps of
sugar and half a
carton of cream.

Horace has a cup.
He takes 2 lumps of sugar and
a full carton of cream.

a) Whose coffee is sweeter, Midge's or Horace's?

b) Whose coffee is creamier, Midge's or Horace's?

2 The next time
Midge has a cup

. . . and Horace has a mug.

a) How much sugar
and how much cream
should Midge have?

b) How much sugar
and how much cream
should Horace have?

Think it through

3 This is what some of Basil's friends take in their coffee.

Meg

2 lumps
$1\frac{1}{2}$ cartons
of cream

Lenny

No sugar
2 cartons
of cream

Fiona

3 lumps
$\frac{1}{2}$ carton
of cream

Midge

2 lumps
$\frac{1}{2}$ carton
of cream

Horace

2 lumps
1 carton
of cream

Winston

4 lumps
No cream

Lou

4 lumps
$2\frac{1}{2}$ cartons
of cream

Maisie

No sugar
$1\frac{1}{2}$ cartons
of cream

a) Who has sweeter coffee than Meg? b) Who has creamier coffee than Meg?
c) Who has the sweetest coffee? d) Who has the creamiest coffee?

4 a) Basil often uses this recipe for coffee.
   He wants to make enough for 16 cups.
   How many coffee bags does he need?

   b) How many bags does he need for 16 mugs?

Basilian Coffee

Pour 4 cups of boiling water onto 3 coffee bags.

5 Basil is making four cups of coffee.
   Another friend arrives.

   He adds

   > another cup of water
   > and another coffee bag.

   Is the result

   > weaker than usual,
   > the same as usual,
   > or stronger than usual?

6 Basil and all his friends
   turn up at Midge's house
   . . . for coffee.
   Midge makes a large jugful.
   She uses this recipe.

   Does it make

   > weaker coffee than Basil's,
   > the same strength coffee as Basil's,
   > or stronger coffee than Basil's?

To make a jugful

Pour 12 cups of boiling water onto 10 coffee bags.

With a friend

7   Here are some more recipes for coffee.

a) Decide between you which ones
make stronger coffee than Basil's.

Basilian Coffee
4 cups of water
3 coffee bags

A
4 cups of water
5 coffee bags

B
3 cups of water
3 coffee bags

C
8 cups of water
7 coffee bags

D
40 cups of water
39 coffee bags

E
20 cups of water
12 coffee bags

F
2 cups of water
3 coffee bags

G
7 cups of water
6 coffee bags

b) Which recipes make weaker coffee than Basil's?

c) Write a recipe of your own.
It should make stronger coffee than Basil's,
but weaker than recipe **C**.
You can make it for as many cups of
coffee as you like.

4 cups; 3 coffee bags

8 cups; 6 coffee bags …

Take note

Coffees which have the same strength
use the same **proportion** of water and coffee.

In proportion

B 1   1 cup holds   25 cl of coffee.
2 cups hold   50 cl.
5 cups hold 125 cl.
12 cups hold   ?

How many centilitres do 12 cups hold?

Take note

The number of cups and the amount of coffee are **in proportion**.

2   10 cm of copper tubing weighs 220 g.
15 cm weighs 330 g.
30 cm weighs 660 g.
40 cm weighs   ?

a) How many grams does 40 cm weigh?

b) Copy and complete the **Take note**.

Take note

The _____ and the mass of the tubing are **in proportion**.

3   At 1 year old Ben weighed 11 kg.
At 2 years old he weighed 14 kg.
At 3 years old he weighed 18 kg
At 4 years old he weighed   ?

a) Why can't we fill in what is missing?

b) Copy and complete the **Take note**.

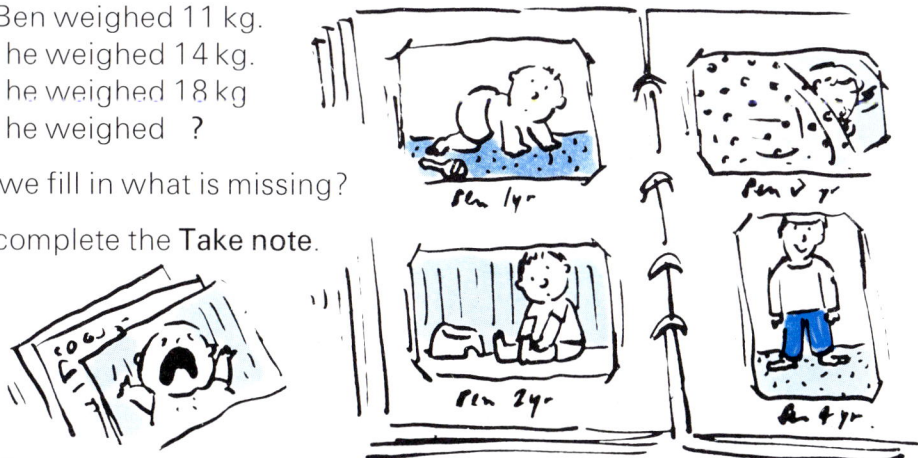

Take note

Ben's _____ and his _____ are **not** in proportion.

4 How many bottles of Makes 5 do you need for

   a) 10 pints of milk?

   b) 20 pints of milk?

   c) 35 pints of milk?

5 Glenda made 2½ pints of milk.
   How much Makes 5 did she use?

6 Here are the directions for use:

DIRECTIONS FOR USE

To make one pint

- add 2 oz (57 g or approx 4 heaped tablespoons)
  to a little cold water.
- Whisk until dissolved and
  make up to one pint with more water.

Once made up,

- store and use in the same way as fresh milk.

ounces

   a) How many oz of Makes 5 do you need for
      (i) 2 pints of milk?
      (ii) 5 pints of milk?

   b) How many grams of powder do you think the bottle holds?

   c) About how many heaped tablespoonfuls of Makes 5 do you
      need for 5 pints?

7 Horace made 4 pints of milk.
   He used 12 oz of powder.
   Was his milk too weak or too strong?

8 Basil made 6 pints of milk.
   He used 400 g of powder.
   Was his milk too weak or too strong?

■ 175
▲ 177
● Next chapter

Mixtures

A 1 Glenda and Ben are mixing paint.
They are painting Glenda's house . . . light blue.

Glenda mixes
2 tins of white
with 5 tins of blue.

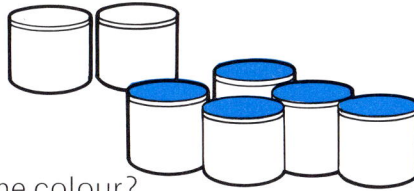

Ben mixes
1 tin of white
with 3 tins of blue.

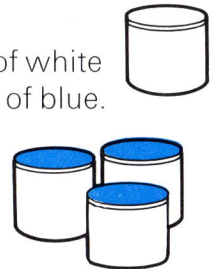

a) Will they get the same colour?
b) What can Ben do now to get the same colour as Glenda?

2 a) Two of these mixtures will give the same colour as each other.
Which are they?

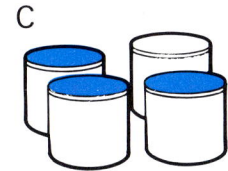

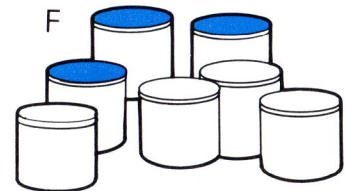

A B C

D E F

b) Two mixtures will give a darker colour than **D**.
Which are they?

Think it through

c) You want mixture **C** to have the same shade as **E**.
What is the smallest number of tins you must add?
Say what colour they should be.

3 Which pancakes will taste more strongly of egg?

A

Pancakes

For 6 pancakes

$\frac{1}{2}$ cup pancake mix
$\frac{1}{2}$ cup milk
1 egg

B

Pancakes

For 12 pancakes

1 cup pancake mix
1 cup milk
1 egg

In proportion . . . or not?

B 1 Basil takes 2 lumps of sugar and 1 teaspoon of dried milk in a cup of tea.
He is making up a flask.
It holds 7 cups.

a) How much sugar should he use?

b) How much milk should he use?

c) When he makes tea in his pint mug he uses 5 lumps of sugar.
How much milk do you think he uses?

2 Three cassettes cost £4.50.

a) How much do 12 cost?

b) Ben paid £13.50 for cassettes.
How many did he buy?

3 Which are **true** and which are **false**?

A The weight of a baby is in proportion to its age.

B The total number of horses' legs is in proportion to the number of horses.

C The number of hairs on your head is in proportion to the number of days you have lived.

D The weight of water is in proportion to its volume.

4 Write down your own example of two things which

a) are in proportion.

b) are not in proportion.

5 Copy and complete:

3 tins of paint cover	18 m²
6 tins cover	?
? tins cover	24 m²
7 tins cover	?

▲177
● Next chapter

Estimating costs

C 1 **Do these in your head.**

Write down your answers.

a)
ROSE BUSHES

£2.96 each

About how much would you expect to pay for 30 bushes?

*Think:
About £3 each.
That's about . . .*

b)
ORANGES

5 for 93p

About how much would you expect to pay for 20 oranges?

c)
TAPES

£1.47 each

THAT'S ABOUT £9 FOR SIX.

Is Midge's price an overestimate or an underestimate?

d)
STICKERS

BIG DADDY   I'VE GOT THE BUG   MY GEAR

3 for £2.42

THAT'S ABOUT £4.80 FOR SIX.

Is Basil's price an overestimate or an underestimate?

e)
BREAD ROLLS

5 for 48p

(i) About how much is this each?

(ii) About how much would you expect to pay for 12 rolls?

Thinking proportionally

D *With a friend*

Do these questions together.
Each of you write down your answers.

1 Is Sam's offer really a bargain?
Explain why.

2 Three of Glenda's jam tarts weigh 85 g altogether.

a) Explain how you would work out how much 12 weigh.

b) About how much do 12 weigh?

3 A recipe for devilled lamb sauce says:

> Use 3 tablespoons of French mustard
> to 120 g of brown sugar.

a) Explain how to find out how many tablespoons of French mustard
you need for 200 g of brown sugar.

b) How many do you need?

4 If you owned the garden shop, how much would you charge for

a) 100 roses?

b) 10 rose bushes?

c) 12 plant pots?

● Next chapter

20 Using letters

A 1 Ben has designed some garden steps.
This is his design.

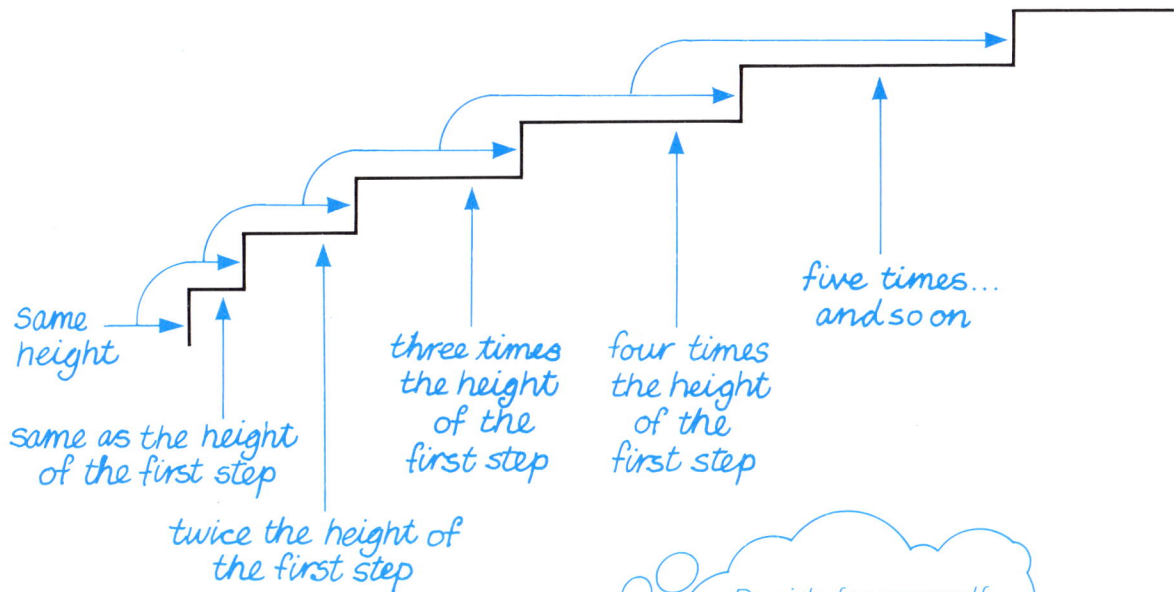

same
height

same as the height
of the first step

twice the height of
the first step

three times
the height
of the
first step

four times
the height
of the
first step

five times...
and so on

*Decide for yourself
how high to make
the first step.*

*Do you remember . . .?
We often use
letters until
we decide which
number to use.*

a) Use squared paper.
Make a drawing which obeys Ben's instructions.

b) The second step is twice as long as the first.
What can you say about the sixth step and the first?

c) Glenda copies Ben's design.
She gives the same instructions . . . using letters.

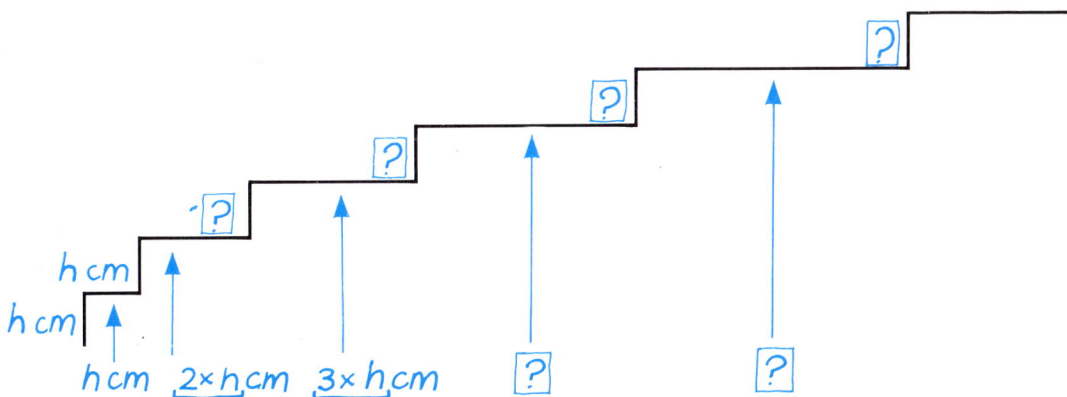

h cm

h cm

h cm   2×h cm   3×h cm

Copy and complete Glenda's instructions on the drawing you made.

d) Glenda chooses h to be 20.
How many **metres** long is the sixth step?

Challenge

2   Ben likes to design zigzags.
These zigzags are all drawn
from the same instructions.
Here are the first two parts
of the instructions.

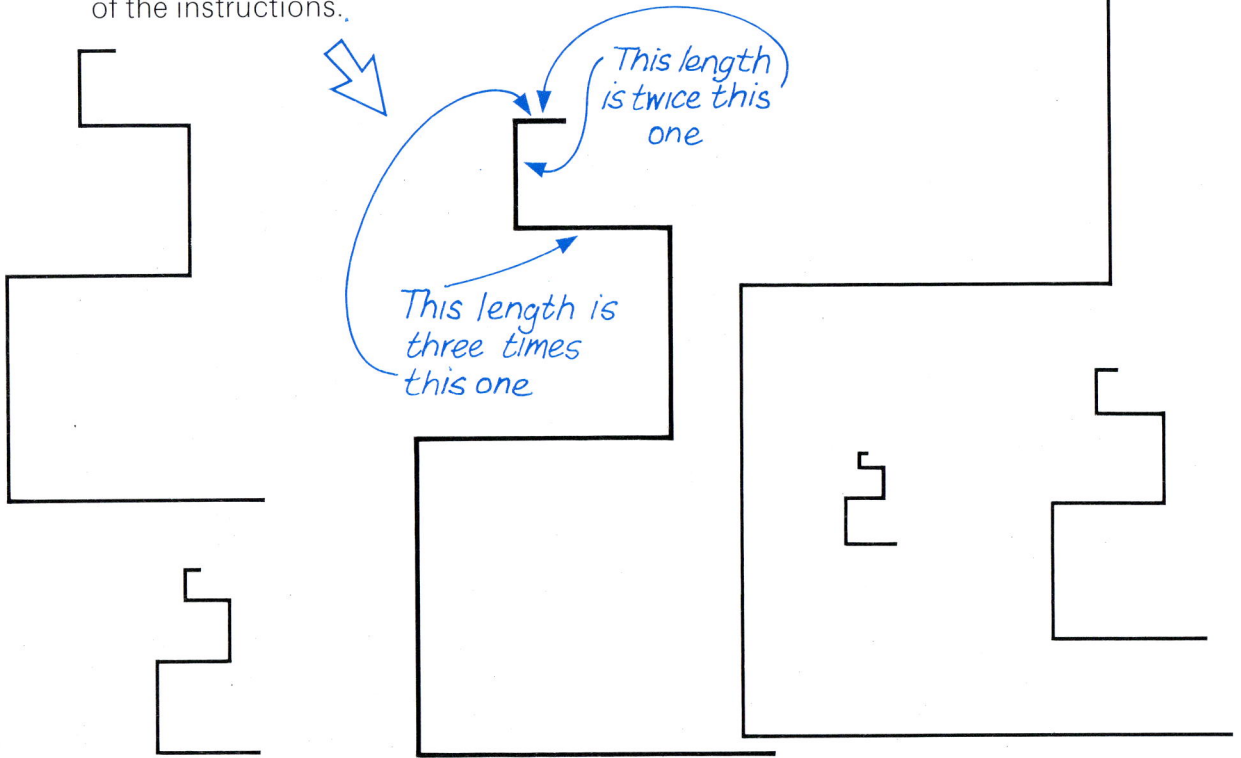

This length is twice this one

This length is three times this one

a) Use squared paper.
Make two accurate copies of the largest zigzag.
On one copy, complete the written instructions.
Use words, not letters.

b) On the other copy, write the instructions again.
This time, use letters.
Start like this.

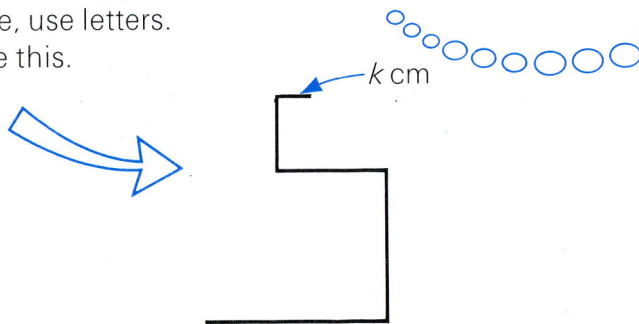

k cm

*Remember:
Your instructions
must work for all
the zigzags.*

c) Which did you find easier, using letters or using words?

d) Design your own zigzag.
Use letters to write the instructions.

20

180 ●

Take note

Letters help us to write instructions in a simple way.

3 You need squared paper.

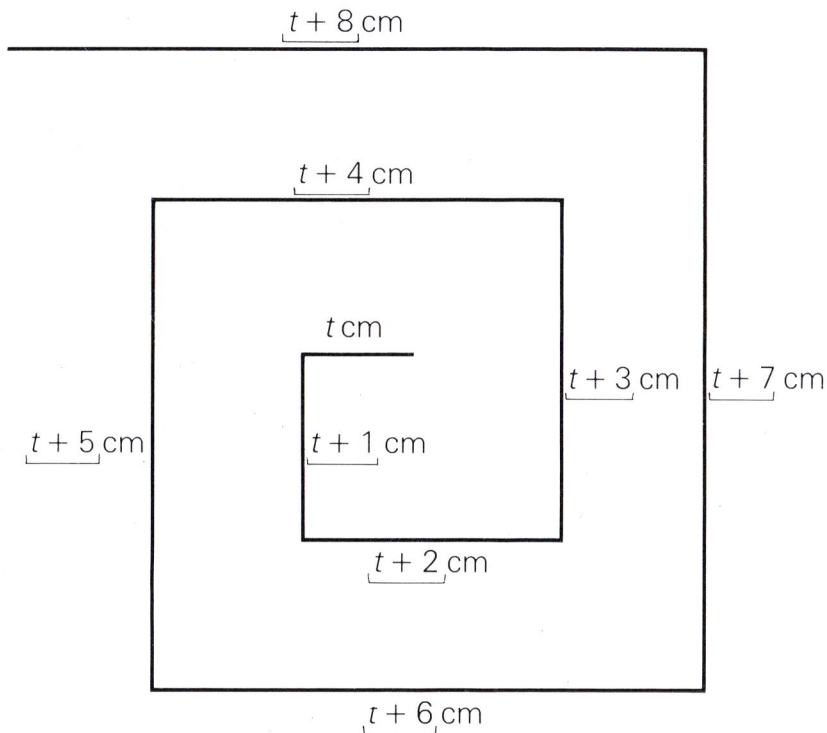

$t + 8$ cm
$t + 4$ cm
t cm
$t + 3$ cm $t + 7$ cm
$t + 5$ cm
$t + 1$ cm
$t + 2$ cm
$t + 6$ cm

Draw this spiral when t is chosen to be

a) 1   b) 2.

4 Draw two versions of this zigzag.
For the first version, choose k to be 1.
For the second, choose k to be 3.

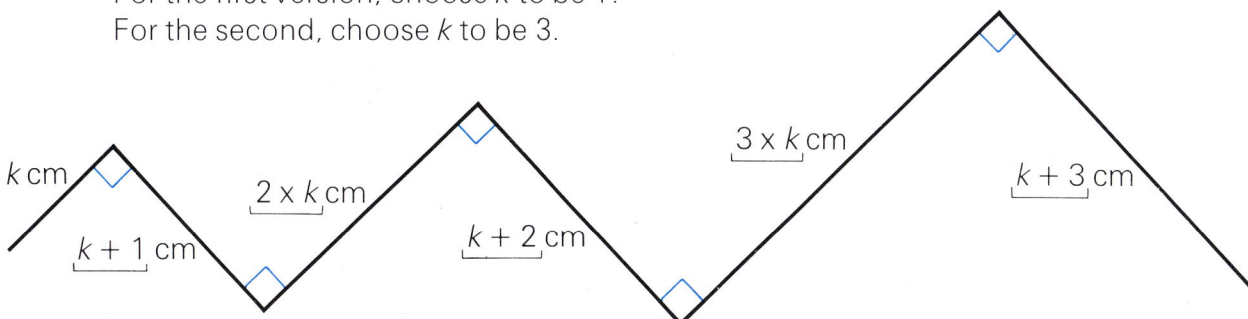

k cm
$k + 1$ cm
$2 \times k$ cm
$k + 2$ cm
$3 \times k$ cm
$k + 3$ cm

5 Ben wrote these instructions for drawing rectangles.
Sketch each rectangle.
Rewrite the instructions using letters.
Choose your own letter each time.

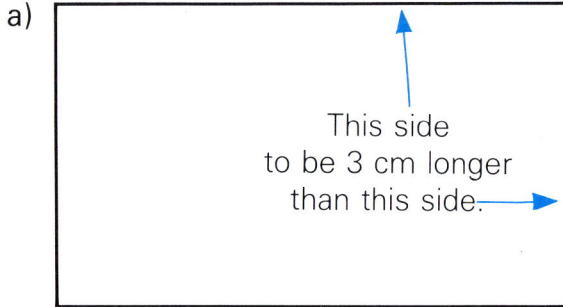

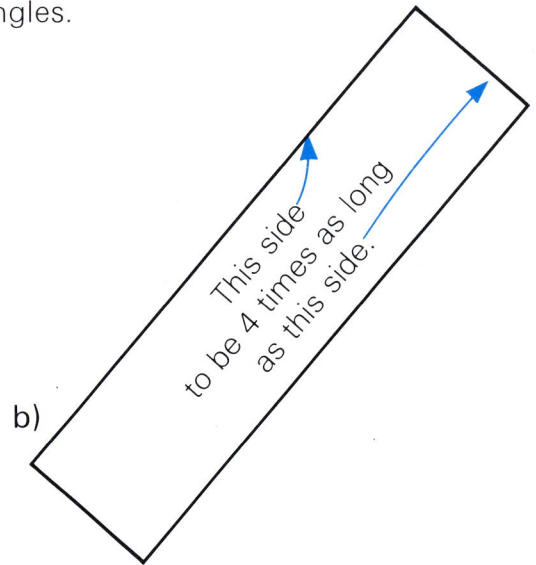

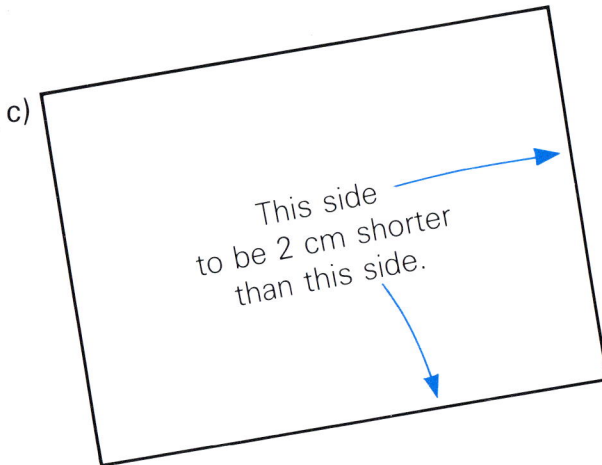

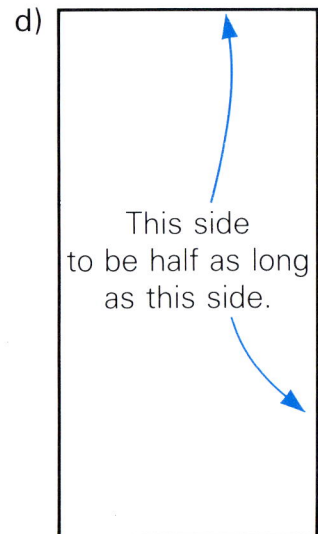

a)

This side
to be 3 cm longer
than this side.

b)

This side
to be 4 times as long
as this side.

c)

This side
to be 2 cm shorter
than this side.

d)

This side
to be half as long
as this side.

6 George wants to build a swimming pool.
He wants it to be 5 m
longer than it is wide.

Sketch the pool.
Write instructions for the
lengths, using letters.

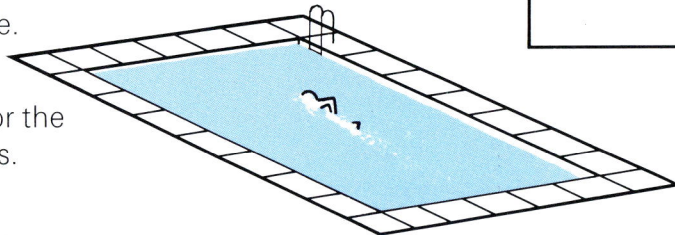

Challenge

7 Ben uses the instructions for the rectangle in question **5c)**.
He draws a rectangle.
He draws another rectangle for question **5d)**.
The rectangles turn out to be the same shape and size.
Draw Ben's rectangle.
Mark in the lengths of its sides.

8 Draw two different rectangles which agree with these instructions.
For each rectangle, write down the number you choose for d.

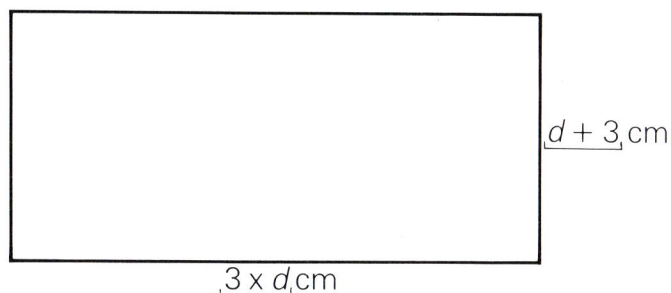

$d + 3$ cm

$3 \times d$ cm

9 Use a protractor.
Draw two different triangles which agree with the instructions.
For each triangle, write down the number you choose for b.

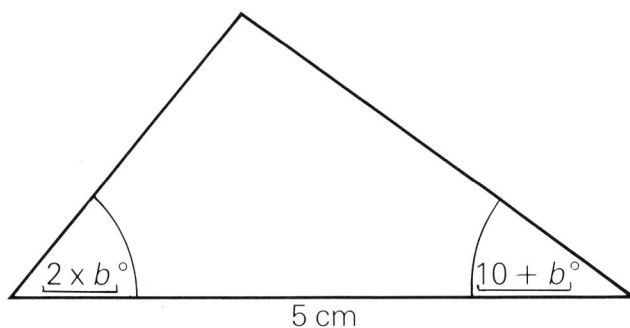

$2 \times b°$

$10 + b°$

5 cm

Challenge

10 These rectangles are drawn from the same instructions.

a) Show the instructions on a drawing of a rectangle.

b) Draw a rectangle, different from those shown here, which agrees with the instructions.

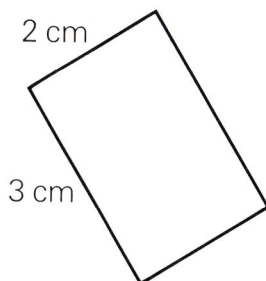

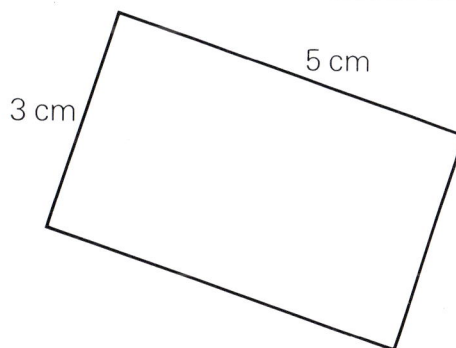

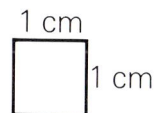

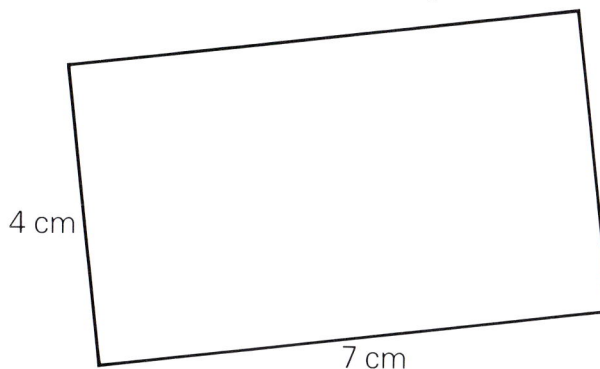

5 cm

3 cm

1 cm

1 cm

2 cm

3 cm

4 cm

7 cm

Making ripples

B 1 This 'ripple' of circles is produced from these instructions, by choosing different numbers for *k*.

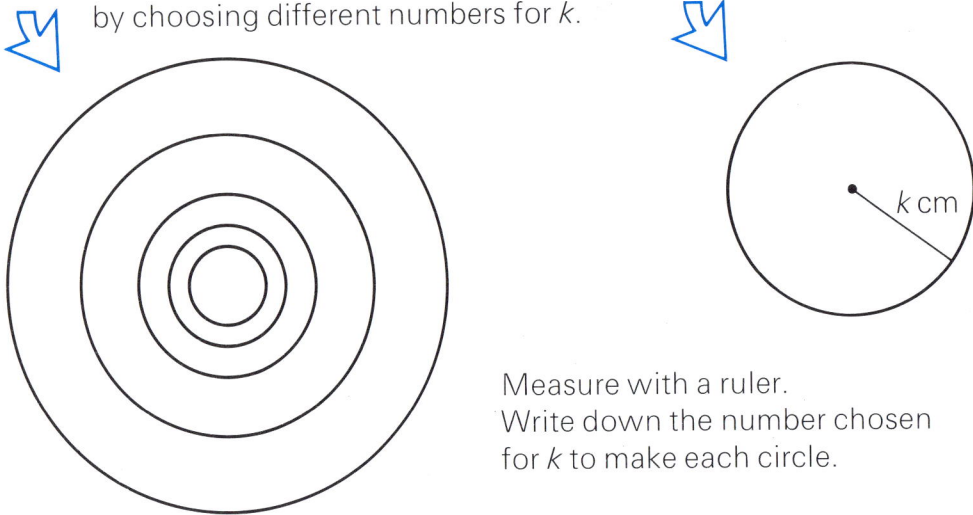

k cm

Measure with a ruler.
Write down the number chosen
for *k* to make each circle.

2 This 'ripple' of rectangles is produced using these instructions, by choosing different numbers for *t*.

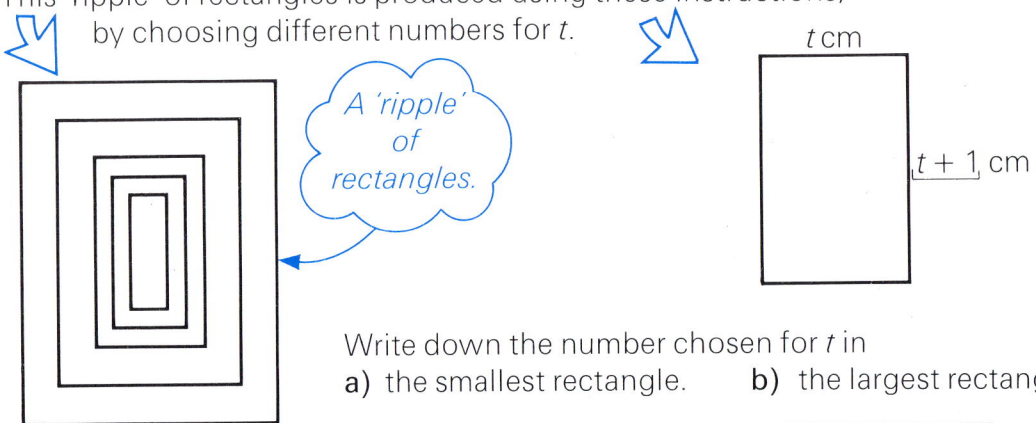

A 'ripple' of rectangles.

t cm

t + 1 cm

Write down the number chosen for *t* in
a) the smallest rectangle.   b) the largest rectangle.

3 a) Draw your own 'ripple'
from these instructions.
Choose numbers for *t* from 2 to 5.

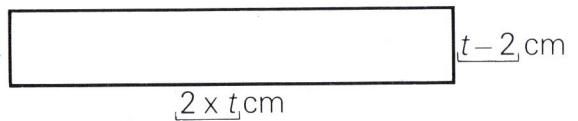

t − 2 cm

2 × *t* cm

b) As *t* gets larger, do the ripples

become more like this

this

or this   ?

*If you choose t to be 2,
your first ripple will
not be a rectangle.*

4 Draw 'ripples' for each of these.
Choose numbers for *t* from 1 to 5.

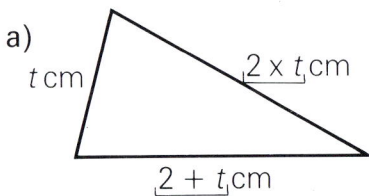

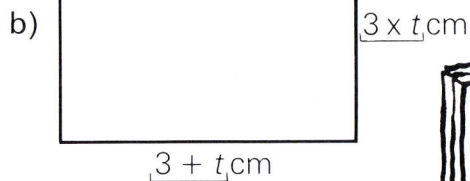

a)
t cm   2 × *t* cm

2 + *t* cm

b)
3 × *t* cm

3 + *t* cm

■ 185
▲ 187
● Next chapter

184 ●

Following instructions . . . to the letter!

1 This rectangle was drawn from these instructions.

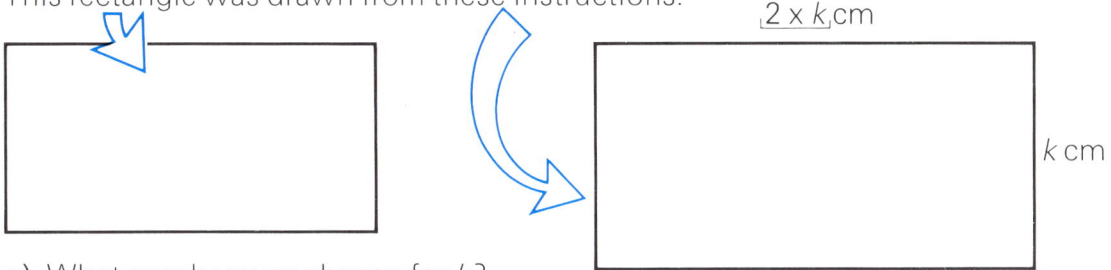

$2 \times k$ cm

k cm

a) What number was chosen for k?

b) Draw a rectangle of your own.
   Make sure it follows the instructions.

c) Write down the number you choose for k.

2 This rectangle was drawn from these instructions.

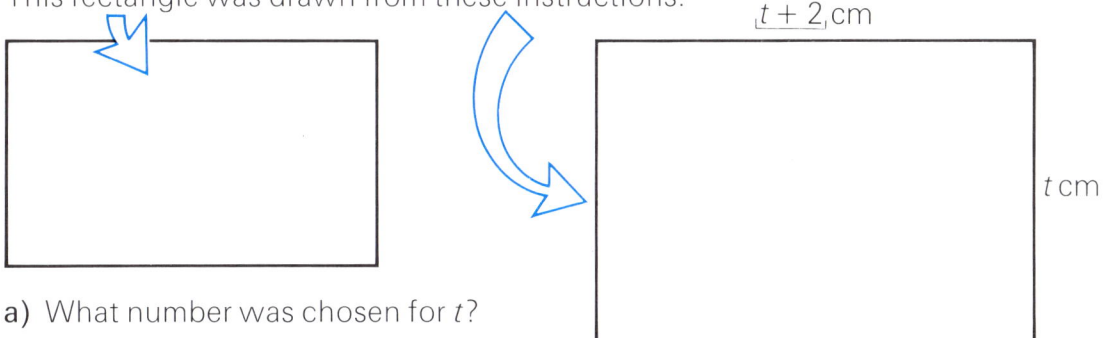

$t + 2$ cm

t cm

a) What number was chosen for t?

b) Draw a rectangle of your own.
   Make sure it follows the instructions.

c) Write down the number you choose for t.

Think it through

3 This triangle was drawn from these instructions.

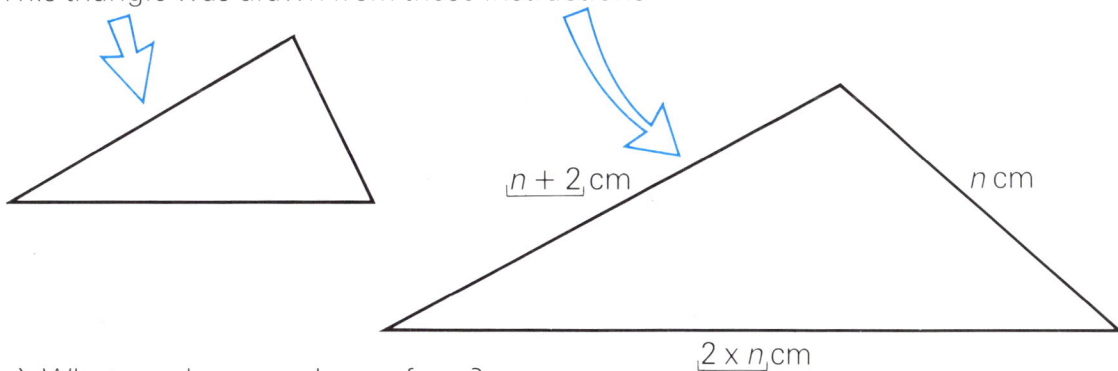

$n + 2$ cm

n cm

$2 \times n$ cm

a) What number was chosen for n?

b) Describe what happens when n is chosen to be 1.

c) n is chosen to be 6.
   What is the perimeter of the triangle?

A drawing might help.

185

4 a) Choose t to be 2.
Draw the ⌐ shape.

b) Now choose t to be 4.
Draw the ⌐ shape.

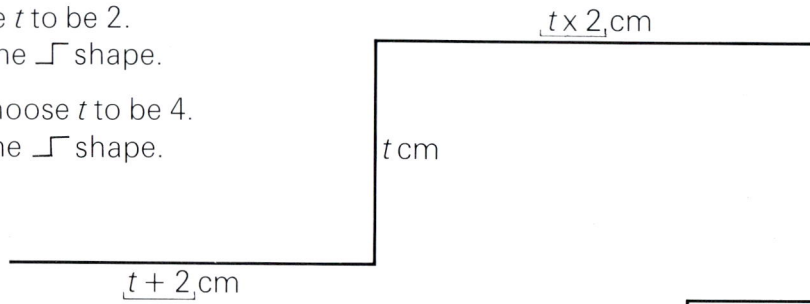

$t \times 2$ cm

t cm

$t + 2$ cm

5 These rectangles were all drawn from these instructions.

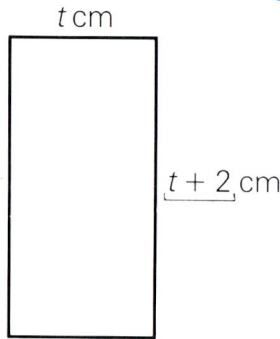

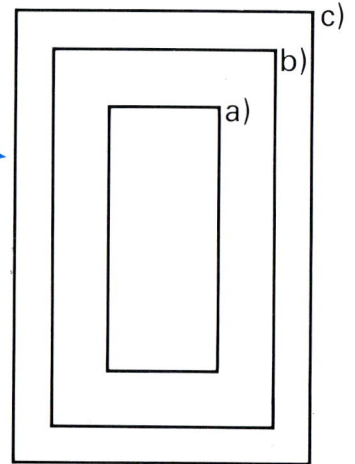

t cm

$t + 2$ cm

a)

b)

c)

What number was chosen for t each time?

6 Sketch this ⌐ design.

a) On your sketch, write the instructions using letters. Start like this:

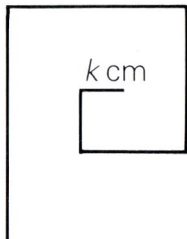

k cm

b) Choose a number for k.
Draw the shape accurately. Write down the number you choose for k.

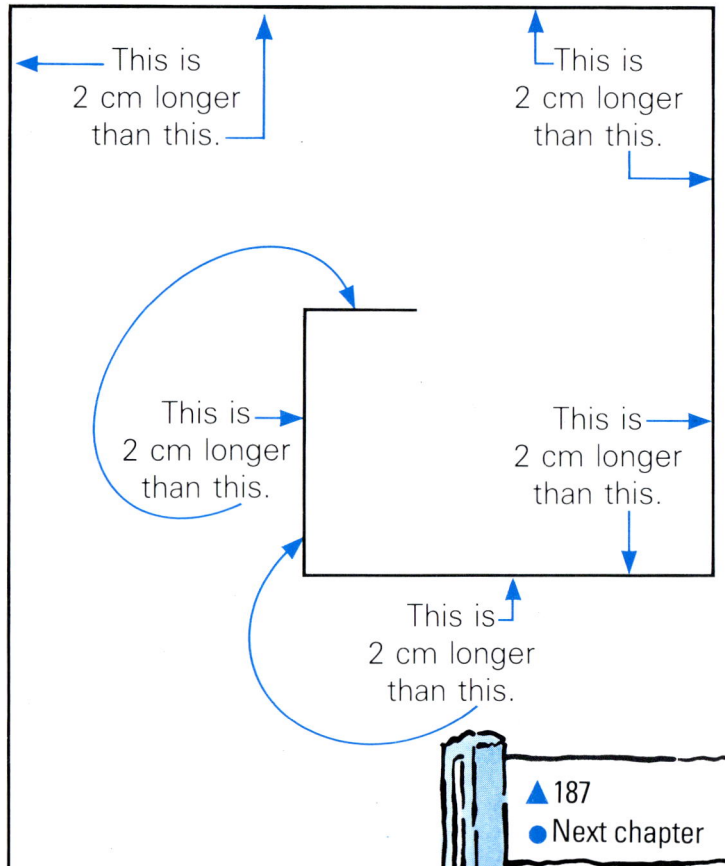

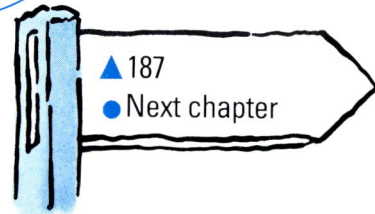

This is 2 cm longer than this.

This is 2 cm longer than this.

This is 2 cm longer than this.

This is 2 cm longer than this.

This is 2 cm longer than this.

▲ 187
● Next chapter

Prisms and rice

C   *Activity*

1   You need some stiff paper or thin card, and some sticky tape.

a) Make two different prisms from these instructions.
One of your prisms must be 9 cm tall.

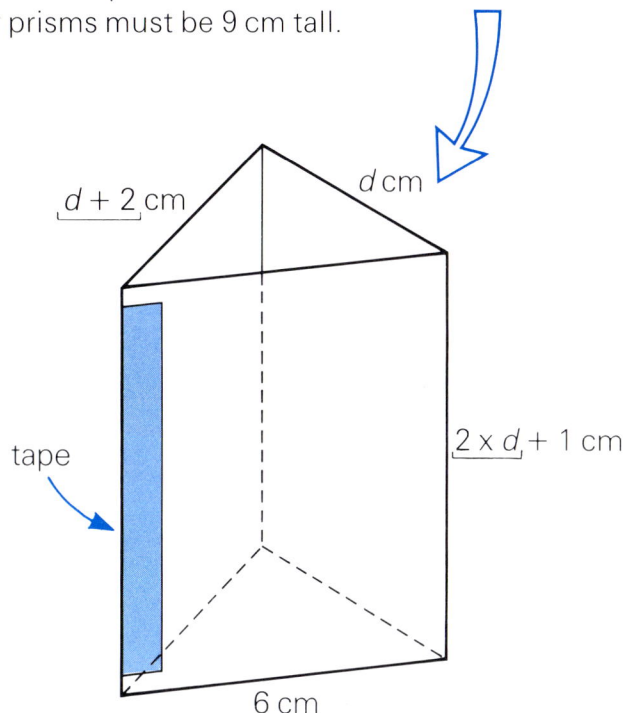

$d + 2$ cm

d cm

$2 \times d + 1$ cm

tape

6 cm

b) On each prism, write down
(i) the number you choose for d.
(ii) the height, and the lengths of all the edges of the prism.

c) Explain what happens when you choose d to be 1. ○○○○

A drawing might help.

d) What is the smallest number you can choose for d, if your result is to be a prism?

e) Is there a largest number you can choose for d?
If you say **yes**, what is the largest number?

f) There are two different numbers you can choose for d which will make the end an isosceles triangle.
What are the two numbers?

Challenge

g) Can the instructions produce a prism whose height is greater than the perimeter of the triangular end?
Explain what you decide.

Challenge: *The rice board*

2   The rice board is designed to challenge you.
The instructions tell you how many grains of rice you must place in each square.

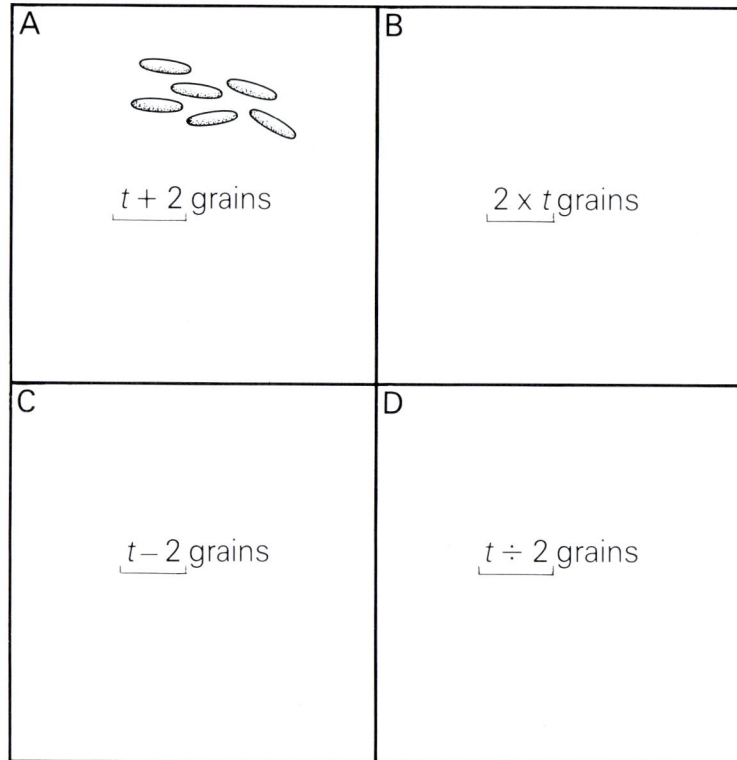

A	B
$t + 2$ grains	$2 \times t$ grains

C	D
$t - 2$ grains	$t \div 2$ grains

a)   There are six grains in **A**.
What number has been chosen for t?

b)   How many grains must be placed in the other squares?

c)   What is the smallest number which can be used for t?

> *Grains of rice cannot be chopped up,
> but you are allowed to have 0 grains in a square.*

d)   How many grains are there in each square for this number?

e)   What can you say about the number chosen for t if there
are more grains in **C** than in **D**?

f)   You use 36 grains of rice on the rice board.
How many are there in each square?

● Next chapter

21 Dealing with information

A 1 On average, how many hours do you sleep each night?
Give a rough estimate.

2 Think about how much you used to sleep when you were 8 or 9 years old.
Would you say you sleep
more hours now,
fewer hours now,
or about the same number of hours now?

3 Think about the people you know who are much older than you.
On average, do you think they sleep
more hours than you do,
fewer hours than you do,
or about the same number of hours as you do?

4 Look at the bar chart.
It shows that an average 4-year-old sleeps about 12 hours each night.
About how many hours does

a) a 10-year-old sleep?

b) a 12-year-old sleep?

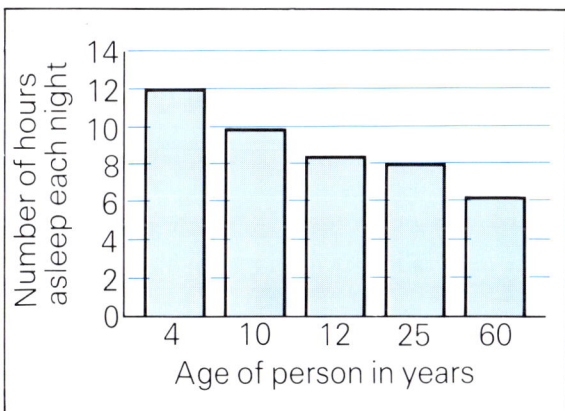

Think it through

5 Use the bar chart to help you.

a) Estimate how many hours these people sleep each night:

(i) a 7-year-old   (ii) an 11-year-old

b) Write one or two sentences to explain how you made each estimate.

ASSIGNMENT

6   Here are **two** assignments.
Your teacher might ask you to choose **one**.

Assignment A

a) Ask some pupils from each Year of your school:

> 'How many hours do you sleep each night?'

You will need to
- decide how many pupils to ask in each Year
- make a chart or table to record the answers.

b) From your results, draw a bar chart
or a pictogram.
It should show how many hours' sleep
pupils in each Year have
(on average) each night.

c) Write one or two short paragraphs to explain
how you worked out the number of
hours' sleep for each Year group.

Assignment B

a) Ask some pupils from each Year of your school:

> 'How many fillings do you have in your teeth?'

You will need to
- decide how many pupils to ask in each Year
- make a chart or table to record the answers.

b) From your results, draw a bar chart
or pictogram.
It should show how many fillings
(on average) pupils have
in each Year group.

c) Write one or two short paragraphs
to explain how you worked out
the number of fillings for
each Year group.

7 **a)** One of these is the number of people in the UK.
   Which do you think it is?
   Have a guess.   A About 56 million   B About 5.6 million   C About 560 million

   b) One of these is the number of people in the world.
   Which do you think it is?
   Have a guess.   A About 4500 million   B About 45 million   C About 450 million

   c) Look at the bar charts.

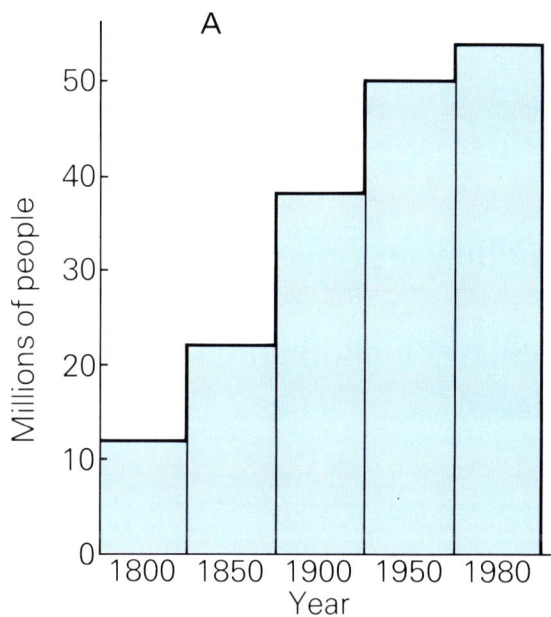

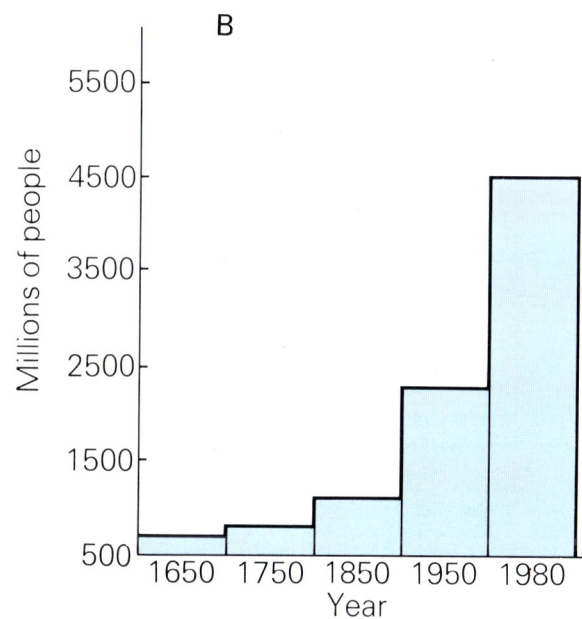

A — Millions of people vs Year (1800, 1850, 1900, 1950, 1980)

B — Millions of people vs Year (1650, 1750, 1850, 1950, 1980)

One shows the changes in the UK population.
The other shows changes in world population.
Which is which?
(You might also like to rethink your guesses in **(a)** and **(b)** now.)

   d) About how many people do you think there will be in the UK
   in the year 2000?
   Use the bar chart to help you to make the estimate.

   e) About how many people do you think there will be in the
   world in the year 2000?
   Use the bar chart to help you to make the estimate.

   f) Write one or two paragraphs to explain how the UK population
   has changed since 1800.
   Do you think the birth rate is speeding up or slowing down?

   g) Write one or two paragraphs to explain how the world
   population has changed since 1650.

● **191**

8   Look at the pictogram.
     It tells us how aircraft changed between 1922 and 1970.

Historical development of aircraft

1922 De Havilland 34
♦♦♦ 8 Seating capacity
⤻ 105 Cruising speed in miles per hour (mph)

1928 Zeppelin L2127
♦♦♦♦♦♦♦♦♦♦♦♦♦ 50   ▬▬▬ 80

1938 'Empire' flying boat
♦♦♦♦♦ 17
⤚ 160

1946 Douglas DC4
♦♦♦♦♦♦♦♦♦♦♦ 44
⤚ 239

1952 Comet
♦♦♦♦♦♦♦♦♦♦♦ 40
⤚ 500

1959 Boeing 707
♦♦♦ 180
⤚ 544

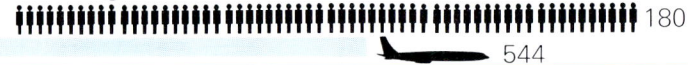

1970 'Jumbo' 747
♦♦ 300
⤚ 600

1970 Concorde
♦♦♦♦♦♦♦♦♦♦♦♦♦♦♦♦♦♦♦♦♦♦♦♦♦♦♦♦♦♦♦♦♦♦♦♦ 140
1500

a) Find the 'Empire' flying boat.
    What was its cruising speed?

b) Which could carry more passengers,
        the Comet
        or the Zeppelin?

c) Which has the higher cruising speed,
        the 'Jumbo'
        or the Boeing 707?

Think it through

d) Use the pictogram to compare the Comet and the Concorde.
    Write down what you find out.

e) In 1971 Brian Bulley flew a Comet from New York
    to London in about 6 hours.
    About how many hours do you think the Concorde would
    take for the same journey?
    Explain how you made your estimate.

Grouped information

B 1 The chart shows how many people in the UK died of TB between 1851 and 1967.

tuberculosis

Deaths from tuberculosis (figures in thousands)

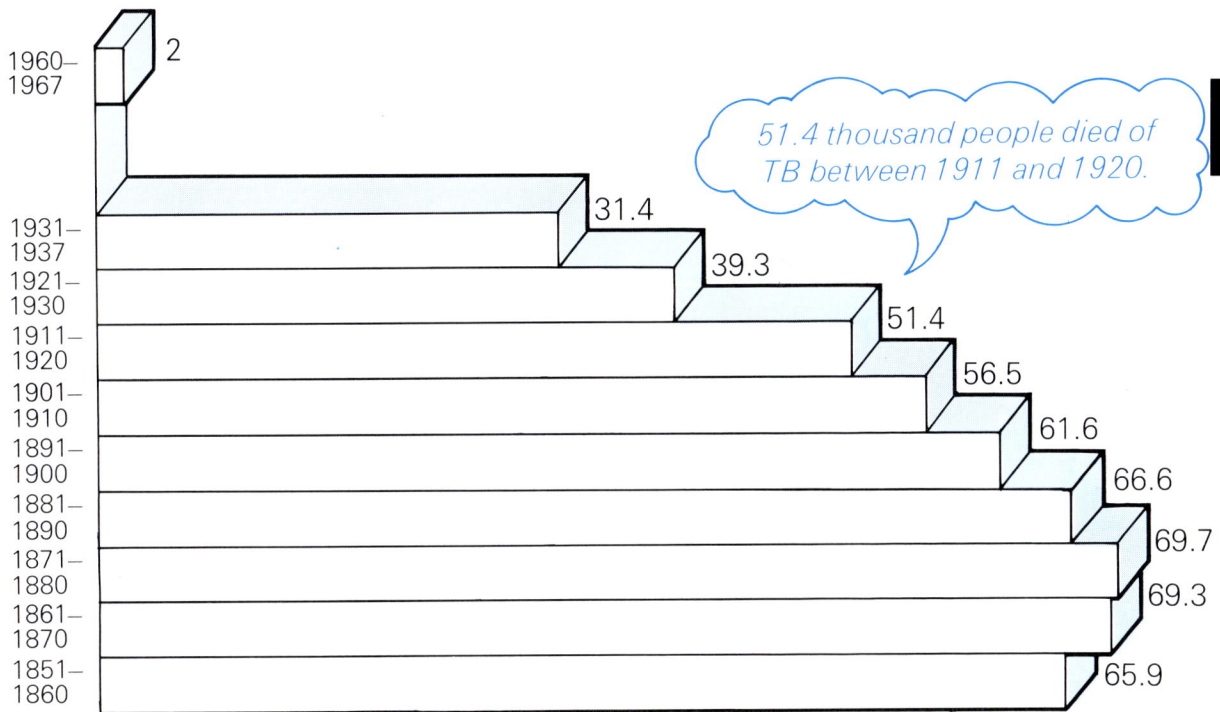

51.4 thousand people died of TB between 1911 and 1920.

Period	Deaths
1960–1967	2
1931–1937	31.4
1921–1930	39.3
1911–1920	51.4
1901–1910	56.5
1891–1900	61.6
1881–1890	66.6
1871–1880	69.7
1861–1870	69.3
1851–1860	65.9

a) How many people died of TB between 1881 and 1890?

b) How many people died of TB between 1851 and 1900?

c) Penicillin is an antibiotic.
   It can cure TB.
   It was first used by doctors during one of these periods:

A Between 1851 and 1861

B Between 1931 and 1941

C Between 1967 and 1977

Which one do you think it is?
Explain why you made this choice.

Find out for yourself

d) Find out who discovered penicillin;
   how and where it was discovered;
   how it is made.

2 The bar chart shows the heights of young trees in a pine forest.

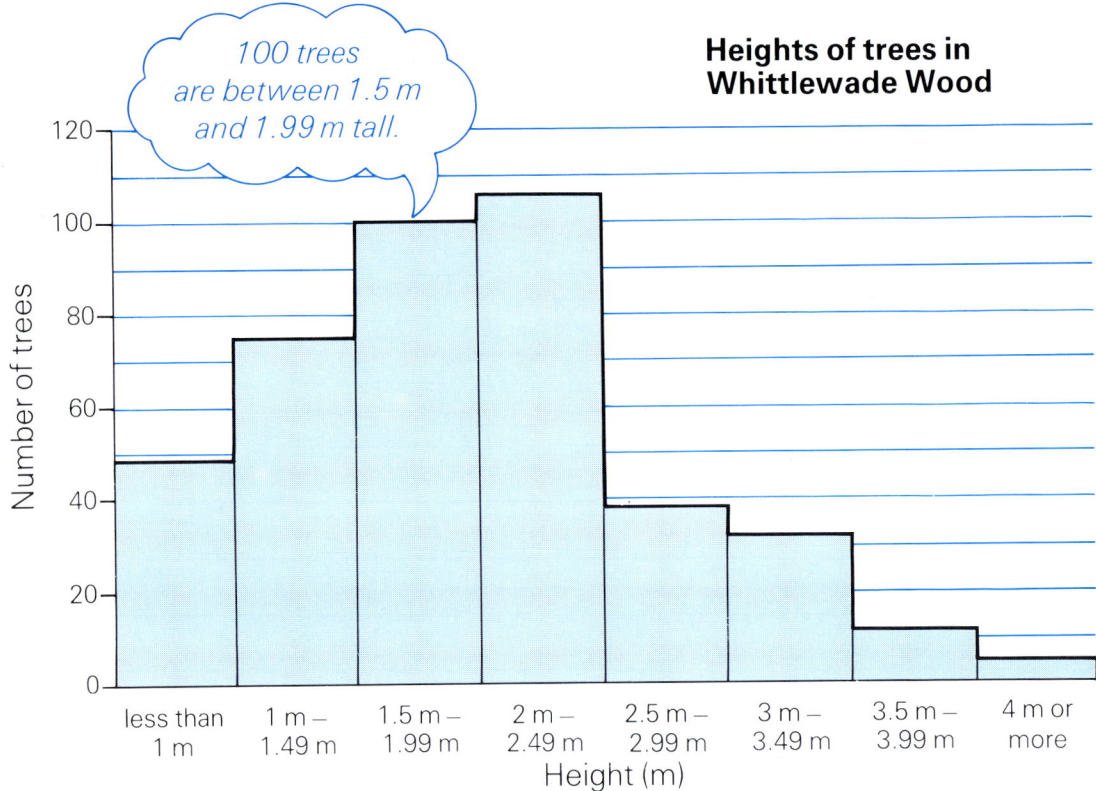

100 trees are between 1.5 m and 1.99 m tall.

Heights of trees in Whittlewade Wood

a) Check from the chart that only four trees are 4 m tall or taller.

b) How many trees are between 2.5 m and 2.99 m tall?

c) How many trees are 3 m tall or taller?

d) How many trees are 2 m tall?
If you think it is not possible to tell from the chart, explain why.

e) The heights of the trees have been grouped.
Draw your own chart of the heights . . .
but use these groupings:

2 m – 2.49 m, . . .

less than 1.5 m   1.5 m – 2.49 m   2.5 m – 3.49 m   3.5 m or more

Think it through

f) When the pines are planted they are 0.5 m tall.
On average, pines grow about 0.5 m each year during the first 5 years.
How long do you think most trees have been in the forest?
Explain how you arrived at your result.

■ 195
▲ 197
● Next chapter

21

Changing the groups

A
B
1 Each week *The Chronicle* publishes a pictogram like this.
It shows the range of secondhand bikes which are
advertised in the paper.

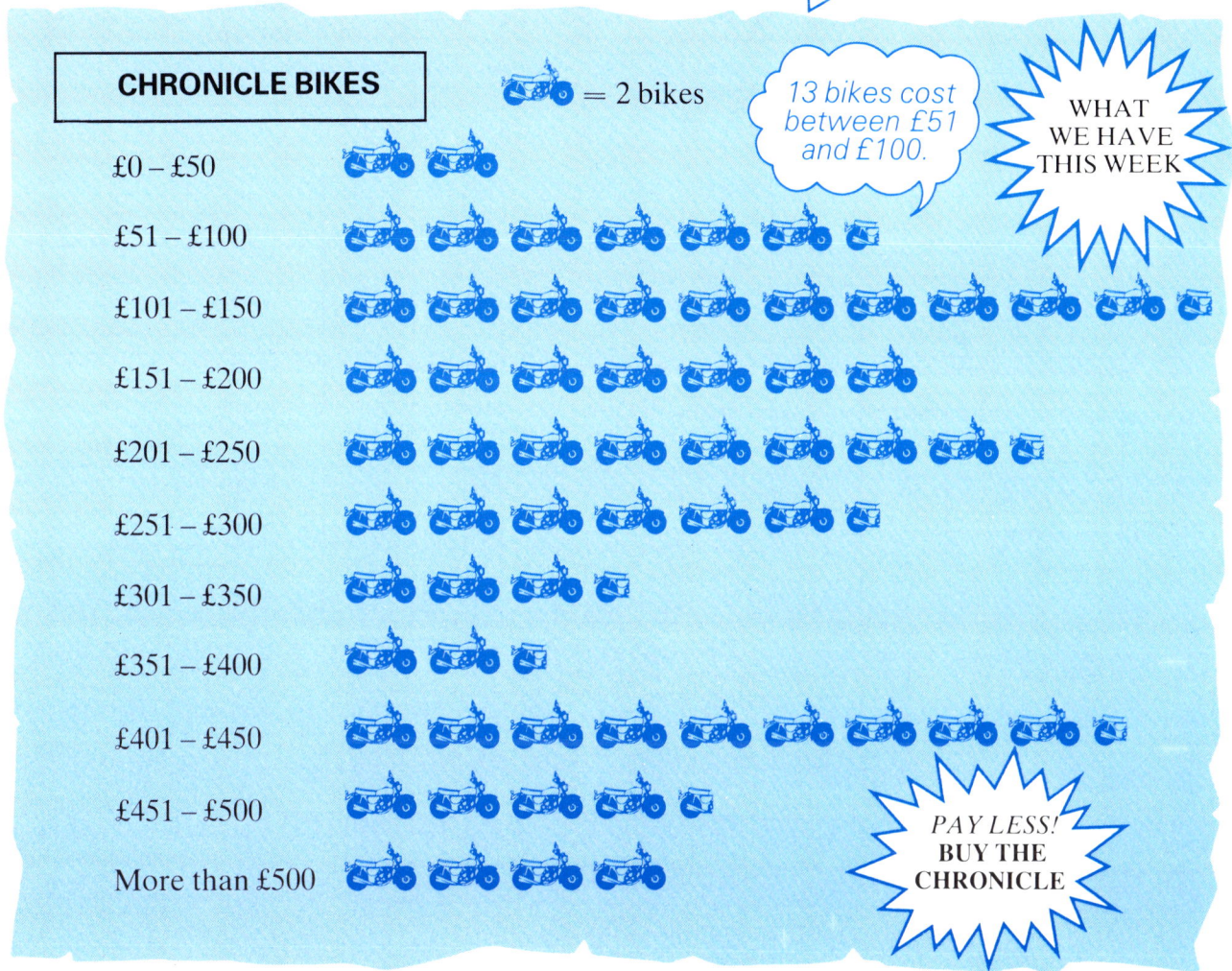

CHRONICLE BIKES 🏍 = 2 bikes

13 bikes cost between £51 and £100.

WHAT WE HAVE THIS WEEK

£0 – £50	
£51 – £100	
£101 – £150	
£151 – £200	
£201 – £250	
£251 – £300	
£301 – £350	
£351 – £400	
£401 – £450	
£451 – £500	
More than £500	

PAY LESS! **BUY THE CHRONICLE**

a) Check from the chart that four of the bikes advertised cost £50 or less.

b) How many bikes cost between £201 and £250?

c) Jemima has £200 to spend on a bike.
How many bikes can she choose from?

She might want to buy a cheap one and save some money.

Think it through

d) Draw your own pictogram for the bikes ...
but use these price groupings.

£100 or less	£101–£200
£201–£300	... and so on

2 This bar chart has been drawn to show the birth weights of 209 babies.
They were all born in one hospital during one year.

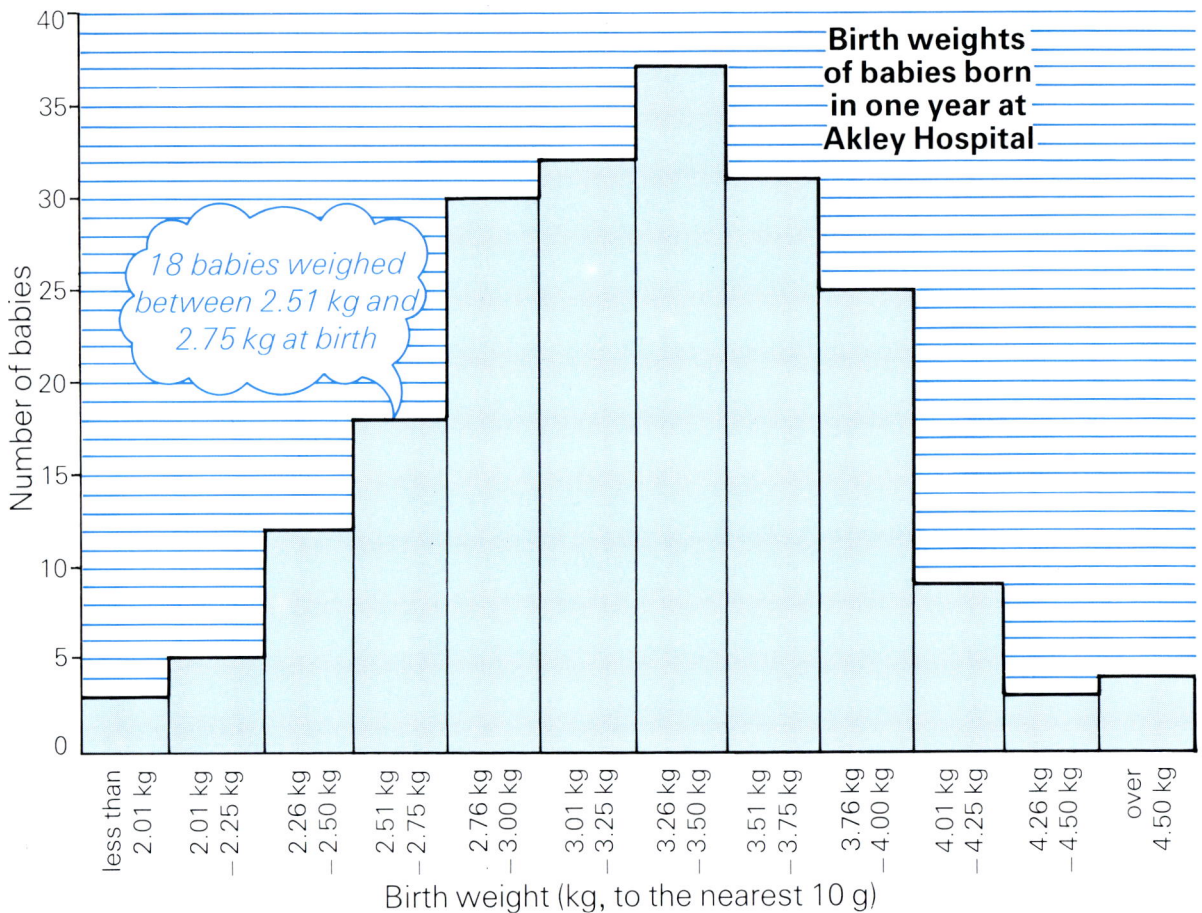

Birth weights of babies born in one year at Akley Hospital

18 babies weighed between 2.51 kg and 2.75 kg at birth

Number of babies

Birth weight (kg, to the nearest 10 g)

less than 2.01 kg / 2.01 kg – 2.25 kg / 2.26 kg – 2.50 kg / 2.51 kg – 2.75 kg / 2.76 kg – 3.00 kg / 3.01 kg – 3.25 kg / 3.26 kg – 3.50 kg / 3.51 kg – 3.75 kg / 3.76 kg – 4.00 kg / 4.01 kg – 4.25 kg / 4.26 kg – 4.50 kg / over 4.50 kg

a) Check from the chart that only three babies weighed less than 2.01 kg.

b) How many babies weighed between 2.26 kg and 2.50 kg?

c) How many babies weighed more than 4.5 kg?

d) How many kilograms would you say the 'average' baby weighs?
Explain why you chose this weight.

e) How many babies weighed 3 kg?
If you think it is not possible to tell from the chart, explain why.

Think it through

f) In the bar chart the weights have been grouped. *2.51 kg–2.75 kg . . .*
Draw your own bar chart of the weights . . .
but use these groupings:

less than 2.5 kg, 2.51 kg – 3.00 kg,
3.01 kg – 3.50 kg, 3.51 kg – 4.00 kg, over 4.00 kg.

▲ 197
● Next chapter

Comparing information

C 1 Look at the charts.
They show the rainfall
        and the temperature
in two American cities.

a) Look at the rainfall chart for city A.
   It shows that most rain falls in
   November, December and January.
   Which three months have most
   rain in city B?

b) Look at the temperature chart for city A.
   July is the hottest month.
   Which is the coolest month?

c) One of the cities has over 280 mm
   of rain during the first
   three months of the year.
   Which is it?

d) Describe how the weather changes in
   city B during the year.
   Write about the rain and
   the temperature.

e) One of the cities is Buenos Aires.

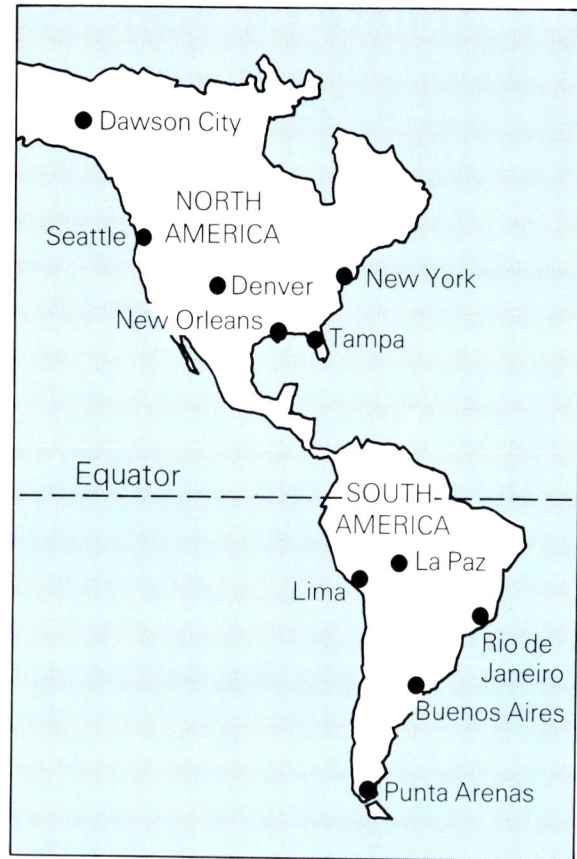

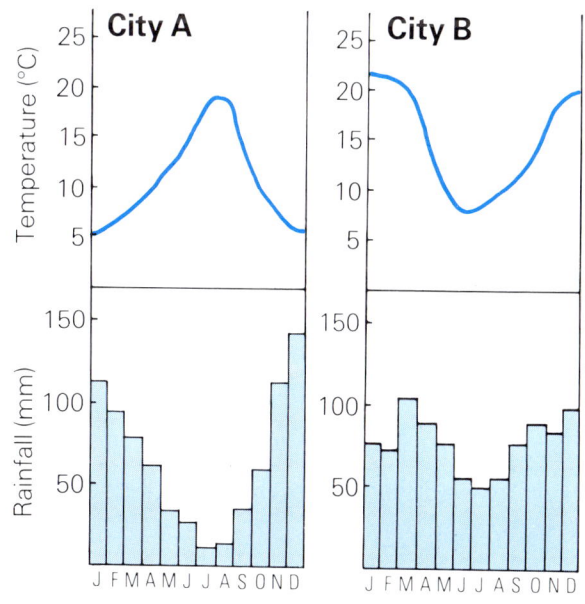

The other is Seattle.

Which chart do you think belongs to which city?
Explain how you made your choice.

Think it through

2   You need squared paper.
The bar charts show the results of a mathematics exam for two schools.
The pupils in both schools sat the same exam.

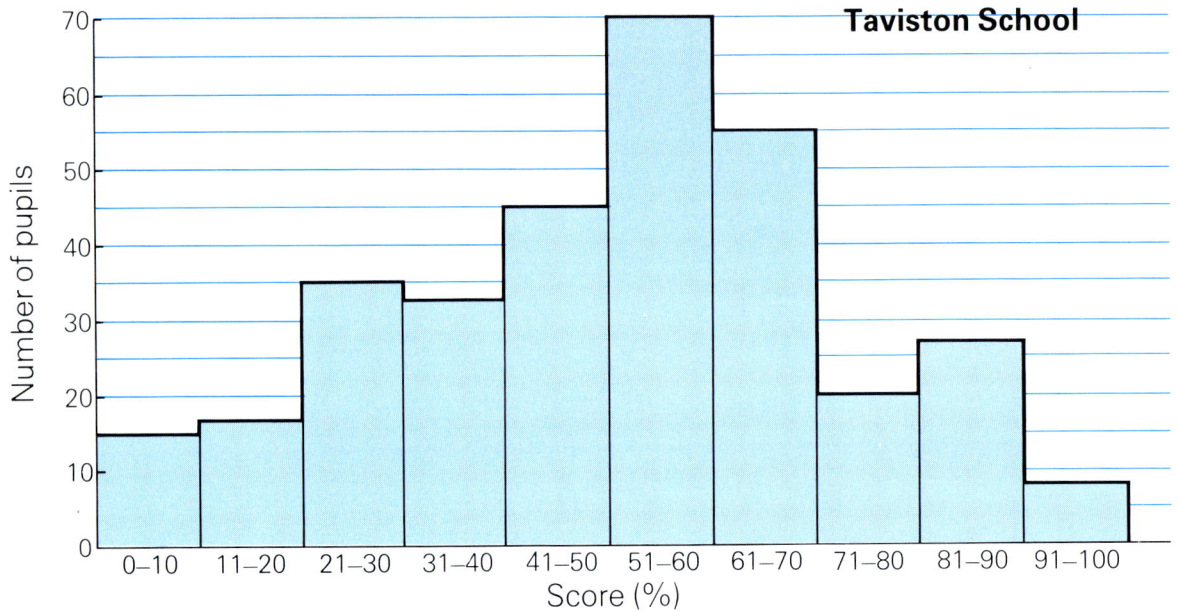

Taviston School

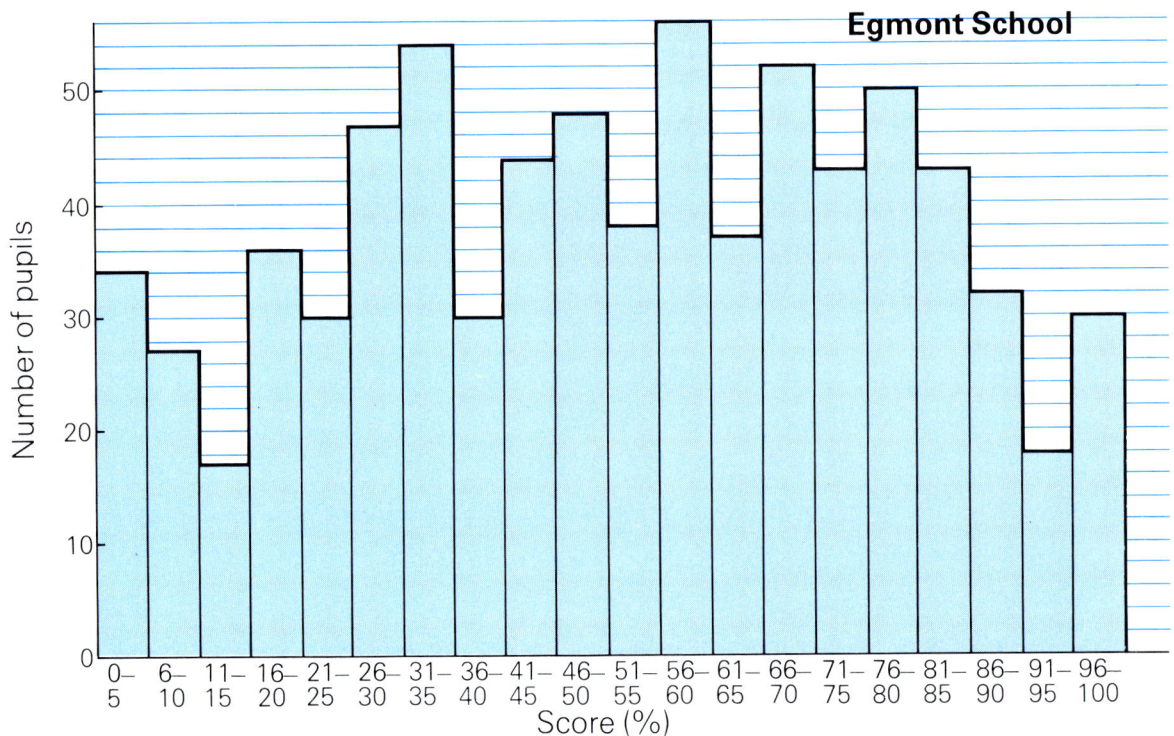

Egmont School

a) Redraw one of the charts so that it is easier to make a comparison.

b) Which school would you say had the better results?
Write one or two sentences to explain why you think this.

● Next chapter

22 Nets and solids

A 1 Phil's latest hobby is stamp-collecting.
(He's a philatelist.)
He keeps his stamps in little boxes.

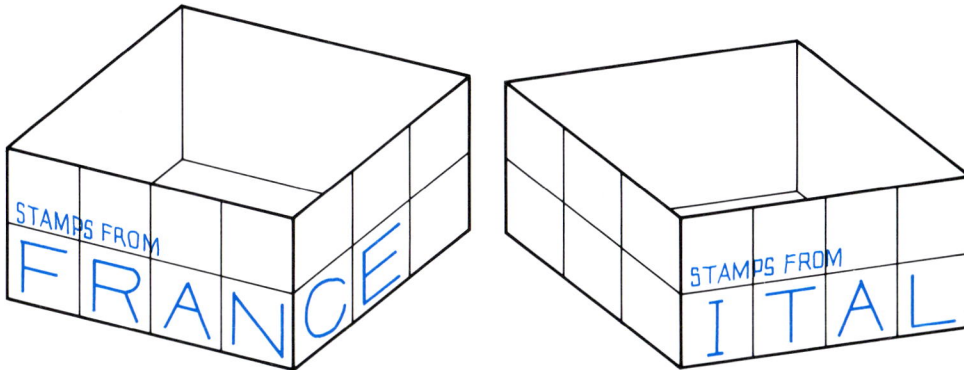

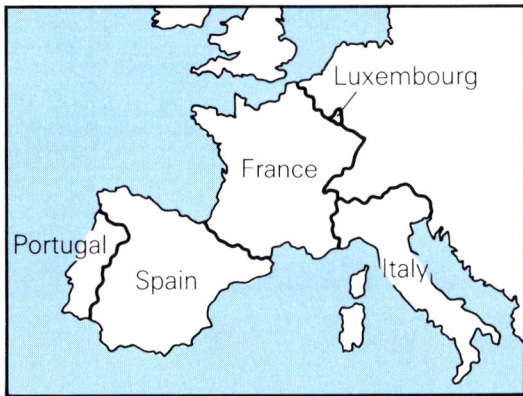

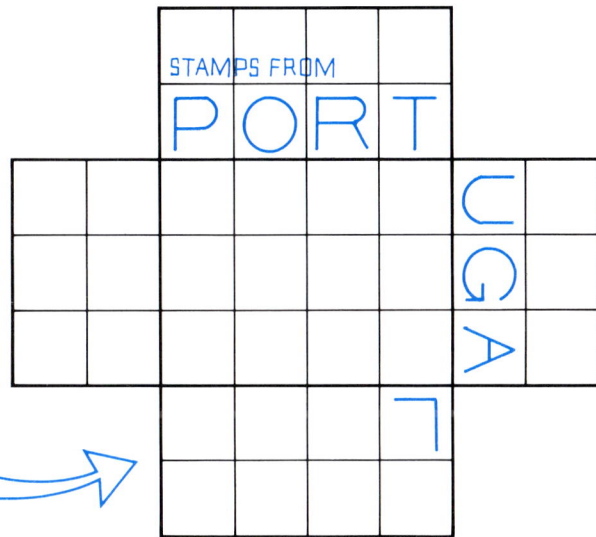

Here is a **net** for one of Phil's boxes.
Which country's stamps will
he keep in it?

2 Here is a net of another box.
It is for stamps from Luxembourg.

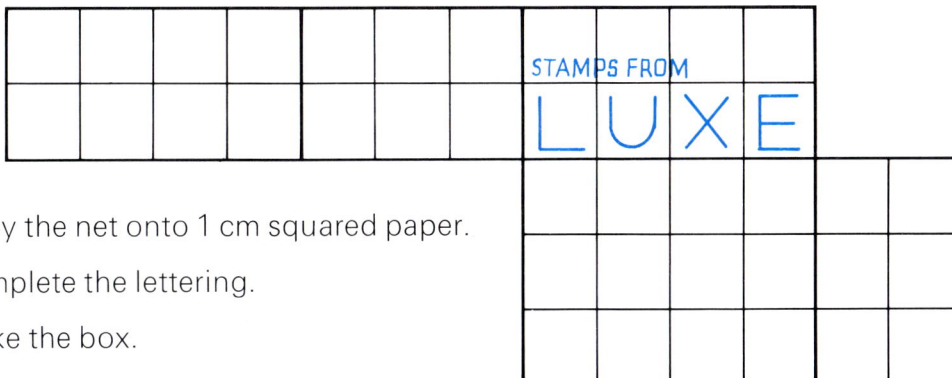

a) Copy the net onto 1 cm squared paper.

b) Complete the lettering.

c) Make the box.

● **199**

3 Phil is making this box from this net.

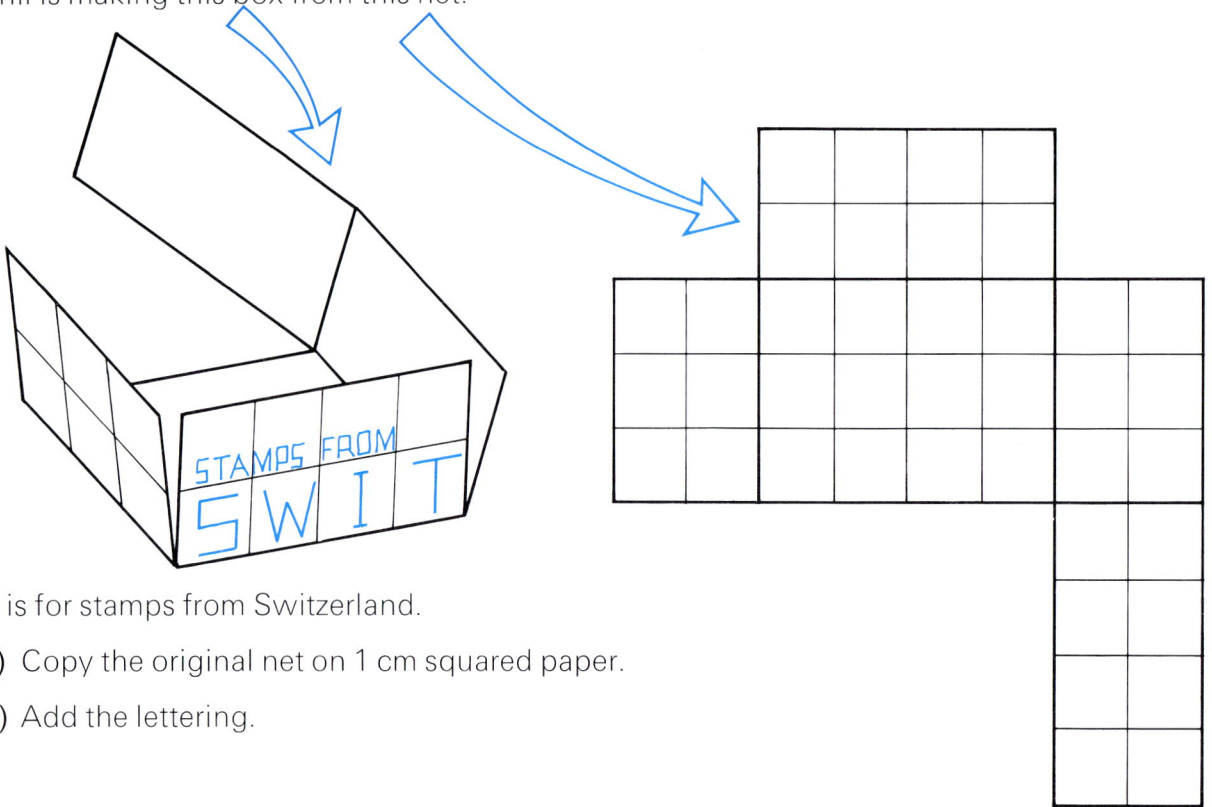

It is for stamps from Switzerland.

a) Copy the original net on 1 cm squared paper.

b) Add the lettering.

4 This box will be for stamps from Liechtenstein.

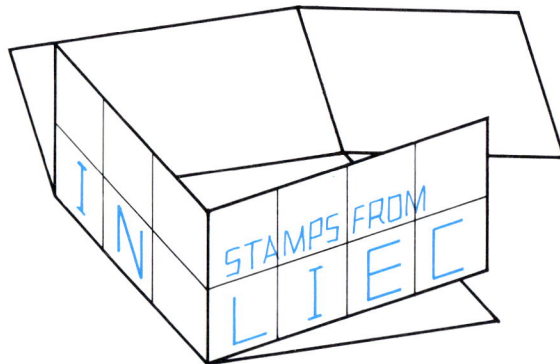

a) Draw the net on 1 cm squared paper.

b) Add the lettering.

5 Phil is making a box for stamps from Great Britain. Unfortunately, one face has come off.

a) Draw the net for the rest of the box. Include the lettering.

b) Add the face that has fallen off.

c) In how many ways can the face be added?

Compare your net with your friends'.

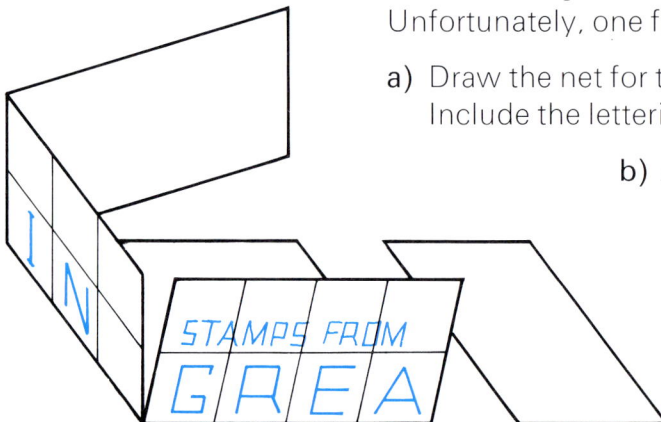

Nets for different boxes

B 1   Phil is given some stamps like this from Cameroun.

He makes an open box for them.

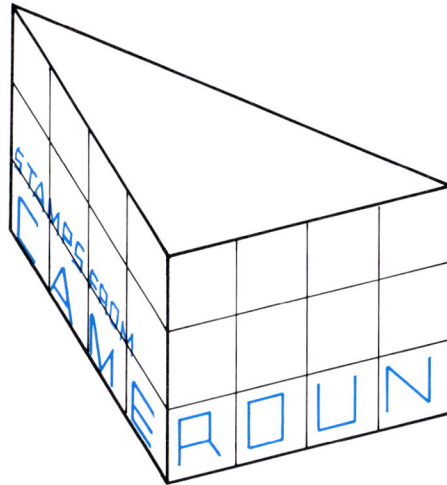

a) Phil uses this net.
Copy it onto squared paper.
Cut it out.
Check that it makes the box.

b) Draw a different net for the box.
Cut it out.
Check that it makes the box.
Stick both nets in your book.

2   Phil uses this net to make a pyramid.

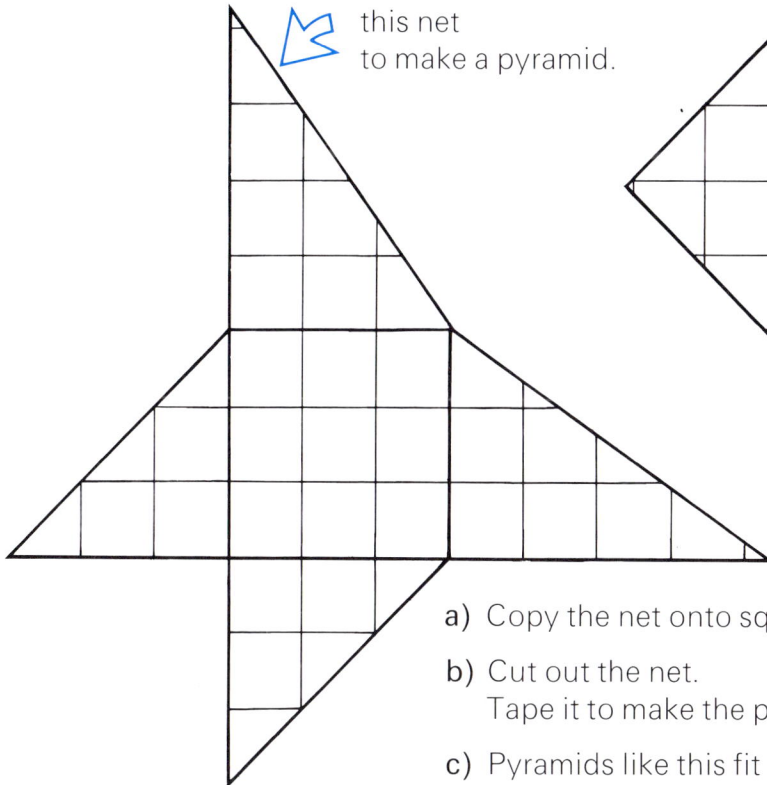

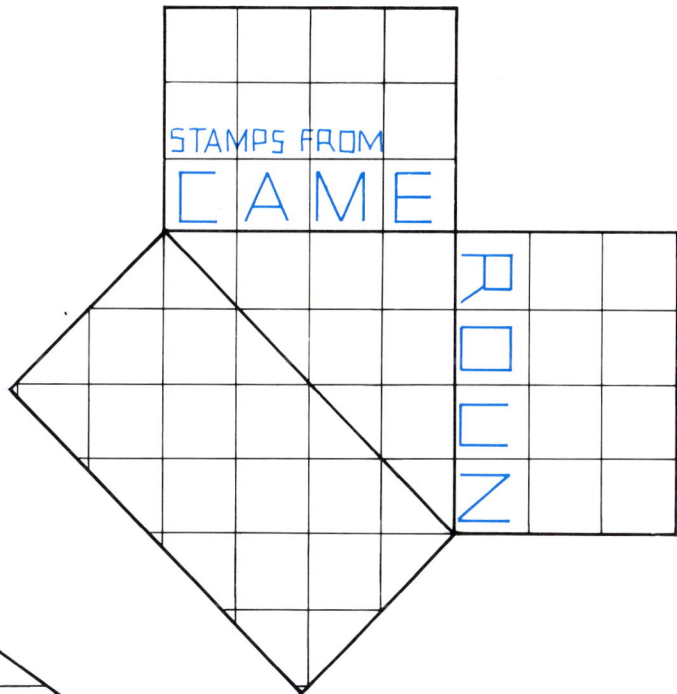

a) Copy the net onto squared paper.

Measure the lengths carefully.

b) Cut out the net.
Tape it to make the pyramid.

c) Pyramids like this fit together to make a cube.
How many do you think are needed?
Check your answer by borrowing some of your friends' pyramids.

3 Phil decides to make a taller
square-based pyramid.
Here is part of its net.

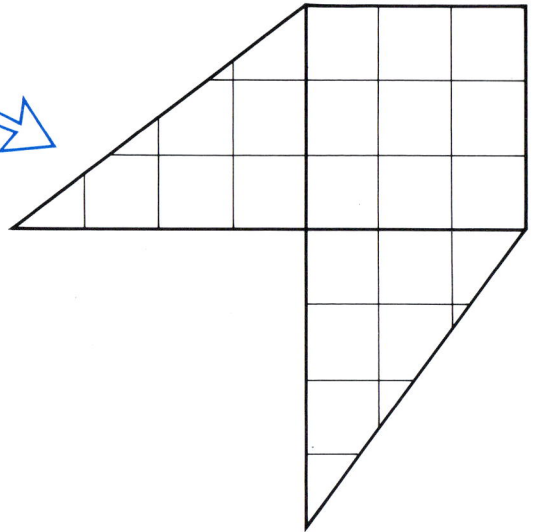

  a) Copy and complete the net.

  b) Cut it out.

  c) Check that it makes a pyramid.

  d) Stick the net in your book.

With a friend

4 You need some 1 cm isometric paper and some sticky tape.

  a) Each of you draw four equilateral
  triangles of this size.

  b) Cut out your four triangles.

  c) Tape them together to make
  a triangular pyramid.

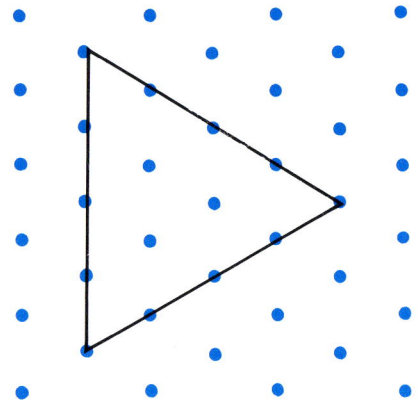

a tetrahedron

5 a) Open and flatten out your pyramid by cutting the tape.
  Make as few cuts as possible.

  b) You should now have a net of the pyramid.
  Compare your net with your friend's.
  Stick it in your book.

6 a) Draw as many different nets of the pyramid as you can.

  b) Cut them out; check that they make pyramids.

  c) Stick the nets in your book.

7 How many different nets did you find?

22

Nets for cylinders

C 1   Liz has some strips of copper.
She wants to make a ring for this finger.
Her finger is about 2 cm wide.

4 cm strip

6 cm strip

8 cm strip

10 cm strip

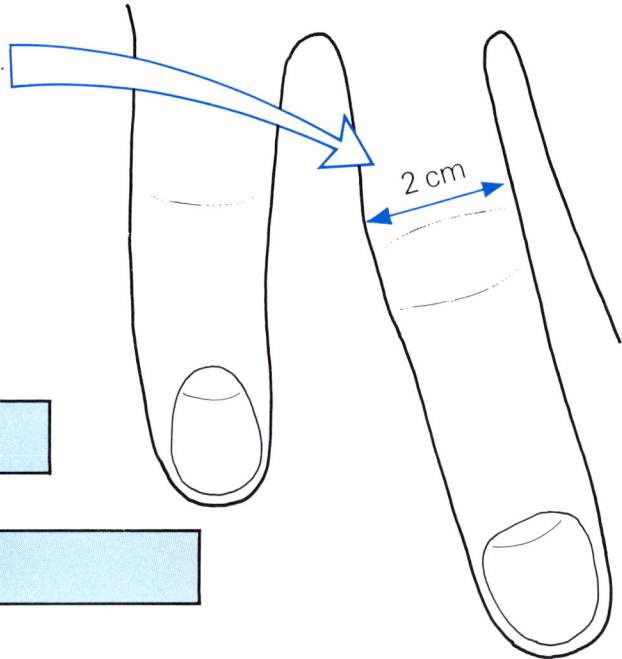

a) Which strip would make the best-fitting ring?
Write down your guess.

b) Make paper strips the same length as Liz's.
Wrap them around your finger.
Which is the best fit?

2   This label has come off a bottle of correcting fluid.

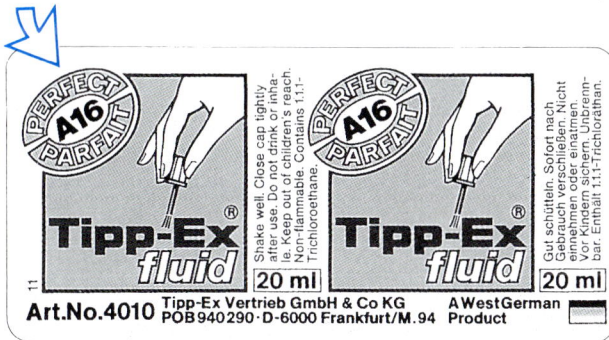

a) What is the diameter of the bottle?
Write down your guess.

b) Check your guess by
- copying the label
- cutting it out
- rolling it into a cylinder
- measuring the cylinder's width.

3   This is a full-size drawing of a standard baked bean tin.

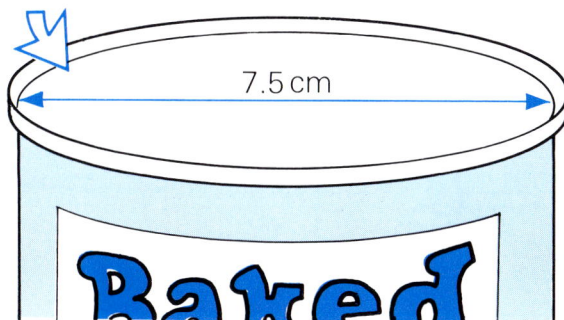

7.5 cm

The diameter of the tin is 7.5 cm.

a) How wide is the label?
Write down your guess.

b) Check it when you get home.

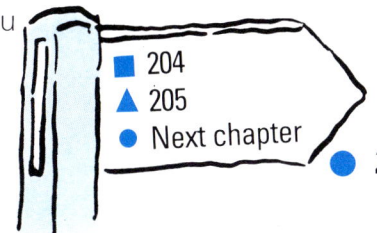
■ 204
▲ 205
● Next chapter

Different nets for the same box

1 Phil makes a box with a lid.

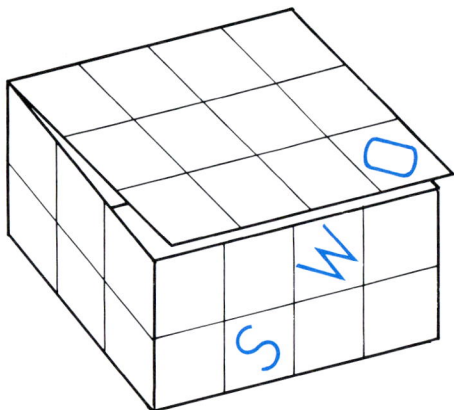

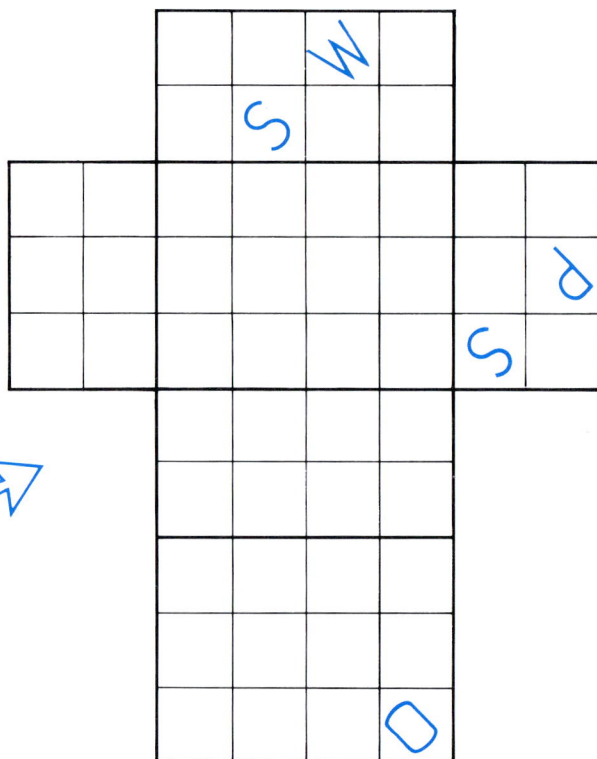

He uses this net.

a) Draw the net on 1 cm squared paper.

b) Cut it out.

c) Fold it into a box.

d) What word is on the box?

2 Phil accidentally cuts off this face.

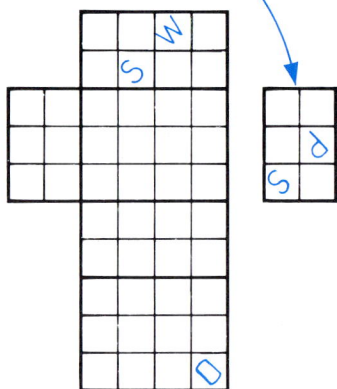

He re-sticks it like this.

a) Cut and re-stick your net in the same way.

Use sticky tape.

b) Fold it.

c) Do you still have the net of a box?

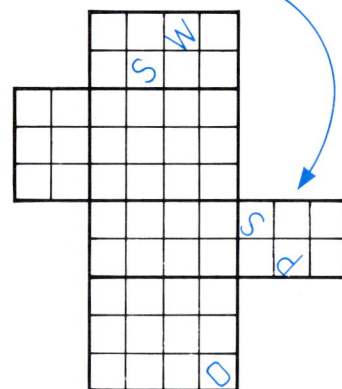

3 Phil could have re-stuck his net in these ways. Which are still nets of a box?

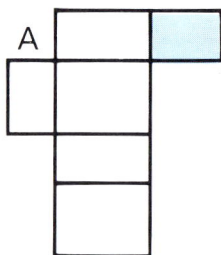

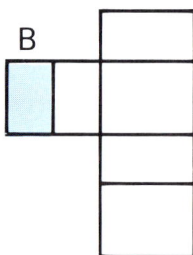

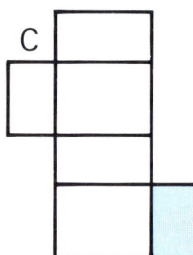

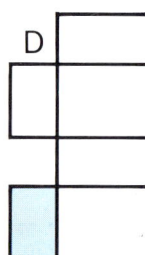

A   B   C   D

▲ 205
● Next chapter

Nets for cones

D 1 a) You need compasses and some plain paper.
Draw a circle with radius 6 cm.

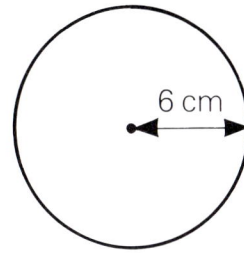

6 cm

b) Keep the arms of the compasses 6 cm apart.
Divide the circle into six equal parts like this.

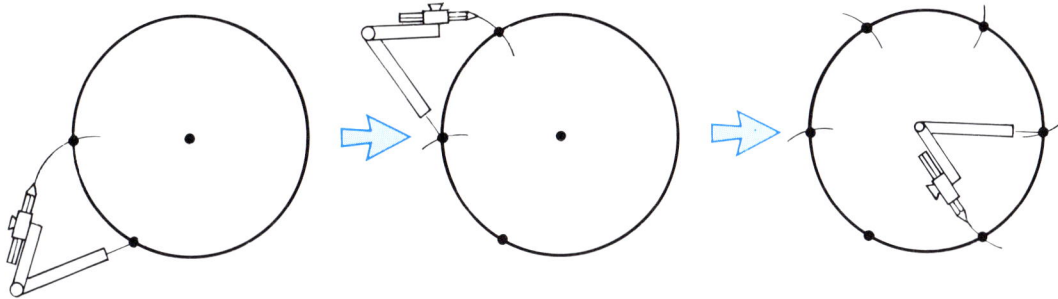

c) Cut out your circle.   Divide your disc into   Separate the two parts.
                                 these two parts.

$\frac{2}{6}$ $\frac{4}{6}$ $\frac{2}{6}$ $\frac{4}{6}$

d) Make this part of   Make these two   Tape the edges together.
your disc into a **cone**.   edges touch.

$\frac{4}{6}$ $\frac{4}{6}$ $\frac{4}{6}$

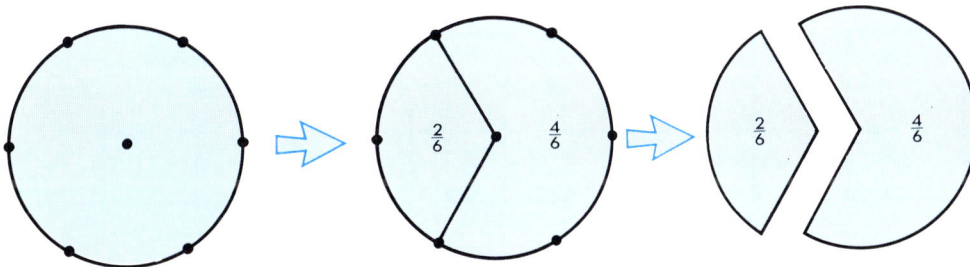

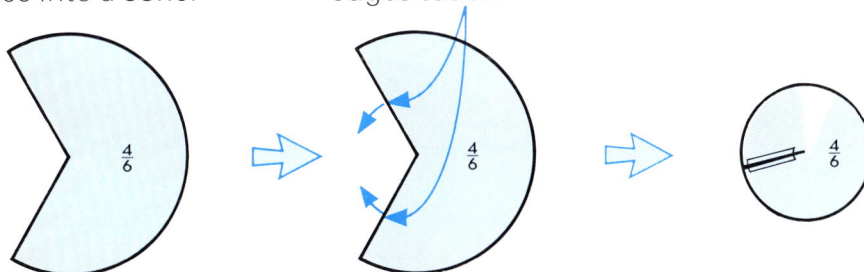

e) Make the other part of your disc into a cone.

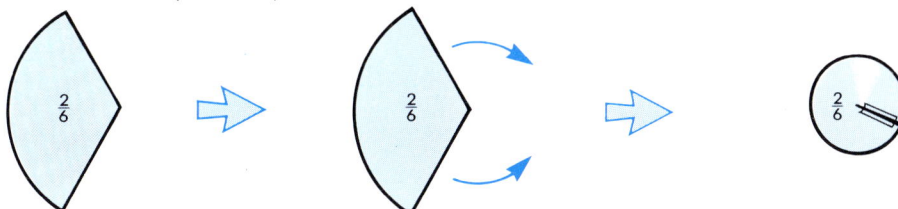

$\frac{2}{6}$ $\frac{2}{6}$ $\frac{2}{6}$

2 a) Stand your cones on a sheet of paper.
Hold them down firmly with one finger.
Draw round each base.

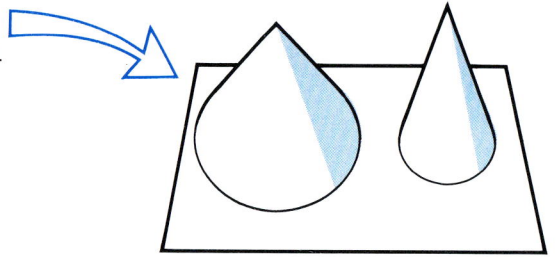

b) What shape is each base?

c) Measure and write down the width
(diameter) of each base.

d) What was the width (diameter) of your original disc?

3 Phil makes cones out of these
two parts of a disc.

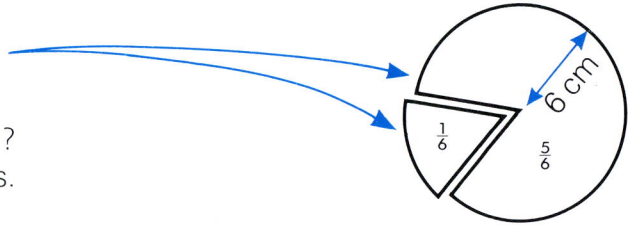

a) How wide will each base be?
Write down your predictions.

b) Make the two cones.
Check your predictions.

With a friend

4 You need paper, scissors and sticky tape.
Phil says a cone can be made like this:

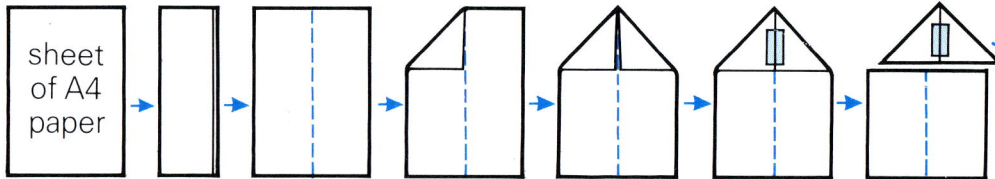

a) Carry out Phil's instructions; make this object.

b) Explain why it is **not** a cone.

c) Flatten your object out again.
Cut it so that it **does**
become a cone.

d) Explain what you have done.

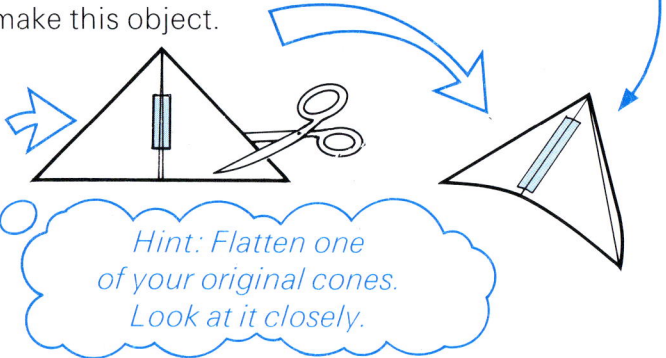

*Hint: Flatten one
of your original cones.
Look at it closely.*

Activity

5 You need paper, scissors and sticky tape.
Try to make a cone which is about 6 cm high and 6 cm across the base.
When you have made it, stick its net in your book.
Underneath, write down what radius circle you used,
and what size wedge you cut away.

*Give the angle of the wedge,
in degrees.*

Exotic nets

E 1 a) Which of these are nets of cones?

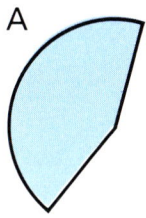

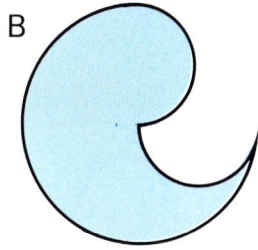

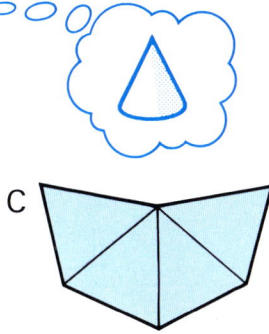

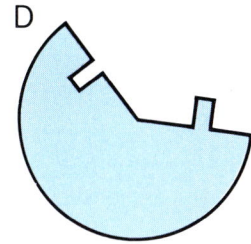

A   B   C   D

b) Draw a cone net of your own.

2 a) Which of these are nets of cylinders?

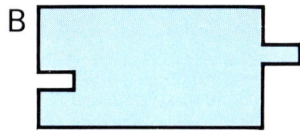

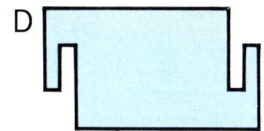

A   B   C   D

b) Draw a cylinder net of your own.

3 a) Which of these are nets of pyramids?

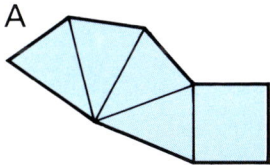

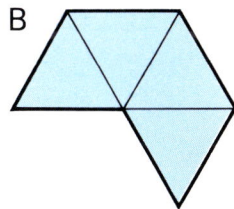

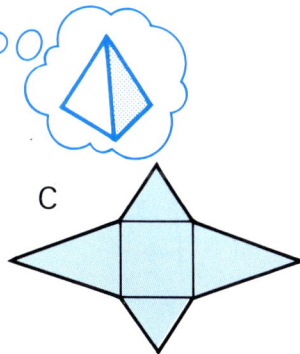

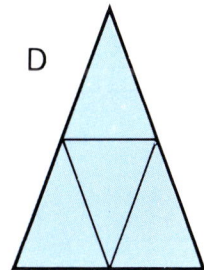

A   B   C   D

b) Draw a pyramid net of your own.

4 For each net, name what solid it makes.

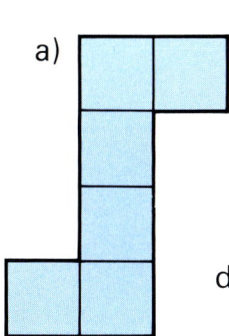

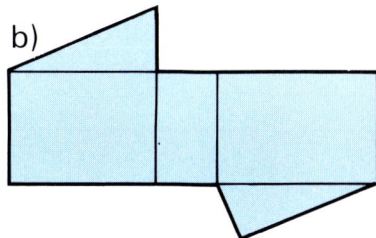

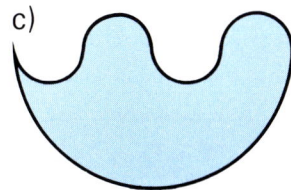

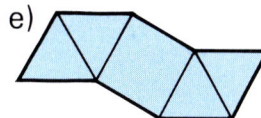

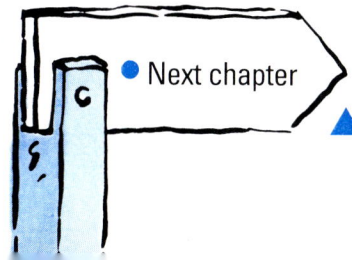

a)   b)   c)

d)   e)

● Next chapter

23 Using percentages

A — With a friend

1. Here are some ways in which **%** or **per cent** is used.
   You will have seen them or heard them said.
   Discuss them with your friend.
   Each of you write down the answers to the questions.

a) This is the label on Rupinder's school blouse.

100% cotton

What does it mean?

b) I'M 100 PER CENT FIT!

Winston

I'M 99% FIT!

Lenny

What do Winston and Lenny mean?

c) SHOE SALE 50% OFF ALL SHOES BUY NOW!

Write what is on the sale card in another way.
Do not use '%' or 'per cent'.

d) Did you know . . . ?
About 60% of you is water.

(i) How much of you is not water?
(ii) About how many kilograms is this?

Take note

100% of something means **all** of it.

one hundred per cent

1% of something means $\frac{1}{100}$ of it.

2 Copy and complete each of these.
The picture of the glass of orange will help.

a) 100% of something is _____ of it.

b) ▢ % of something is $\frac{1}{2}$ of it.

c) 25% of something is _____ of it.

d) ▢ % of something is $\frac{3}{4}$ of it.

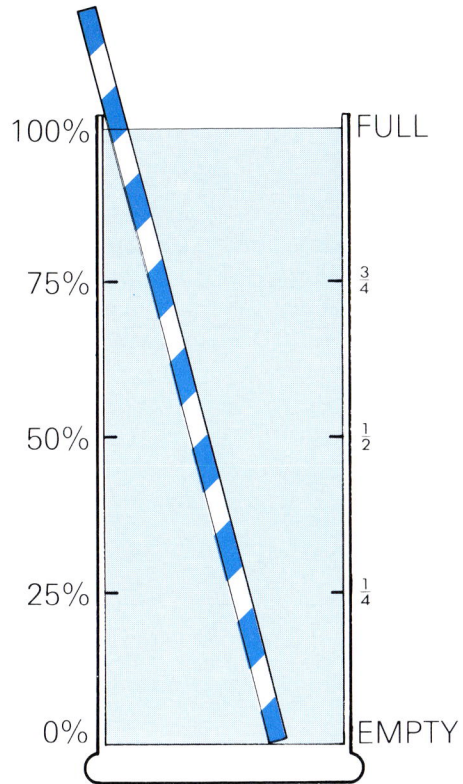

3 Copy and complete each of these.
The picture of the glass of orange will help.

a) 10% of something is $\frac{1}{\Box}$ of it.

b) 20% of something is $\frac{\Box}{10}$ of it.

c) 30% of something is $\frac{\Box}{10}$ of it.

d) 60% of something is $\frac{\Box}{10}$ of it.

e) ▢ % of something is $\frac{7}{10}$ of it.

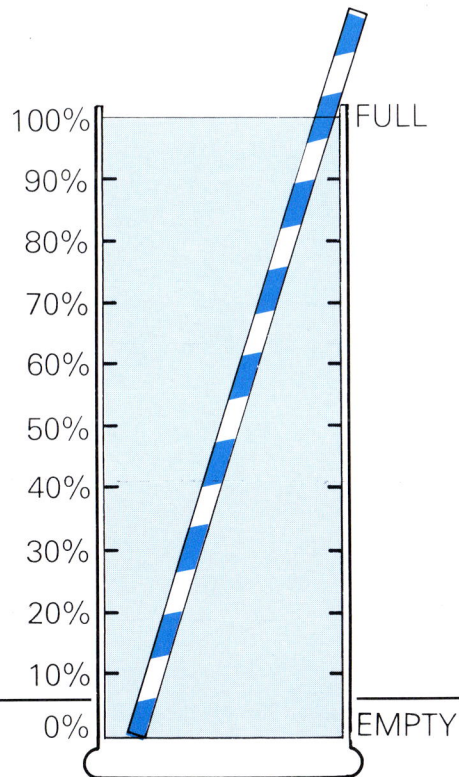

Think it through

4 Copy and complete:

a) ▢ % of something is $\frac{1}{5}$ of it.

b) 80% of something is $\frac{\Box}{5}$ of it.

5 Copy and complete each of these.
The picture of the petrol tank gauge will help.

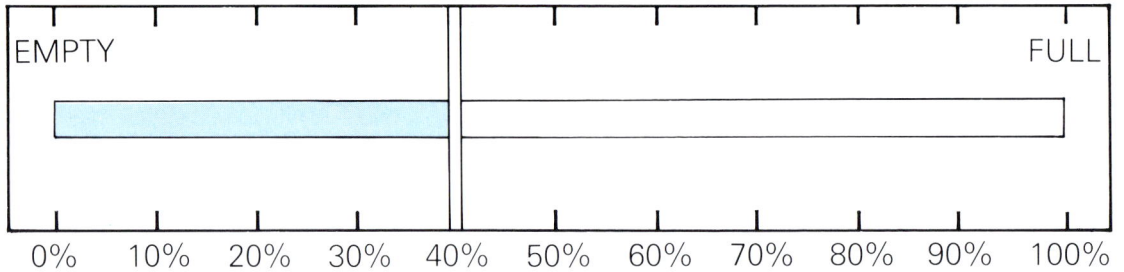

EMPTY · FULL

0% 10% 20% 30% 40% 50% 60% 70% 80% 90% 100%

a) ☐ % is a full tank.

b) 0% is an _____ tank.

c) 10% full is only $\frac{\square}{\square}$ full.

d) ☐ % full is only $\frac{1}{5}$ full.

e) 60% full is $\frac{\square}{5}$ full.

f) 1% full is only $\frac{\square}{\square}$ full.

g) 33% full is only $\frac{\square}{100}$ full.

h) ☐ % full is $\frac{3}{4}$ full.

i) 99% full is $\frac{\square}{\square}$ full.

j) ☐ % full is only $\frac{1}{20}$ full.

k) 15% full is only $\frac{\square}{20}$ full.

l) 95% full is $\frac{\square}{20}$ full.

Think it through

6 Copy and complete:

$\frac{2}{5}$ full is $\frac{\square}{10}$ full ... which is $\frac{\square}{20}$ full ... which is ☐ % full.

7 The English test has been marked.
The teacher is working out the percentages.

Meg Matthews $\frac{12}{50}$ (24%) (18%)

Lenny Smith $\frac{17}{50}$

Horace Hornblower $\frac{24}{50}$

Dilip Patel $\frac{27}{50}$

Midge Williams $\frac{30}{50}$

Winston Meredith $\frac{35}{50}$

Fiona Brown $\frac{43}{50}$

Ron Maddox $\frac{50}{50}$

a) Meg scored $\frac{12}{50}$ ... which is $\frac{24}{100}$ ... which is 24%.
Write down the percentage mark for each of the other friends.

b) Alan Bean scored 18%.
How many marks out of 50 did he score?

c) Sandra Bailey scored $\frac{26}{50}$.
Karen McCulloch scored 54%.
Who had the better mark?

Surveys

B 1

7/10 of all cats prefer PURR!

a) The Purr advertisement says that seven out of every ten cats prefer Purr. How many is this in every 100?

b) What percentage is this?

2 Match each sentence with one of these percentages.

a) One in every two cars is foreign.

b) 76 out of every 200 cars are faulty.

c) Four out of five people prefer butter to margarine.

d) One girl in every five is married before she is 18.

e) Last year, the police solved 9 in every 20 major crimes.

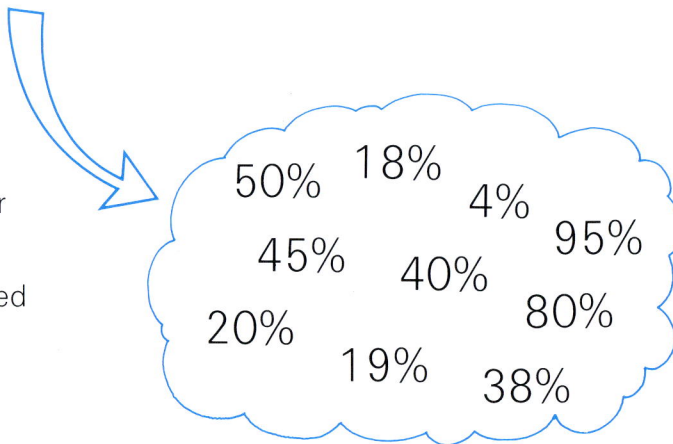

50% 18% 4% 95% 45% 40% 20% 80% 19% 38%

3 Copy and complete:

a)
5% means
5 in every 100
1 in every ☐
4 in every ☐
☐ in every 200

b)
20% means
20 in every ☐
10 in every ☐
☐ in every 25
25 in every ☐

c)
75% means
☐ in every 100
☐ in every 4
☐ in every 20
12 in every ☐
☐ in every 1000

Estimating

C 1 a) About how much would you save on the bike? Write down a rough answer.

  b) Jim uses his calculator to find the exact saving. He presses

  $\boxed{C}\ \boxed{2}\ \boxed{5}\ \boxed{\div}\ \boxed{1}\ \boxed{0}\ \boxed{0}\ \boxed{\times}\ \boxed{1}\ \boxed{9}\ \boxed{7}\ \boxed{=}$

   (i) Will this give him the correct answer? Why?

   (ii) Calculate the exact saving yourself.

CYCLES

SAVE 25%

WAS £197

12% is about 10%.

2 These bicycles are all reduced by 12%.

  a) **Estimate** how much you would save on each one.

  b) Use your calculator to work out the exact saving.

SALE - 12% off all bikes

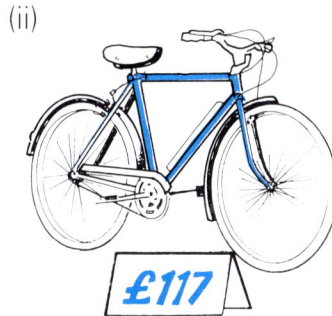

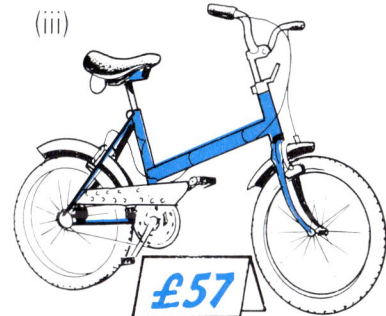

(i) £255

(ii) £117

(iii) £57

Challenge

3 These computers are all reduced by 33%.

  a) **Estimate** how much you would save on each one.

  b) Use your calculator to work out the saving to the nearest 1p.

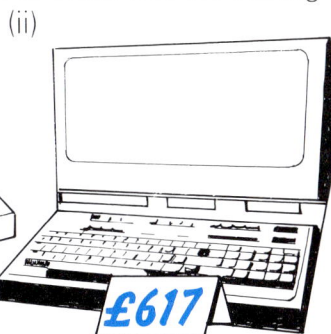

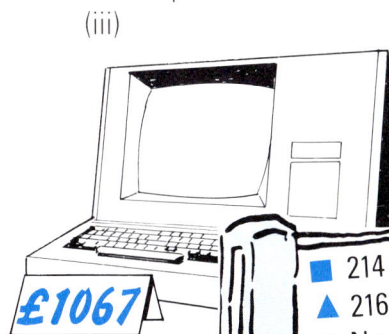

(i) £240

(ii) £617

(iii) £1067

■ 214
▲ 216
● Next chapter

Fractions and percentages

23

A
B 1 Copy and complete the table:

Fraction	$\dfrac{?}{100}$	Percentage
$\dfrac{1}{10}$	$\dfrac{\ }{100}$	%
$\dfrac{1}{2}$	$\dfrac{\ }{100}$	%
$\dfrac{1}{5}$	$\dfrac{\ }{100}$	%
$\dfrac{3}{4}$	$\dfrac{\ }{100}$	%
$\dfrac{7}{10}$	$\dfrac{\ }{100}$	%

2 a) Copy the table.
    Try to complete it by doing the calculations in your head.

Amount	10%	5%	15%	20%	25%	50%	75%
500 kg	50 kg						
450 m							
£15							
£95							

5% is a half of 10%.

10% of 500 kg is one-tenth of 500 kg.

b) Check your work with your calculator.

3 These are Meg's autumn and summer term test scores.
   In which term did she do better

   a) in Mathematics?

   b) in English?

   c) in Home Economics?

   d) in Technology?

Autumn term marks

Subject	Mark
Mathematics	$\dfrac{17}{20}$
English	$\dfrac{35}{50}$
Home Economics	$\dfrac{22}{25}$
Technology	$\dfrac{30}{40}$

Summer term marks

Subject	%
Mathematics	90
English	80
Home Economics	80
Technology	78

Estimating percentages

C Do not use a calculator on this page.

1 Write down a rough answer to each of these.

a)

12% of £50.00

b)

23% of £8

c)

47% of 36 kg

d)

76% of 12 litres

2 About how much would you save on each pair of shoes in the sale?

SALE SAVE 25% SALE

a)

WERE £15.83

b)

WERE £19.36

c)

WERE £37.77

With a friend

3 Work together.
Find the original price of each TV.

SALE 10% off all electrical goods

a)

WAS ? NOW £360

b)

WAS ? NOW £126

▲ 216
● Next chapter

Comparisons

D 1 Midge has a dog and a cat. She feeds the cat on Go-Cat and the dog on Frolic.

a) Which has the larger percentage of protein, Go-Cat or Frolic?

b) How many grams of protein are there in
   (i) a pack of Go-Cat?
   (ii) a pack of Frolic?

c) Midge's cat has 85 g of Go-Cat each day. Her dog has 150 g of Frolic. Which has more protein each day, the cat or the dog?

d) Which pet has more
   (i) ash
   (ii) fibre
   each day?

23

To open push here and tear back

MACKEREL & HERRING VARIETY
Go-Cat
'GO-CAT' COMPLETE CAT FOOD
Serve 'Go-Cat' either on its own or together with canned cat food.

INGREDIENTS:
Cereals, vegetable protein extracts, meat and animal derivatives, fish and fish derivatives, minerals, oils and fats, yeast.
Contains permitted antioxidants and preservatives.
Vitamin levels guaranteed until date on top of pack.

NUTRITIONAL ANALYSIS PER KG
Protein   30%
Oil   8%
Fibre   3%
Ash   7.5%
Vitamin A 11000 i.u.
Vitamin D 1000 i.u.
Vitamin E   80 i.u.
Selenium 0.17 mg

Store the opened pack in a cool, dry place.

Carnation
36 Park Street, Croydon, Surrey CR9 1TT

400 g
14.1 oz

Frolic
made with REAL BEEF and LIVER
Frolic
The Complete Food Dogs Adore

500 g

Frolic
Pedigree Petfoods, Shrewsbury Avenue, Peterborough, PE2 0BY.
FROLIC is a complete compound food for dogs

INGREDIENTS
Cereals, meat and animal derivatives, various sugars, vegetable protein extracts, oils and fats, minerals, vitamins A, D and E. Contains permitted preservatives, permitted colourants, permitted antioxidant.
Contents per 100g:
Protein 17g; Oil 6g;
Ash 6g; Fibre 2.5g
Vitamins (I.U. per kg)
Vitamin A   = 3,900
Vitamin D   = 390
Vitamin E   =   40

Vitamin content is guaranteed for 18 months from the date of manufacture shown on the top of the packet.

See back panel for feeding instructions

Frolic is a registered trademark.
© Copyright Pedigree Petfoods 1983

500 g

made with REAL BEEF and LIVER

THE GO...
THE CRI...

INGREDIENTS:
Cereals, vegetable protein extracts, meat and animal derivatives, fish and fish derivatives, minerals, oils and fats, yeast.
Contains permitted antioxidants and preservatives.
Vitamin levels guaranteed until date on top of pack.

NUTRITIONAL ANALYSIS PER KG
Protein   30%
Oil   8%
Fibre   3%
Ash   7.5%
Vitamin A 11000 i.u.
Vitamin D 1000 i.u.
Vitamin E   80 i.u.
Selenium 0.17 mg

INGREDIENTS
Cereals, meat and animal derivatives, various sugars, vegetable protein extracts, oils and fats, minerals, vitamins A, D and E. Contains permitted preservatives, permitted colourants, permitted antioxidant.
Contents per 100g:
Protein 17g; Oil 6g;
Ash 6g; Fibre 2.5g
Vitamins (I.U. per kg)
Vitamin A   = 3,900
Vitamin D   = 390
Vitamin E   =   40

Vitamin content is guaranteed for 18 months from the date of manufacture shown on the top of the packet.

2 Which is the larger amount?

a) $\frac{1}{3}$ of the glass of lemonade, or 30% of it

b) $\frac{3}{4}$ of 80 kg, or 70% of 80 kg

c) $\frac{1}{2}$ of 50 cm, or 55% of 50 cm

d) $\frac{1}{10}$ of 12 l, or 12% of 12 l

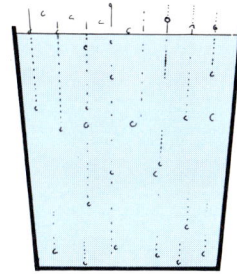

3 Which is the larger amount?

a) $\frac{73}{100}$ of the glass of lemonade, or 72% of it

b) $\frac{40}{100}$ of 50 kg, or 42% of 50 kg

c) $\frac{3}{10}$ of 75 cm, or 28% of 75 cm

d) $\frac{2}{5}$ of 100 m, or 25% of 100 m

e) $\frac{1}{8}$ of 80 l, or 8% of 80 l

f) $\frac{1}{5}$ of 5 m, or 5% of 5 m

4 Sleep-and-Dream labels tell you the proportion of materials used, by mass. Pillowcases weigh 120 g each.

a) How many grams of cotton and how many of linen are needed for each pillowcase?

b) 80 g of cotton are used for each Sleep-and-Dream sheet. What does a sheet weigh?

●Next chapter

24 Fixing positions

A 1 **DO YOU REMEMBER...?**

*The map shows two villages.
Imagine you are at Blagdon.
The position of Neasdon from Blagdon is
(4 km E, 3 km N).*

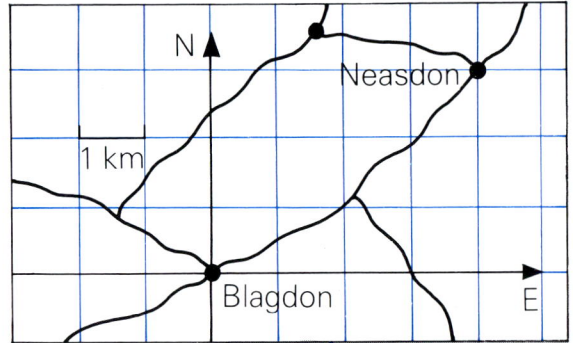

Write down the position of Blagdon from
Neasdon, like this:

(☐ km W, ☐ km S).

2 This is a map of the Penzance area.

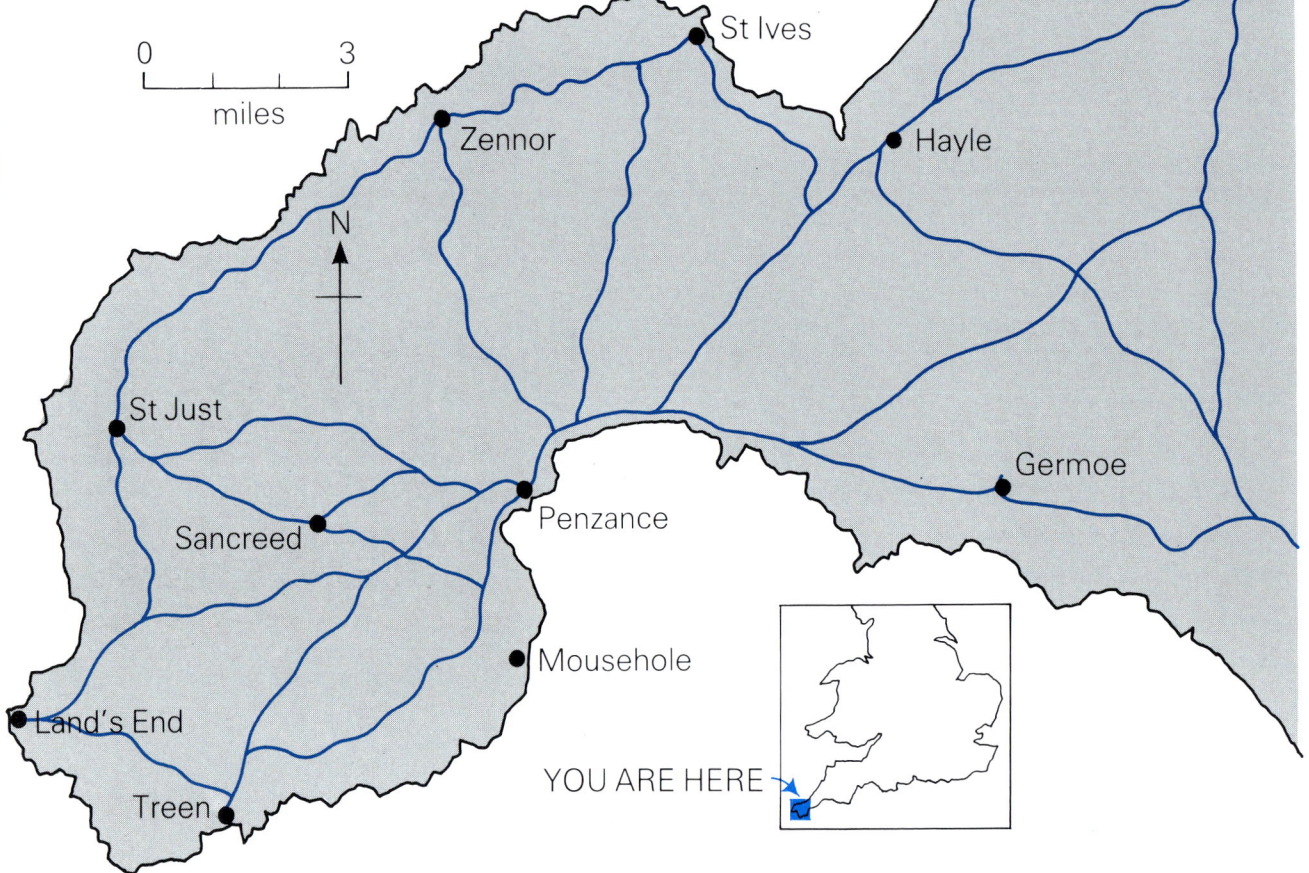

YOU ARE HERE

a) These are the approximate positions of three towns relative
   to Penzance.
   Which towns are they? *from*
   (i) (6 miles W, 1 mile N)
   (ii) (7 miles E, 0 miles N) (iii) (5½ miles E, 5 miles N)

b) What is the approximate position of these places from Sancreed?

   (i) Land's End (ii) Mousehole (iii) St Just

B 1 DO YOU REMEMBER ...?

*We can also fix a position
by using a distance and a direction.
Imagine you are at Hallington.
The position of Bettle
from Hallington is
(10 km, N 80° E).*

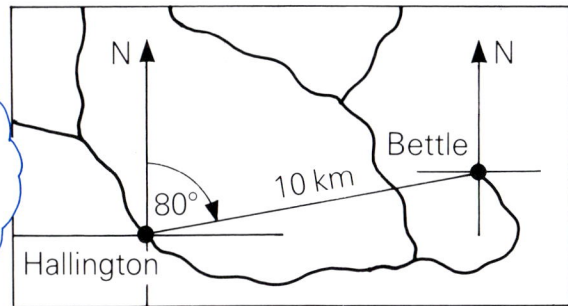

Write down the position of Hallington relative to
Bettle, like this: (☐ km, S ☐° W).

from

2   You need a ruler and a protractor.
    These are the approximate
    positions of three cities
    relative to Birmingham.
    Which cities are they?

    a) (110 km, S 20° W)

    b) (330 km, N 45° W)

    c) (150 km, S 40° E)

3   Write down the
    approximate
    positions of these
    cities relative to
    Manchester:

    a) Birmingham

    b) Aberdeen

    c) Bristol

Think it through

4   Write the approximate position of Bristol
    relative to Belfast in these two ways.
    a) (☐ km ☐, ☐ km ☐)

    b) (☐ km, ☐☐°☐)

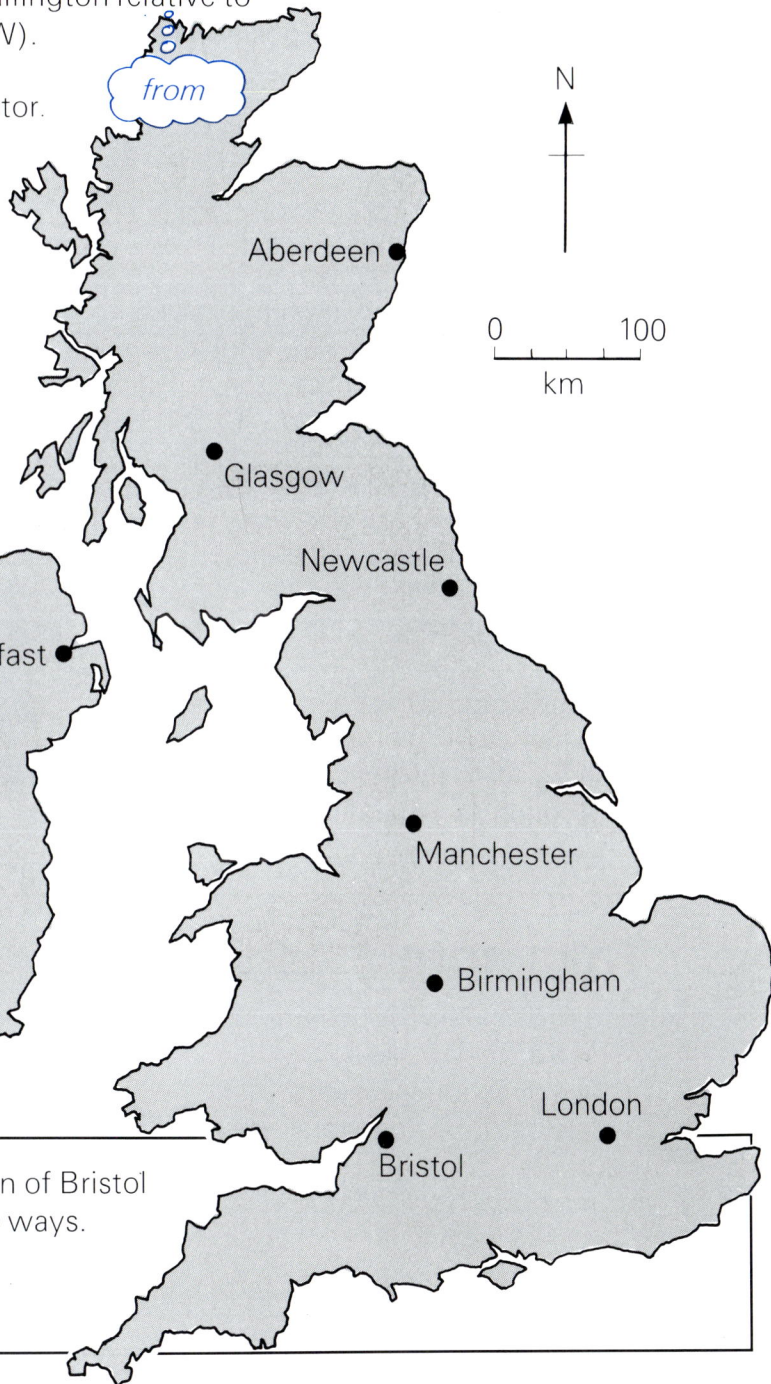

● 219

With a friend

5 Discuss each of these questions with your friend.
Each of you write down what you decide.

a) Glenda and Duke are on an outward bound course.
They radio their position to the leader at base camp.
The leader writes their position like this:
(7 km, 060°).
Write down what you think she means.

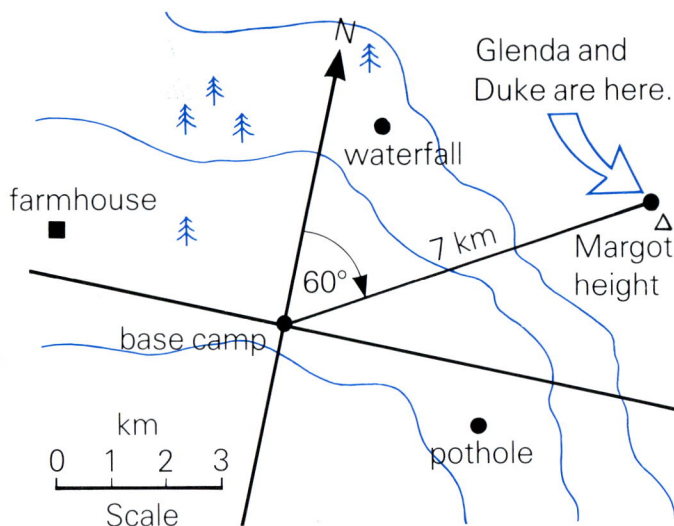

b) Jamal and Rupinder are at one of the places marked on the map.
They radio their position to base camp.
The leader writes: (4 km, 115°).
Where do you think they are, . . .
the farmhouse,
the waterfall,
or the pothole?

c) Harry and Ben are at the waterfall.
They radio their position.

(4 km, N 15° E)

Write down what you think the leader writes.

d) This is how the leader writes the position of Ralph and Melanie: (8 km, 200°).
Write their position like this: (☐ km, S☐° W).

Take note

We sometimes write positions using
● a distance
and ● a direction measured clockwise from north.

The position of Exon from Tulse is (7 km, 105°).

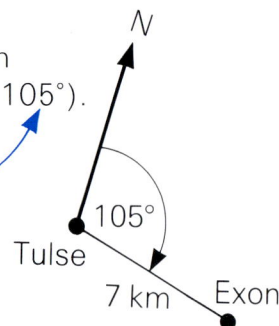

Take note

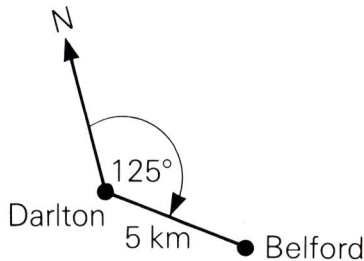

Relative to Darlton, Belford is at the position (5 km, 125°).
We call a direction written like this a three-figure **bearing**.
We say the **bearing** of Belford from Darlton is 125°.

Measure from the north. Always measure clockwise. Always write three digits ... 015° not 15°.

6   The map shows some airports in England and Wales.

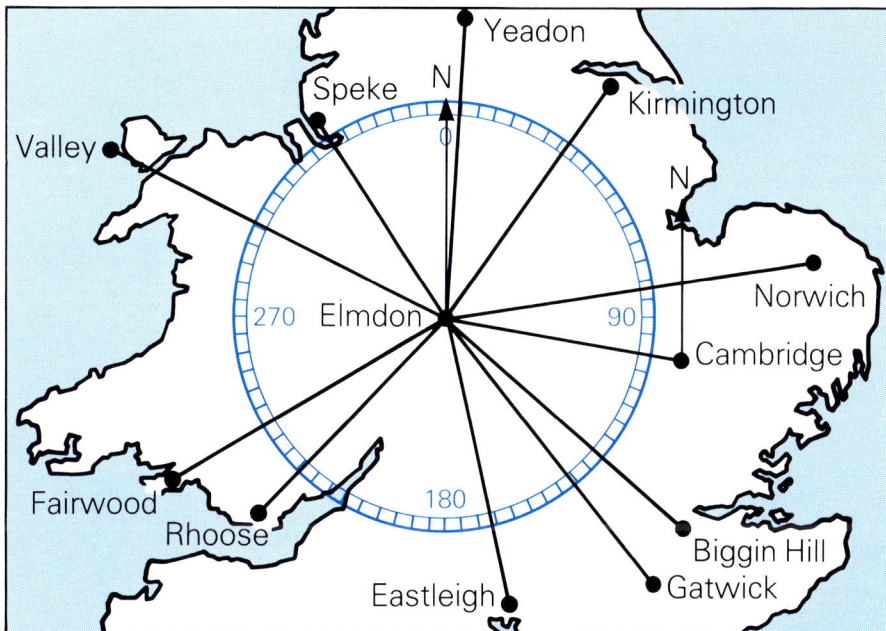

a) Which airport is almost due south of Elmdon?

b) From Elmdon which airport is on a bearing of about
   (i) 100°   (ii) 225°   (iii) 035°?

c) Write down the approximate bearings of these airports from Elmdon:
   (i) Kirmington   (ii) Gatwick   (iii) Valley   (iv) Rhoose

   Remember to use three digits.

d) Use a circular protractor.
   Measure the bearings of these airports from Cambridge:
   (i) Norwich   (ii) Elmdon   (iii) Fairwood

24

7 The map shows some mountains in the Yorkshire dales.

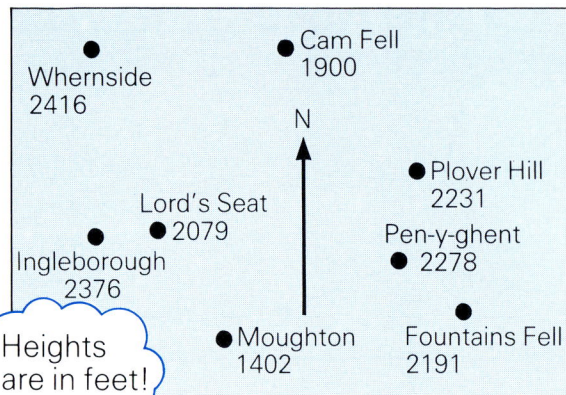

Cam Fell 1900

Whernside 2416

N

Lord's Seat ● 2079

Plover Hill 2231

Ingleborough 2376

Pen-y-ghent ● 2278

Heights are in feet!

● Moughton 1402

Fountains Fell 2191

These pictures are taken from the summit of Moughton.

a)

Looking on a bearing of 309°.

b)

Looking on a bearing of 065°.

Looking on a bearing of 335°.

c)

d)

For each picture, decide from the map which mountain it shows.
Write down the name of each mountain.

Looking on a bearing of 012°.

■ 223
▲ 225
● Next chapter

24

Positions and bearings

A 1 a) Check on the map that the quarry is 4 km due south of Shady Pond.

b) Is Gray's Farm due east or due west of the church?

c) How many kilometres is it from the church to Gray's Farm?

d) You are at the Old Moat House. Which places on the map are at these positions relative to you?

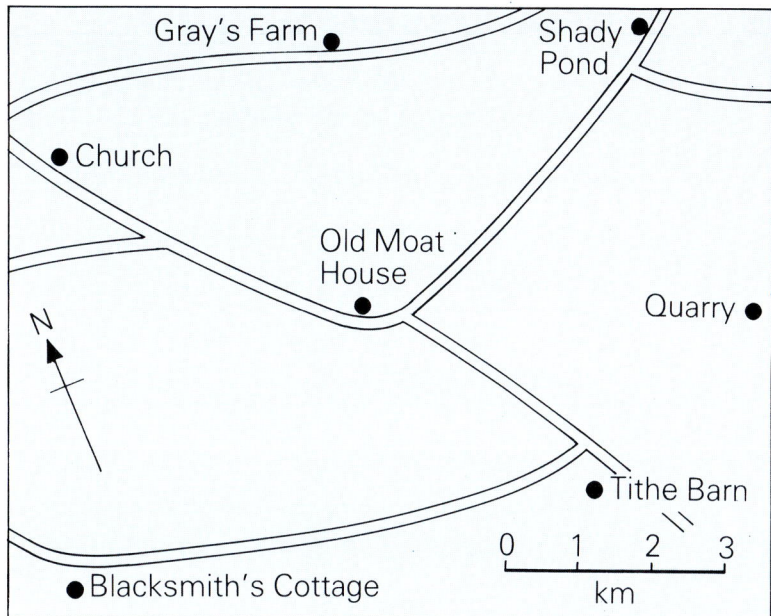

(i) (5 km W, 2 km S)   (ii) (5 km E, 2 km N)   (iii) (2 km E, 3½ km S)

e) What is the approximate position of Shady Pond relative to the church?
Write the position like this:

(☐ km E, ☐ km S).

2 Dove House is not on the map, but its position relative to the Post Office is (3 km E, 2 km S).
What is the position of the Post Office relative to Dove House?

Challenge

3 a) The position of place A relative to place B is (4 km W, 6 km N).
What is the position of place B relative to place A?

b) Suppose you know the position of place X relative to place Y.
How quickly can you write down the position of Y relative to X?
Try some examples of your own.
Write down a simple rule for doing this.

With a friend: Which town?

1 You need an atlas of the British Isles and a protractor.
   Find out the names of these towns:

   a) a port on the south coast on a bearing of 113° from London

   b) a cathedral city near Stonehenge on a bearing of 247° from London

   c) a Welsh city with a famous castle on a bearing of 301° from London

   d) a city with an archbishop on a bearing of 344° from London

The city, not the archbishop!!

2 a) Shannon and Zendar are two oil rigs.
      Their positions are shown on the sketch.
      Write down the bearing of Zendar from Shannon.

   b) Maxim I is another oil rig.
      Its direction relative to Shannon is S 40° E.
      Write this as a three-figure bearing.

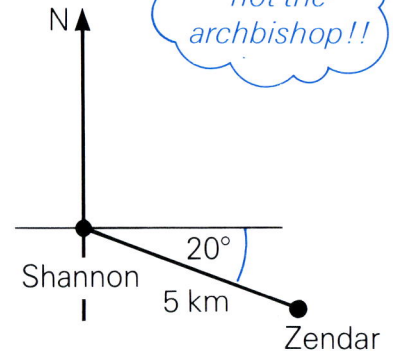

3 Write these directions as three-figure bearings:

   a) N 20° E   b) N 70° E   c) S 50° E

   d) S 10° W   e) east   f) west   g) N 10° W

Challenge

4 The bearing of Barnham
   from Axecliff is 123°.
   What is the bearing of
   Axecliff from Barnham?

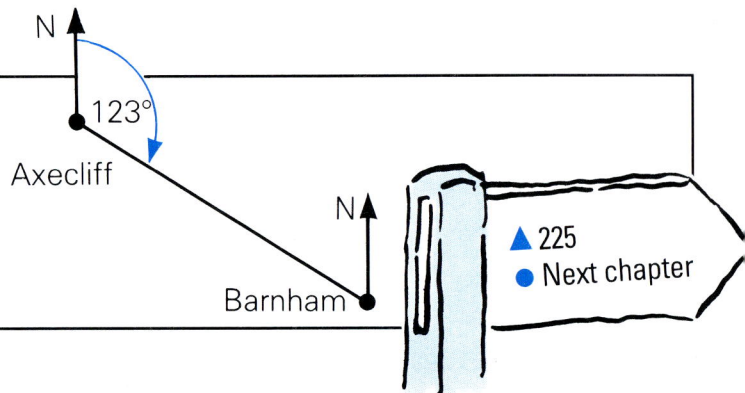

▲ 225
● Next chapter

Bearings, positions and flight

C   Challenge 1

1   You need a protractor.

A helicopter starts at
Aberystwyth.
It visits six of the towns
shown on the map, then
flies back to Aberystwyth.
These are the directions
in which it flies as it
leaves each town:

    stage 1   091°
    stage 2   195°
    stage 3   326°
    stage 4   077°
    stage 5   263°
    stage 6   200°
    stage 7   back to Aberystwyth

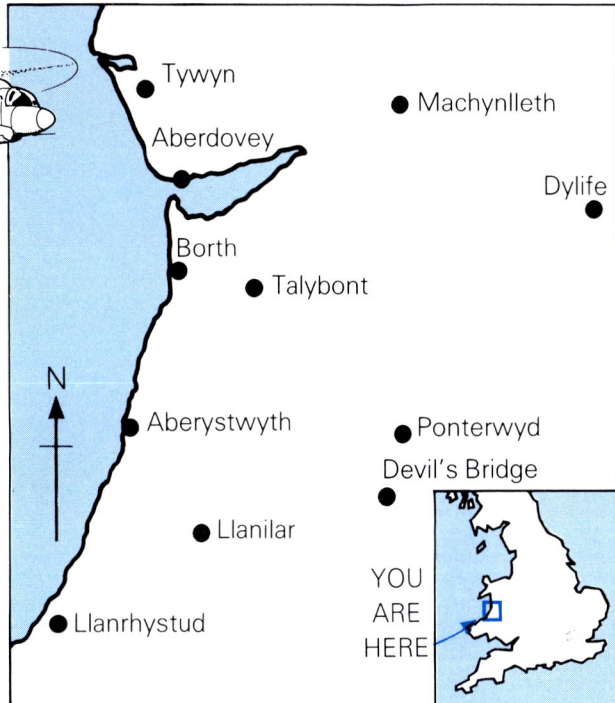

Tywyn
Machynlleth
Aberdovey
Dylife
Borth
Talybont
N
Aberystwyth
Ponterwyd
Devil's Bridge
Llanilar
YOU
ARE
HERE
Llanrhystud

a) Which six towns does it visit?

b) Approximately, on what bearing does it fly during stage 7?

Challenge 2

2   The map shows five oil rigs.
A service helicopter flies
from oil rig Axis on a
bearing of 217°.
It arrives at Sweetholm.
From there it flies on a
    bearing of 302° to Seadove.
From there it flies on a
    bearing of 271° to Target.
From there if flies on a
    bearing of 061° to Harvik.
Which oil rig is which?

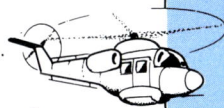

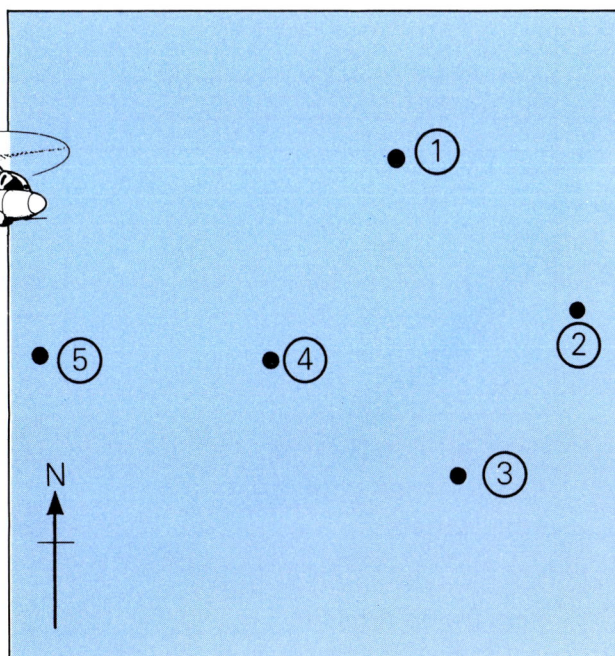

①
②
⑤
④
③
N

3 **a)** Make an accurate
copy of the radar
screen.

b) The coloured line
shows the path of
an aircraft.
At 10:00 it is
at the position
(40 km, 270°).
It is coming
in to land at
Heathrow.

What is its
position at
   (i) 10:03?
  (ii) 10:08?
 (iii) 10:11?

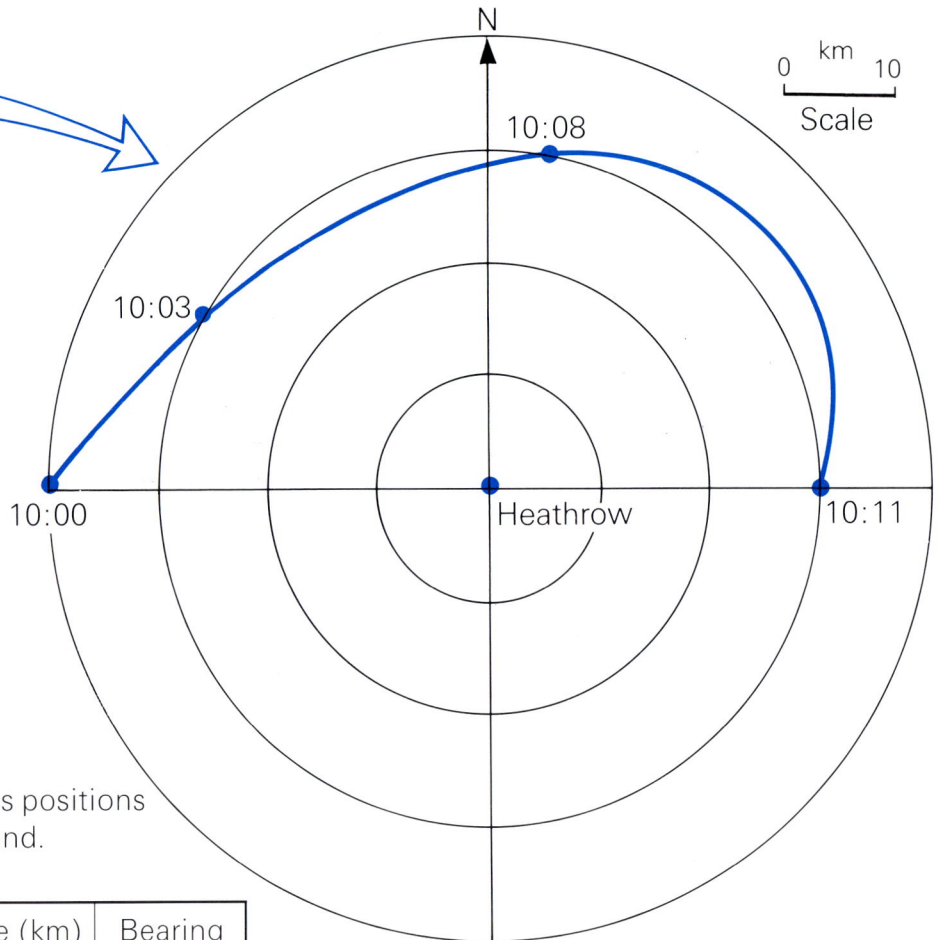

N

10:08

0 km 10
Scale

10:03

10:00

Heathrow

10:11

c) The table shows its positions
as it comes in to land.

Time	Distance (km)	Bearing
10:00	40	270°
10:03	30	300°
10:08	30	010°
10:11	30	090°
10:13	20	200°
10:15	20	260°
10:16	10	330°
10:17	5	030°
10:18	0.5	230°

Plot the flight path of
the aircraft on your
radar screen.

d) In which direction do you
think it was travelling
as it touched down?

● Next chapter

25 Working with letters

A

DO YOU REMEMBER ...?

We can use letters to give instructions.

Instructions in words

The length is twice the width.

Instructions using letters

$2 \times k$ cm

k cm

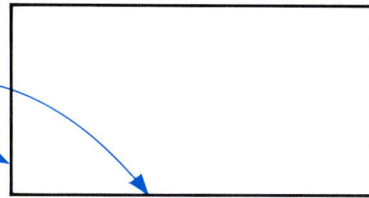

1 Phil found this new design for a commemorative stamp.

By choosing different numbers for k the designer gets different sized stamps.

18p

WILD LIFE YEAR

k cm

$2 \times k$ cm

a) What number is chosen for k to give this version?

b) What number is chosen to give this version?

18p

WILD LIFE YEAR

18p

WILD LIFE YEAR

c) One version of the stamp has a perimeter of 12 cm.
   (i) What number is chosen for k?
   (ii) What size is the stamp?

distance around the edge

d) One of these versions does not obey the instructions. Which one?

A

B

C

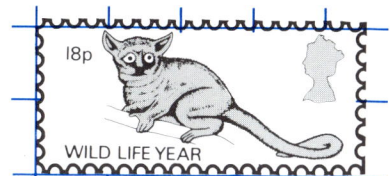

2 Ben drew this 'ripple' of rectangles.

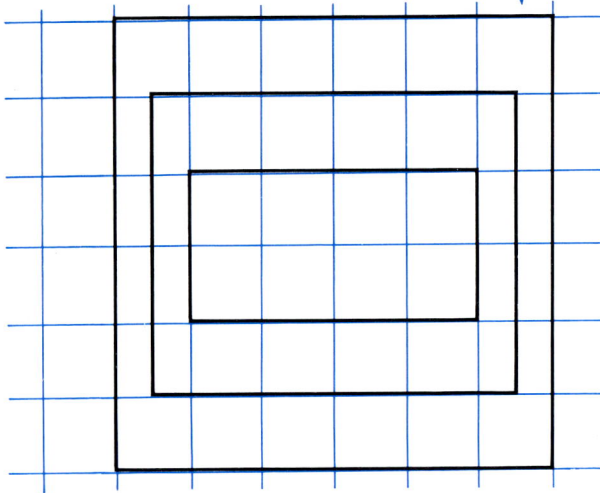

He used these instructions:

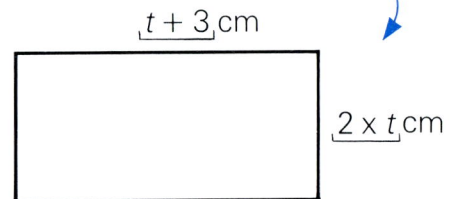

$t + 3$ cm

$2 \times t$ cm

a) Look at Ben's smallest rectangle.
   What number did Ben choose for t?

b) What number did he choose for t for the next rectangle?

c) One choice for t gives a square.
   What number is it?

d) Copy Ben's 'ripple'.
   Draw the rectangle when t is chosen to be
   (i) 4   (ii) 5   (iii) 6.

e) Think about choosing larger and larger numbers for t.
   Does the rectangle become

   long and thin, like this

   or tall and thin, like this

   or square-shaped, like this

   ?

Challenge

3 Look at the 'ripple' of rectangles.
   Copy and complete these
   instructions for it.

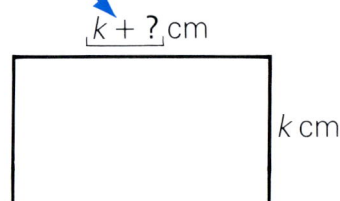

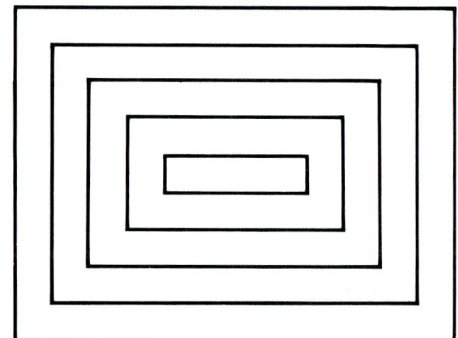

   $k + ?$ cm

   k cm

B *Think it through*

1 Here are seven sets of instructions.
   Three of them give a set of square 'ripples'.
   Which ones?

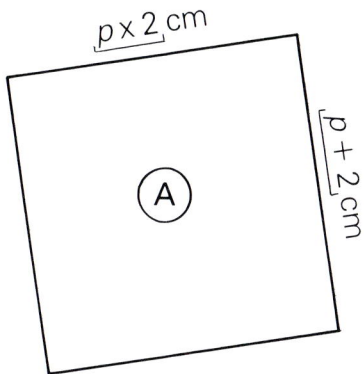

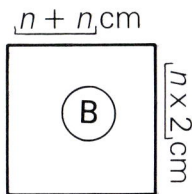

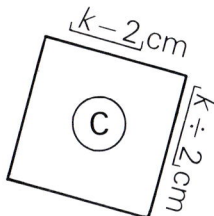

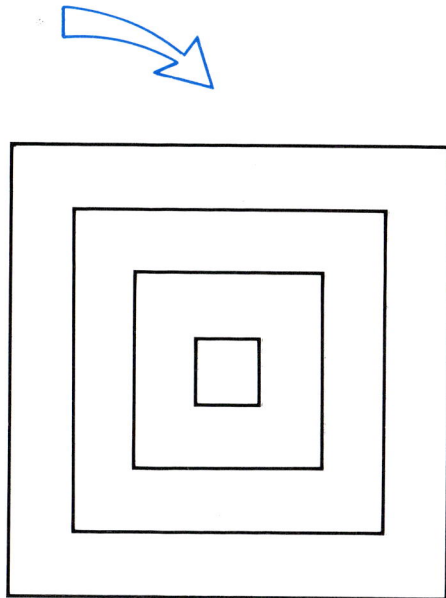

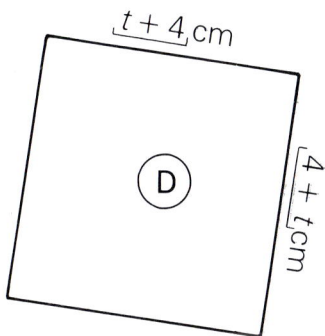

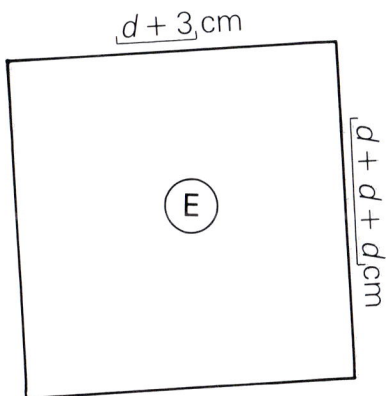

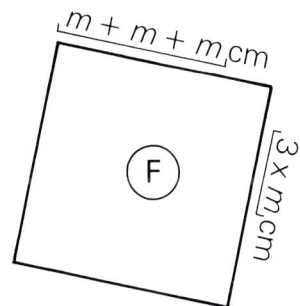

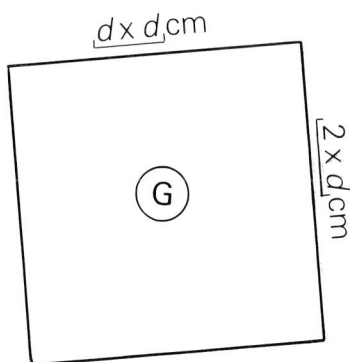

$p \times 2$ cm

$p + 2$ cm

A

$n + n$ cm

$n \times 2$ cm

B

$k - 2$ cm

$k \div 2$ cm

C

$t + 4$ cm

$4 + t$ cm

D

$d + 3$ cm

$d + d + d$ cm

E

$m + m + m$ cm

$3 \times m$ cm

F

$d \times d$ cm

$2 \times d$ cm

G

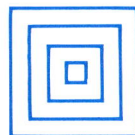

We get a square set of 'ripples'.
This shows that $n + n$ cm and $n \times 2$ cm are the same distance, no matter what number we choose for n.

2 Look carefully at your result for **B**.
   It should show that $n + n$ can be written as $n \times 2$.
   Write each of these using $\times$:

   a) $d + d + d$ b) $k + k$ c) $b + b + b + b$

3 **a)** Check that these two expressions stand for the same number when p is chosen to be 3.

$$2 \times p \qquad p + 3$$

b) Check that they stand for different numbers when p is chosen to be 2.

c) Can you find another choice for p which makes them stand for the same number?
If so, write it down.

d) Here are some more pairs of expressions.
For each pair, choose numbers for the letter.
Try to make the two expressions stand for the same number.
  • For some you will find only one possibility.
  • For some you will find none.
  • For others you will find many.
Write down what you find out for each pair.

(i)
$$t - 7 \qquad t + 1$$

(ii)
$$3 \times k \qquad k + k + k$$

(iii)
$$2 \times m \qquad m + 4$$

(iv)
$$k \div 2 \qquad 2 \div k$$

(v)
$$k \div 2 \qquad k - 3$$

(vi)
$$n \times 15 \qquad 15 \times n$$

(vii)
$$12 - t \qquad 3 \times t$$

(viii)
$$d + d \qquad d \times d$$

25

4 Here are some pairs of expressions.
In some pairs, the two expressions stand for the same number
for all numbers you might choose for the letter.
Which pairs are they?

A $p + p$ | $2 \times p$

B $k \times k$ | $k + k$

C $t + 2$ | $2 + t$

D $m \div 4$ | $4 \div m$

E $n+n+n$ | $3 \times n$

F $4 \times h$ | $h \times 4$

5 Copy and complete the **Take note**.

Take note

$2 \times n$ is a different way of writing $n + \square$ and $n \times \square$.

$3 \times p$ is a different way of writing $\square + p + p$ and $p \times \square$.

$k \times 4$ is a different way of writing $\square + k + k + k$ and $4 \times \square$.

6 a) $2 \times n$ cannot be rewritten as $n + 2$.

Explain why not.

b) $5 + m$ cannot be rewritten as $5 \times m$.

Explain why not.

Challenge

7 a) Write down the perimeter
of the square
   (i) using three $+$s.
   (ii) using one $\times$.

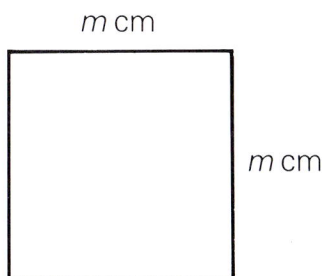

b) Write down the perimeter
of the rectangle
   (i) using three $+$s and two $\times$s.
   (ii) using five $+$s.
   (iii) using one $\times$.

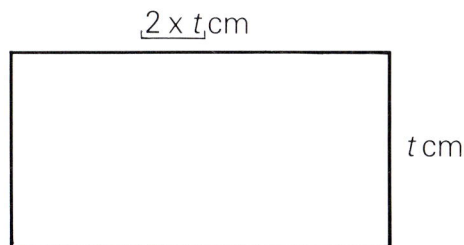

m cm

m cm

$2 \times t$ cm

t cm

C 1 These are designs for shapes to be made from gold wire.

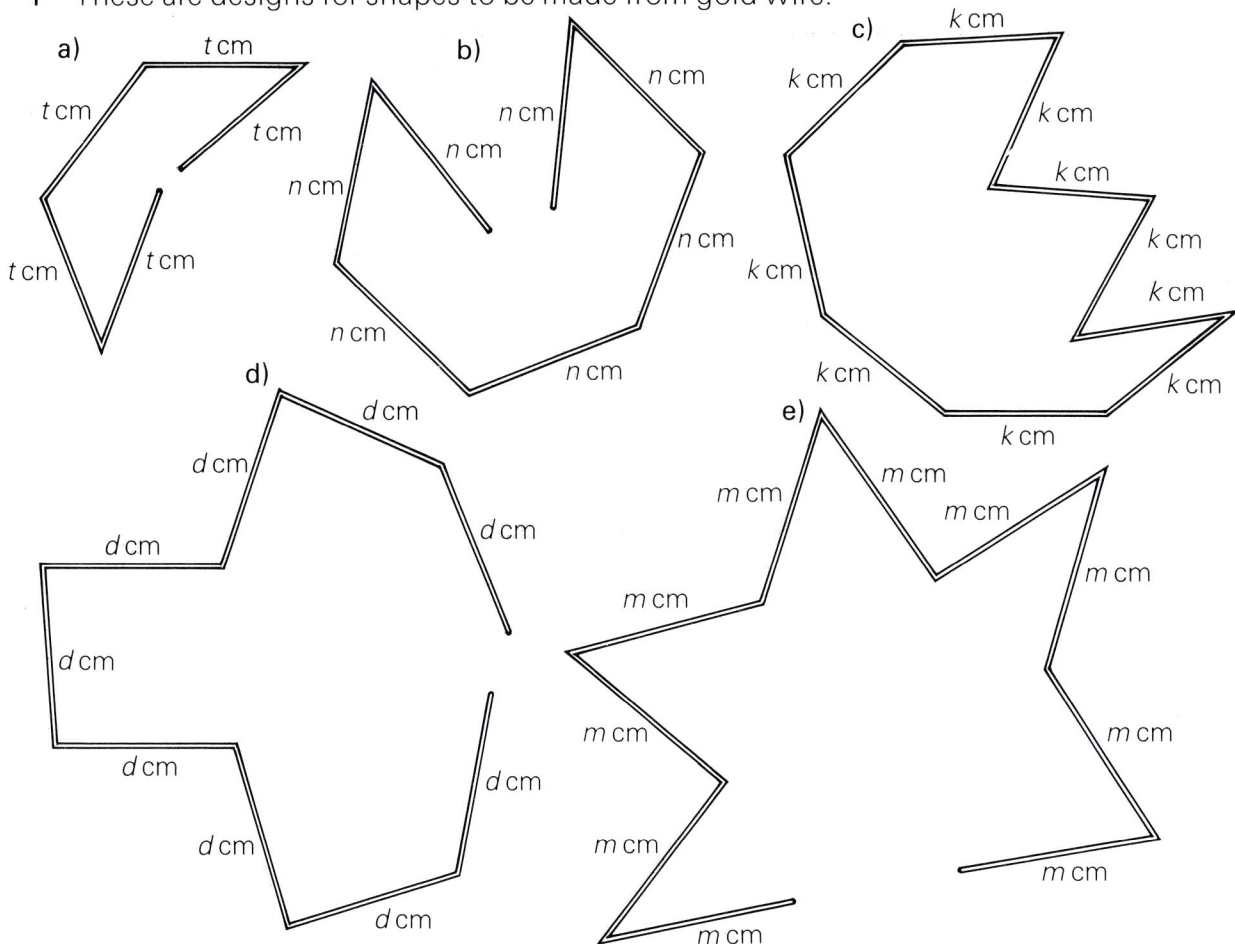

a) *t* cm, *t* cm, *t* cm, *t* cm, *t* cm, *t* cm

b) *n* cm, *n* cm, *n* cm, *n* cm, *n* cm, *n* cm, *n* cm

c) *k* cm, *k* cm, *k* cm, *k* cm, *k* cm, *k* cm, *k* cm, *k* cm, *k* cm, *k* cm

d) *d* cm, *d* cm, *d* cm, *d* cm, *d* cm, *d* cm, *d* cm, *d* cm, *d* cm

e) *m* cm, *m* cm, *m* cm, *m* cm, *m* cm, *m* cm, *m* cm, *m* cm, *m* cm, *m* cm

Write down an expression for the length of gold wire in each shape
(i) using +s   (ii) using a ×.

Challenge

2 The perimeter of each triangle can be written very simply.
Show how for each one.

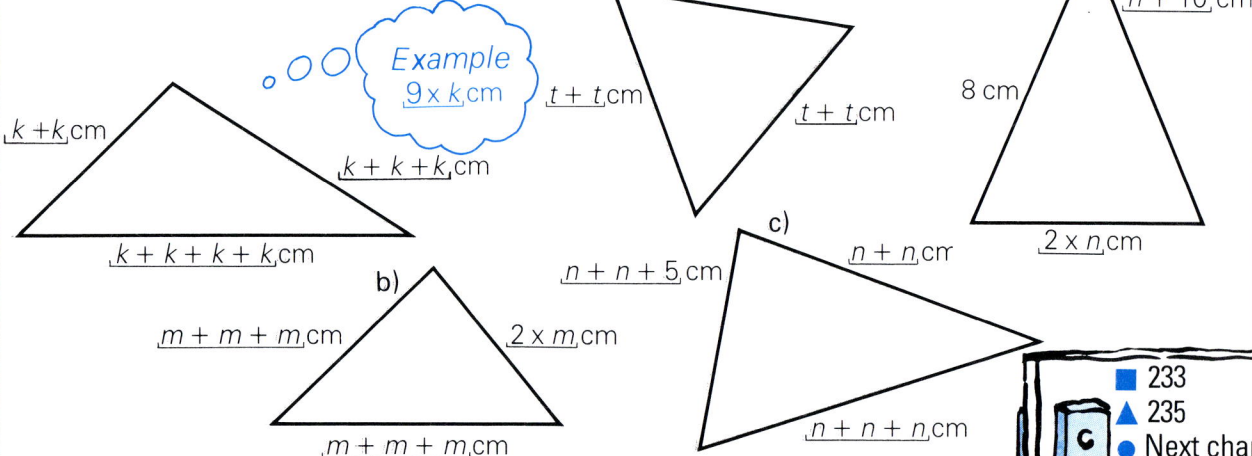

Example
$9 \times k$ cm

$k + k$ cm
$k + k + k$ cm
$k + k + k + k$ cm

a) $t + t$ cm, $t + t$ cm, $t + t$ cm

b) $m + m + m$ cm, $2 \times m$ cm, $m + m + m$ cm

c) $n + n + 5$ cm, $n + n$ cm, $n + n + n$ cm

d) $n + 10$ cm, 8 cm, $2 \times n$ cm

■ 233
▲ 235
● Next chapter

Instructions with letters

A 1 The teacher wants everybody in Horace's class to draw a rectangle.
She writes these instructions
on the board, and says:

DRAW YOUR RECTANGLE ON 1cm SQUARED PAPER. CHOOSE YOUR OWN NUMBER FOR k.

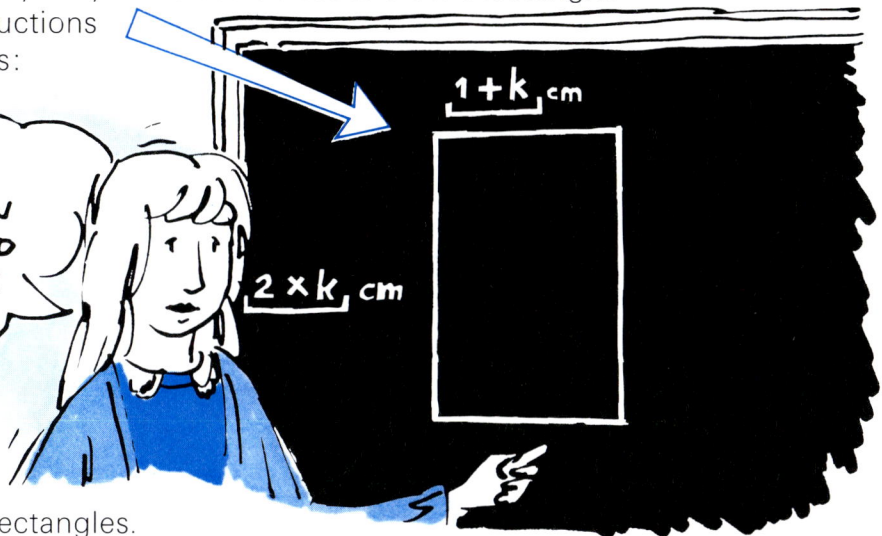

$1+k$ cm

$2 \times k$ cm

Here are some of the rectangles.

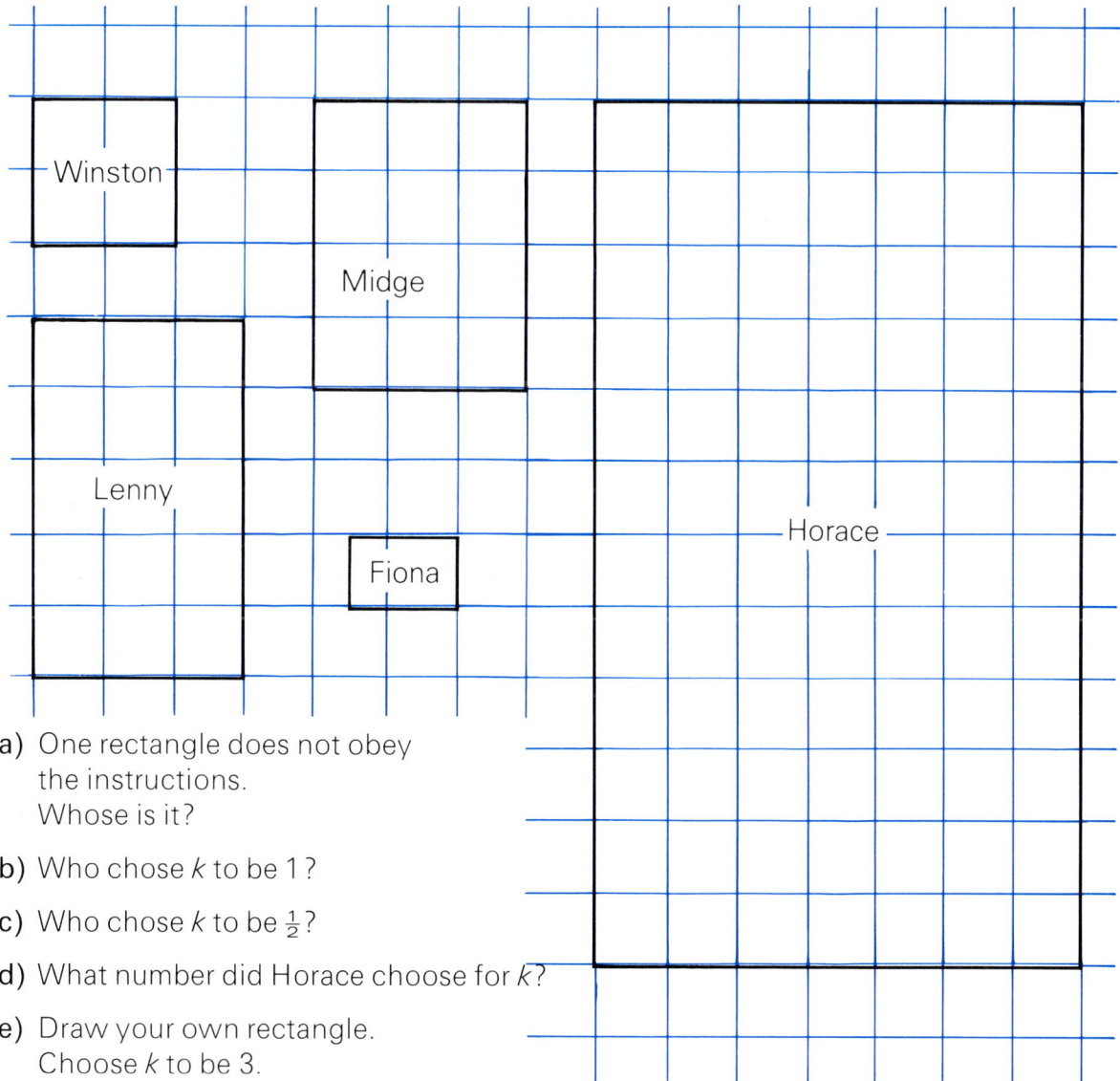

Winston

Midge

Lenny

Fiona

Horace

a) One rectangle does not obey
the instructions.
Whose is it?

b) Who chose k to be 1?

c) Who chose k to be $\frac{1}{2}$?

d) What number did Horace choose for k?

e) Draw your own rectangle.
Choose k to be 3.

Rewriting expressions

B
C

1 Rewrite each of these using × :

   a) $p+p$ b) $k+k+k$ c) $n+n+n+n+n$ d) $m+m+m$

2 Rewrite each of these using +s:

   a) $2 \times n$ b) $3 \times n$ c) $d \times 4$ d) $h \times 2$

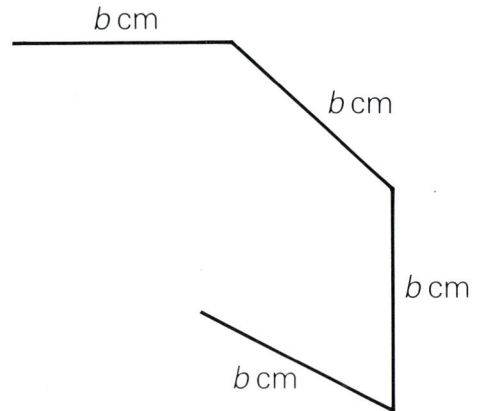

3 Write down the total length of the line

   a) using three +s.

   b) using one × .

4 Write down the perimeter of each shape
   (i) using +s.
   (ii) using × .

25

a)

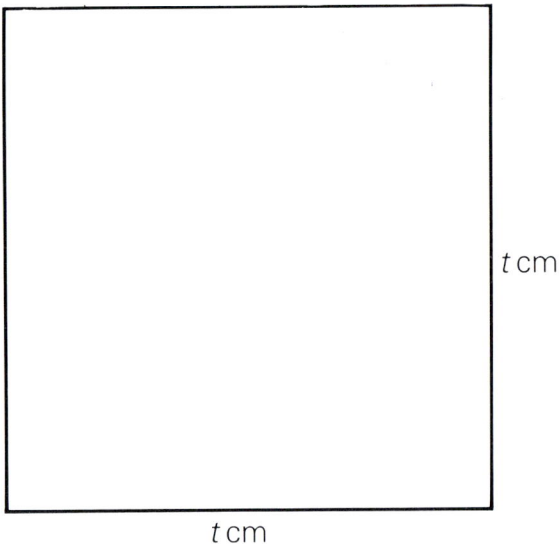

b)

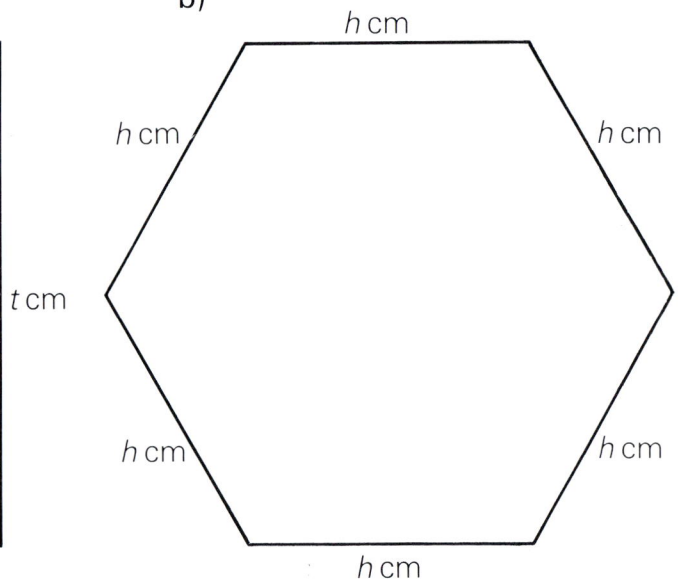

Think it through

5 **a)** Sketch a square.
     Write in lengths for its sides using n, so that its perimeter is $4 \times n$ cm.

   b) Sketch a rectangle.
     Write in lengths for its sides using n, so that its perimeter is $6 \times n$ cm.

   c) Sketch a triangle.
     Write in lengths for its sides using p, so that its
     perimeter is $8 \times p$ cm.

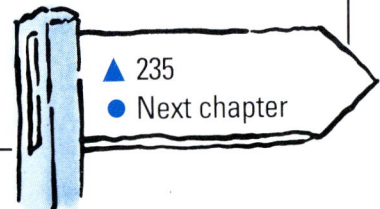

▲ 235
● Next chapter

Solving problems with letters

D 1 The area of a square is $k \times k$ cm².
What is its perimeter?
Write an expression using k.

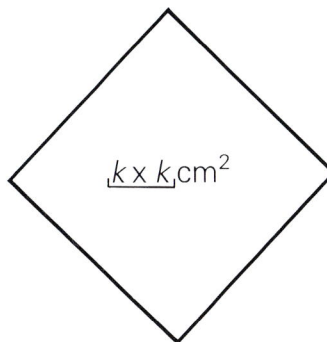

2 The perimeter of a square is $2 \times p$ cm.
How long is each side?
Write an expression using p.

3 The perimeter of a rectangle is $28 \times n$ cm.

a) Write down one possible length and width using n.

b) Write down another possible length and width using n.

4 Copy and complete:
Area of the triangle $= \square \times \square \div \square$ cm².

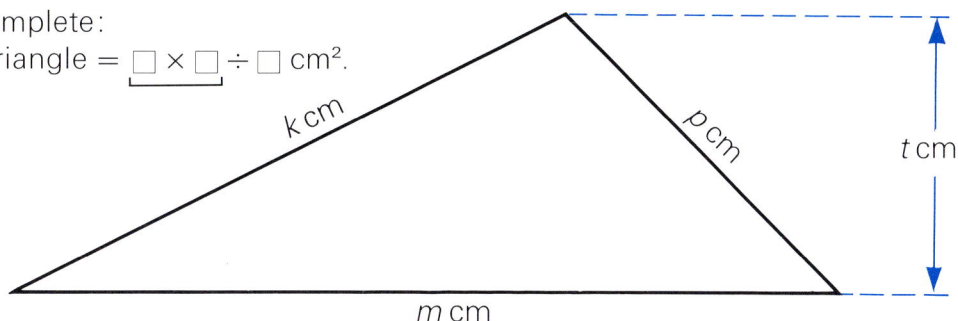

25

5 How many centimetres longer than the width is the length of this rectangle?
Write an expression using m.

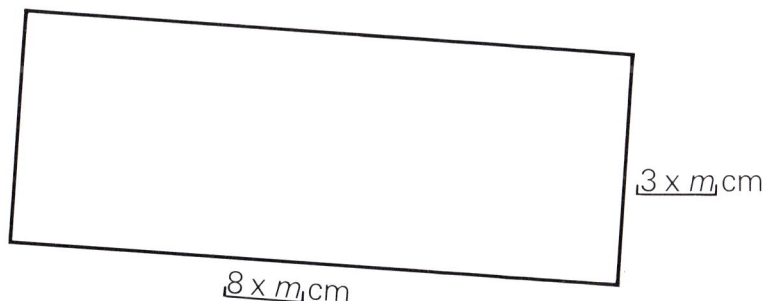

$3 \times m$ cm

$8 \times m$ cm

Challenge

6 What is the perimeter
of this shape?
Write an expression using p.

*You will need to use
one × and one +.*

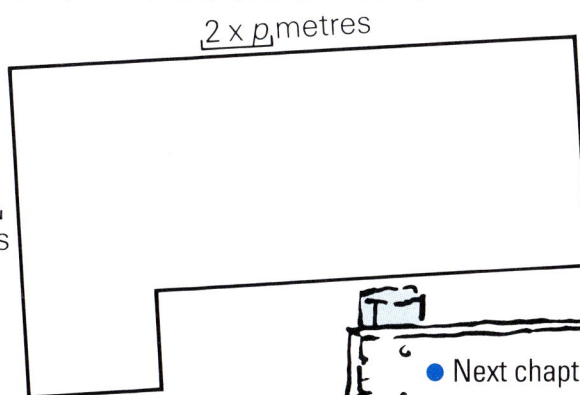

$2 \times p$ metres

$p + 4$
metres

● Next chapter

▲ **235**

26 Using directed numbers

A *With a friend*

1

A

B

C

D

E

F

$\overline{}13°C$ $12°C$ $8°C$ $2°C$ $\overline{}3°C$ $9°C$

It is 1 January: New Year's Day.

The photographs are
of the six places
on the map.

The temperatures
are also for the six places
on the map.

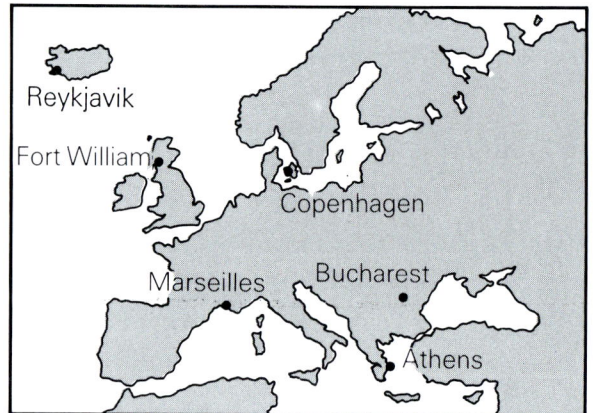

Reykjavik
Fort William
Copenhagen
Marseilles
Bucharest
Athens

Each of you copy this thermometer.
Draw it on squared paper.

freezing point   VERY COLD

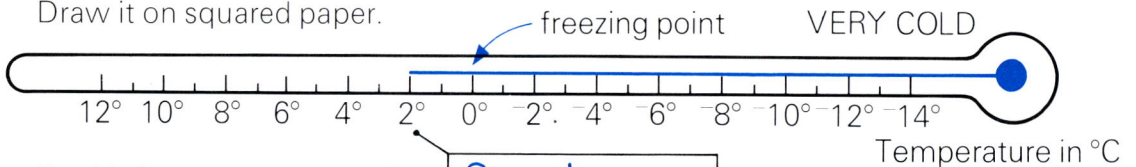

12° 10° 8° 6° 4° 2° 0° ‾2° ‾4° ‾6° ‾8° ‾10° ‾12° ‾14°

Temperature in °C

Decide between you which
places, photographs and
temperatures go together.
Each of you show them like this.

Copenhagen
photograph B

2 At 10:45 the temperature in Fort William is ⁻5°C.
Look at the other time/temperature signs.
They are all for Fort William.

a) Did the temperature in Fort William **rise** or **fall** between 10:45 and 11:00?

b) By how many degrees did the temperature rise between 11:00 and 12:00?

c) When was it colder, at 18:30 or at 11:00?

> 4 JANUARY
> TIME: 10:45
> TEMP: ⁻5°C
>
> FORT WILLIAM
> WELCOMES YOU

●Fort William

> 4 JANUARY
> TIME: 11:00
> TEMP: ⁻4°C
>
> FORT WILLIAM
> WELCOMES YOU

> 4 JANUARY
> TIME: 12:00
> TEMP: ⁻2°C
>
> FORT WILLIAM
> WELCOMES YOU

26

d) Copy and complete this table:

> 4 JANUARY
> TIME: 15:00
> TEMP: 2°C
>
> FORT WILLIAM
> WELCOMES YOU

*⁺1° means a **rise** of 1°C.*

> 4 JANUARY
> TIME: 18:30
> TEMP: ⁻6°C
>
> FORT WILLIAM
> WELCOMES YOU

Temperature changes		To			
		11:00	12:00	15:00	18:30
From	10:45	⁺1°			
	11:00				
	12:00				
	15:00				⁻8°

*⁻8° means a **fall** of 8°C.*

Think it through:

3 At 22:30 the temperature is ⁻4°C.
During the next hour it changes by ⁻2°.

a) What is the temperature at 23:30?

b) At midnight the thermometer shows 2°C.
Describe how the temperature has changed since 23:30.
Write any change in temperature in colour.

> We will write temperature **changes** in colour.

Sentences to describe changes

B 1 The temperature is 23°C.
During the day it rises by 10°C.
Later, during the night
it falls by 20°C.

We can write:

$$23 + {}^{+}10 + {}^{-}20 = \square$$

a 10°C rise

starting temperature

a 20°C fall

finishing temperature

What number is missing from the sentence?

TAMPA FLORIDA
8:00 AM
23° C
HAVE A NICE DAY

starting temperature

$^{+}10$ 33 $^{-}20$ 23   13

26

2 Write a number sentence for each of these.
Write rises and falls in colour.

a)
25   15   12
starting temperature

b)
$^{+}18$ 20 $^{-}8$
starting temperature

c)
35   $^{-}14$ $^{-}10$
starting temperature

d)
5   0 $^{-}4$
starting temperature

e)
3   $^{-}8$ $^{+}3$
starting temperature

f)
$^{+}12$ $^{-}4$ $^{-}3$
starting temperature

3 The diving bell is lowered 20 fathoms.
   Then it is raised 5 fathoms.

   a) Copy this number sentence:

   $$15 + {}^{+}20 + {}^{-}5 = 30$$

   b) Describe in your own words what the
      sentence tells us about
      the diving bell.

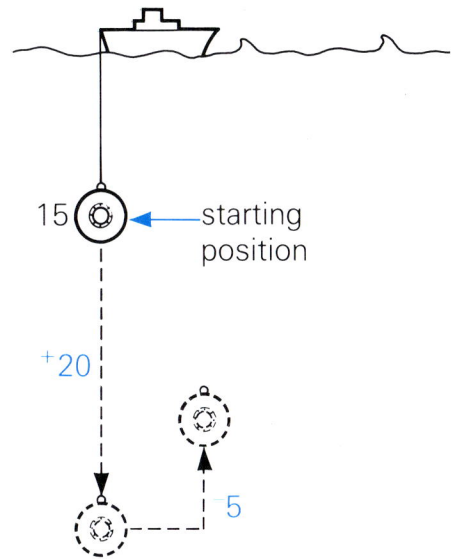

15 ○ ← starting position

${}^{+}20$

${}^{-}5$

4 Annie has £50 in her bank account.
   She withdraws £10 . . .
   and another £20 the next day.
   Describe in your own words what this sentence tells us:

   $$50 + {}^{-}10 + {}^{-}20 = 20$$

26

5 Write a number sentence for this story.
   In your sentence, use 14, 20, ${}^{-}12$ and ${}^{+}6$.

Post Office

Alan has £20 in his Post Office savings.
He pays in another £6.
The next day he withdraws £12.
He now has £14 in his account.

6 Write a number sentence for this story.

The NMP hot air balloon is 10 m above the ground.
It rises 5 m to clear some trees.
Then it descends 4 m to miss a flock of birds.

NMP
UP, UP
AND
AWAY

Your sentence should tell us how
high the balloon is now.
Write rises and falls in colour.

7   Look at the **Take note**.
    Think of another change which we could represent by directed numbers.

    In your example, explain what $^+6$ and $^-10$ would mean.

8   Write **two** stories of your own for this number sentence:

$$4 + {}^+3 + {}^-4 = 3$$

Your first story should be about temperatures.
Choose your own topic for your second story.

26

Find out for yourself

9   Each day, national newspapers print tables like these:

Tuesday

Wednesday

Find out what these columns of numbers mean.
Write about what you find out.

Challenge

10 Copy and complete each sentence:

a) $4 + {}^-3 = \square$

b) $\square + {}^-6 = 2$

c) $2 + \square = 0$

d) $\square + {}^+4 = 7$

e) $^-3 + {}^-2 = \square$

f) $\square + {}^+1 = {}^-4$

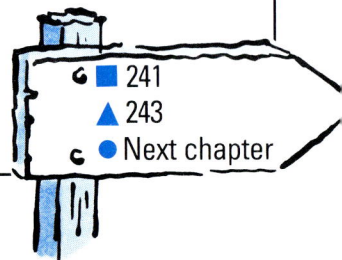

■ 241
▲ 243
● Next chapter

Numbers for a change

A 1 This is a temperature chart for a ski resort:

Temp (°C) at 12 noon	Mon	Tue	Wed	Thu	Fri	Sat	Sun
	2°	0°	3°	0°	⁻1°	⁻3°	
Change from previous day	⁻1°	⁻2°	⁺3°				⁺6°

a) Write down the missing changes for Thursday, Friday and Saturday.

b) What is the temperature at 12 noon on Sunday?

c) What was the temperature at 12 noon on the previous Sunday?

2 In this question, ⁺7 means a change in depth of 7 metres; *7 metres deeper*

⁻4 means a change in depth of 4 metres. *4 metres nearer the surface*

The submarine is at a depth of 20 metres.
This is how its depth changes during
the next 5 minutes:

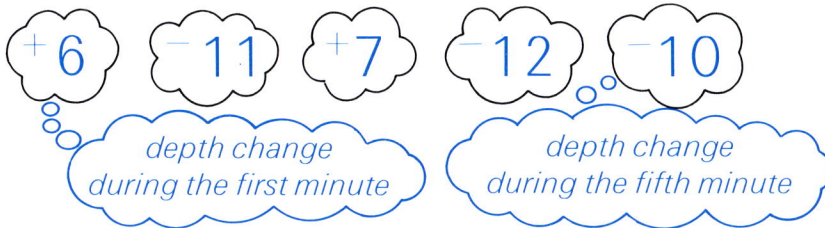

⁺6   ⁻11   ⁺7   ⁻12   ⁻10

depth change during the first minute   *depth change during the fifth minute*

a) How deep is the submarine at the end of each minute?

b) Use 1 mm squared paper.
Copy and complete this record of
the depth of the submarine.

3 Use squared paper.
Copy and
complete the
table to show
the height
differences
between the
camps.

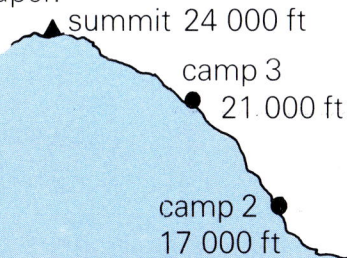

summit 24 000 ft

camp 3 21 000 ft

camp 2 17 000 ft

camp 1 14 000 ft

	Camp 1	Camp 2	Camp 3	Summit
Summit				/////
Camp 3		⁻4000	/////	
Camp 2	⁻3000	/////		
Camp 1	/////	⁺3000		

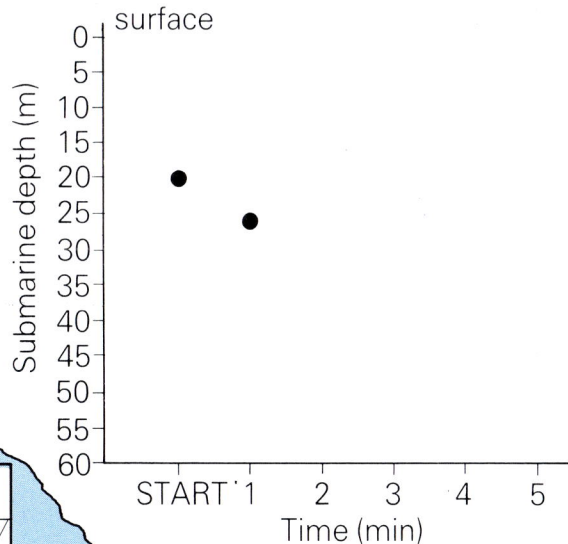

26

B **1** Each drawing represents a change in temperature.

a) Match each thermometer with a number sentence.

(i) $7 + {}^+3 + {}^-6 = \square$

(ii) $4 + {}^+3 + {}^+2 = \square$

(iii) $4 + {}^-3 + {}^+7 = \square$

(iv) $7 + {}^+6 + {}^-3 = \square$

(v) $4 + {}^+2 + {}^-6 = \square$

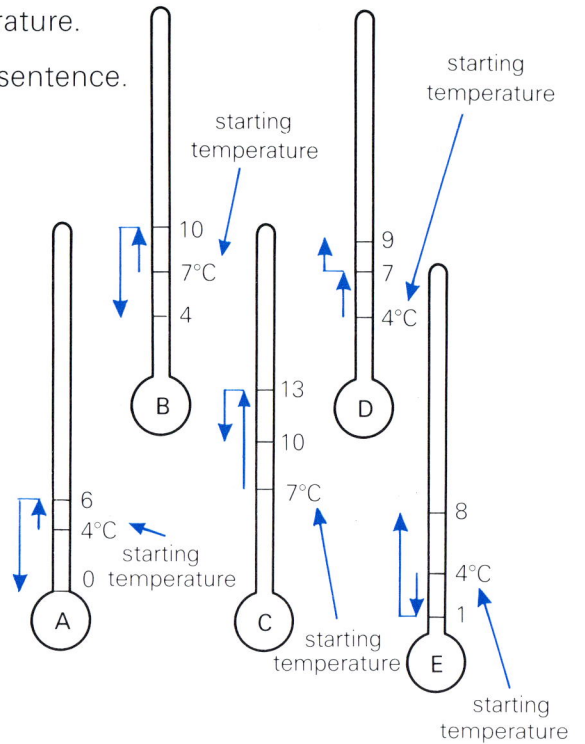

b) Copy and complete each sentence.

2 This is a record for the oil tank.
It shows how much oil is used or added each day.

	Sun	Mon	Tue	Wed	Thu	Fri	Sat
Week 1	$^-20$	$^-10$	$^+15$	$^-20$	$^+10$	0	0
Week 2	$^+15$						

20 litres used on Sunday

15 litres added on Tuesday.

oil tank

a) At the beginning of Week 1 there are 50 litres of oil in the tank.
Explain in your own words what the sentence tells us.

$50 + {}^-20 = 30$

b) This sentence is for Monday of Week 1:
Copy and complete it.

$30 + {}^-10 = \square$

c) Write your own sentences
(i) for Tuesday   (ii) for Wednesday.

d) How many litres of oil are in the tank at the end of Week 1?

3 Copy and complete: a) $4 + {}^-6 = \square$ b) $^-5 + {}^-6 = \square$

c) $7 + \square = {}^-2$ d) $\square + {}^+2 = 0$

e) $\square + {}^+5 = {}^-1$ f) $^-3 + \square = {}^+2$

▲ 243
● Next chapter

26

Movements and positions

Arnold

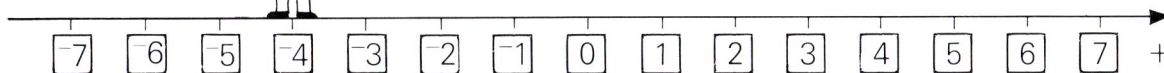

These numbers give positions along the line, to the right and left of $\boxed{0}$.

a) Arnold is at the position $\boxed{^-4}$.

He receives these instructions, one after the other:

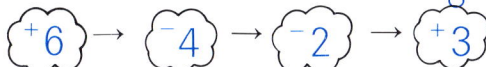

$^+6$ → $^-4$ → $^-2$ → $^+3$

Check that his new position is $\boxed{^-1}$.

These numbers tell us movements along the line.

$^+6$ $^+3$
   $^-4$ $^-2$

b) Meg is at the position $\boxed{5}$.

She receives these instructions:

$^-2$ → $^-3$ → 0 → $^-4$ → $^-20$

What is her new position?

c) Henry is at the position $\boxed{^-6}$.

Explain in your own words what this sentence tells us:

$\boxed{^-6}$ + $^-7$ = $\boxed{^-13}$

d) Hilary is at the position $\boxed{4}$.

She receives five instructions:

$^-3$ → ? → $^-6$ → ? → $^+4$

Her new position is $\boxed{0}$.

Write down **three** possible pairs for the missing instructions.

e) Write down a list of **five** instructions which will move Alex from $\boxed{^-5}$ to $\boxed{5}$.

Try to use five different directed numbers in your list.

26

2

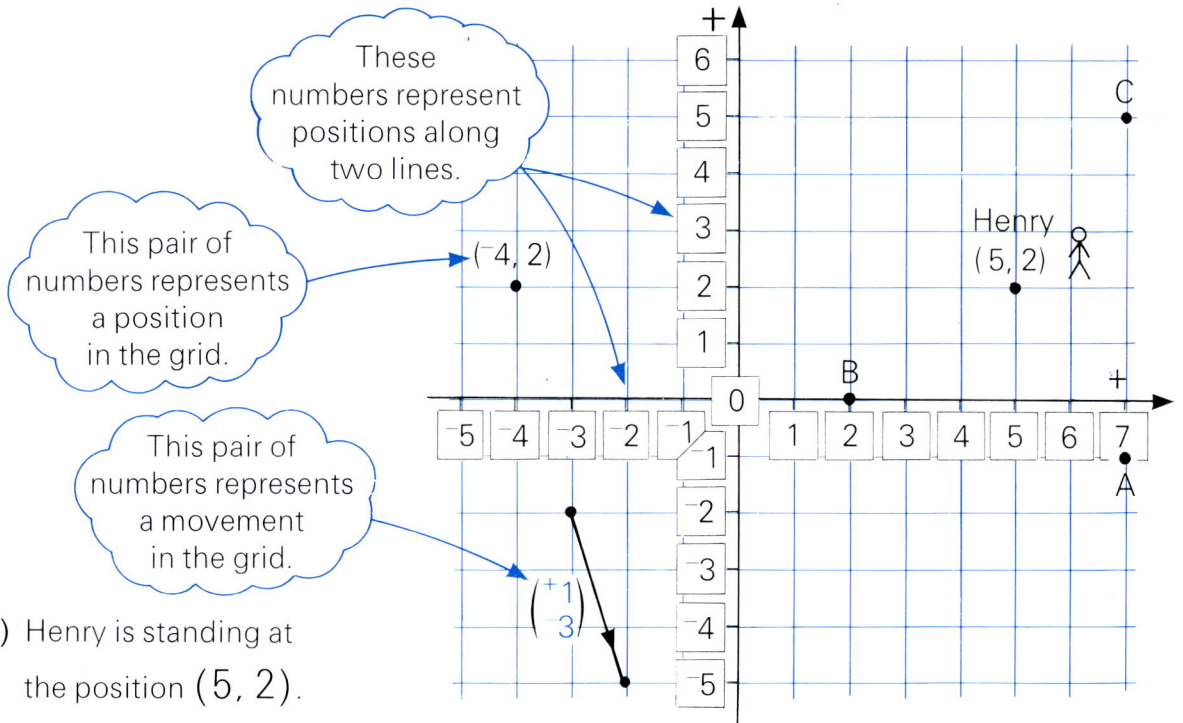

These numbers represent positions along two lines.

This pair of numbers represents a position in the grid.

$(^-4, 2)$

This pair of numbers represents a movement in the grid.

$\begin{pmatrix} ^+1 \\ ^-3 \end{pmatrix}$

Henry $(5, 2)$

a) Henry is standing at

the position $(5, 2)$.

26

He receives this instruction: $\begin{pmatrix} ^+2 \\ ^-3 \end{pmatrix}$

Where should he go, point A,
                  point B
               or point C?

b) When he is at point A, Henry receives this instruction: $\begin{pmatrix} ^-5 \\ ^+1 \end{pmatrix}$

Where should he go, point B
                  or point C?

c) Henry is now at point B.
   What instruction will send him to point C?
   Write your instruction in colour.

Think it through

d) Alice is standing at the point $(^-4, 2)$.
   Explain in your own words what this sentence might mean:

$$(^-4, 2) + \begin{pmatrix} ^-5 \\ ^+1 \end{pmatrix} = (^-9, 3)$$

From / To

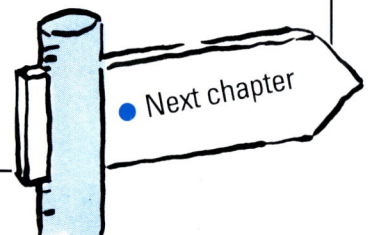

● Next chapter

27 Using graphs

A 1 Horace and his friends have had a medical examination. Here are some of the results:

Name	Height (cm)	Mass (kg)
Horace	150	40
Midge	160	50
Fiona	155	45
Bruce	160	55

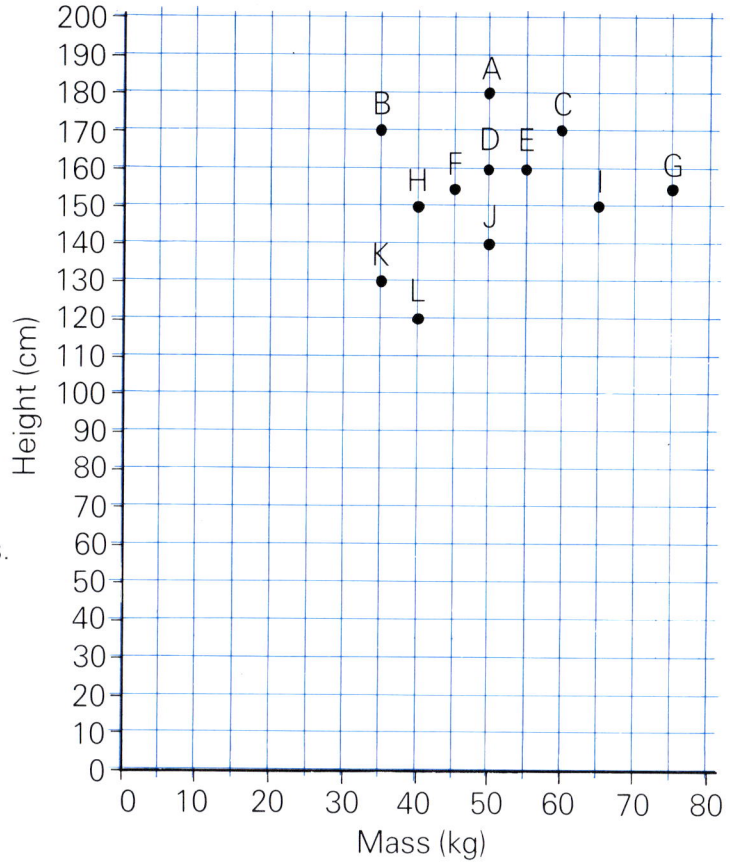

a) Four of the dots on the grid are for Horace and his friends. Write down which dot is for
   (i) Horace.
   (ii) Midge.
   (iii) Fiona.
   (iv) Bruce.

b) Write down which dot is for each of these:

Auntie Flo
170 cm, 60 kg

Cousin Hettie
130 cm, 35 kg

Best friend Quentin
50 kg

Make sensible guesses.

Uncle Ernie

Think it through

2 You need 5 mm squared paper.
Draw your own axes for mass and height.
Mark and label dots for these people.

Height
Mass

Horace's dad
185 cm, 75 kg

Sister Irma
90 cm, 15 kg

Baby Buster
60 cm, 5 kg

Yourself

Your teacher

The heaviest person you know

The diagram in question 1 is called a **graph**.

3   Horace knows lots of people.
The graph shows the ages and heights of some of them.

a) Which dot do you think
is for
   (i)   Baby Buster?
   (ii) Sister Irma?
   (iii) Horace's dad?
   (iv) Best friend Quentin?
   (v)   Horace's grandma?

*Some of these
people are drawn
on page 245.*

b) One dot is obviously in
the wrong place.
Which do you think it is?
Explain why.

c) Three other dots are
probably also in
the wrong place.
Which ones do you
think they are?

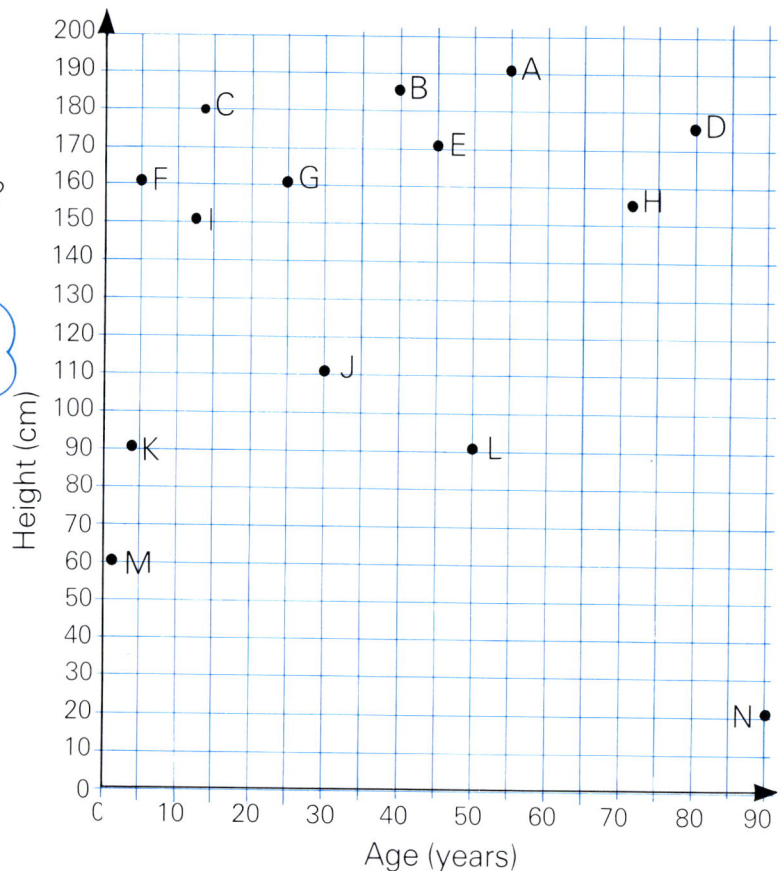

27

Challenge

4   Use 5 mm squared paper.

a) Copy the axes for age and height.
Leave out the dots.
Shade the area on the grid
where you would expect most
normal people's dots to be.

b) Choose three people you know.
Mark a dot for yourself
and a dot for each person.
Estimate their heights and ages if you have to.

*This is not easy!
Think about . . .*

tall people   short people   senior citizens

babies

5 The City Hospital is trying out this new machine.
   It draws a graph to show a patient's temperature.

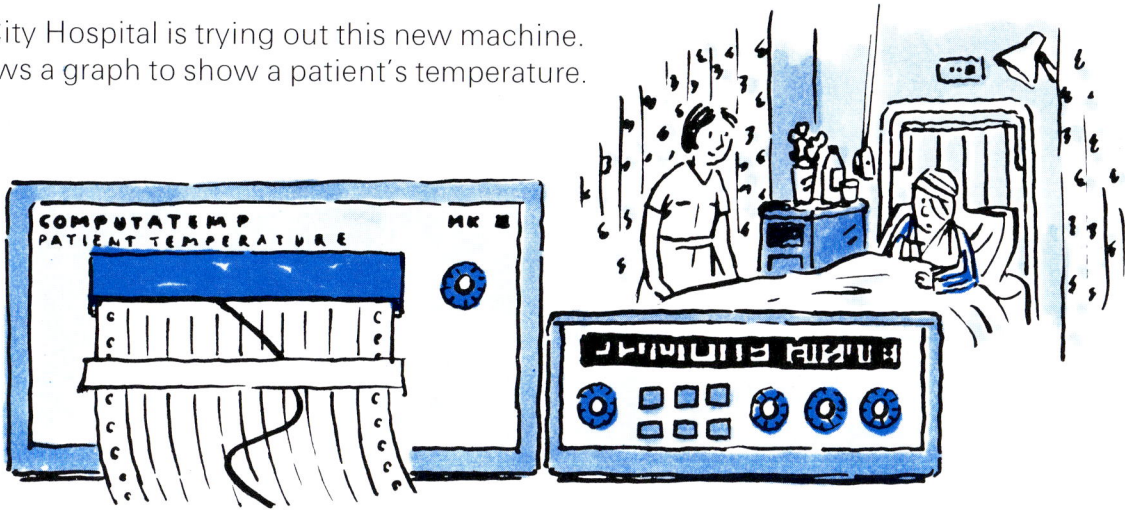

Here is Mary Symes's print-out for Monday.

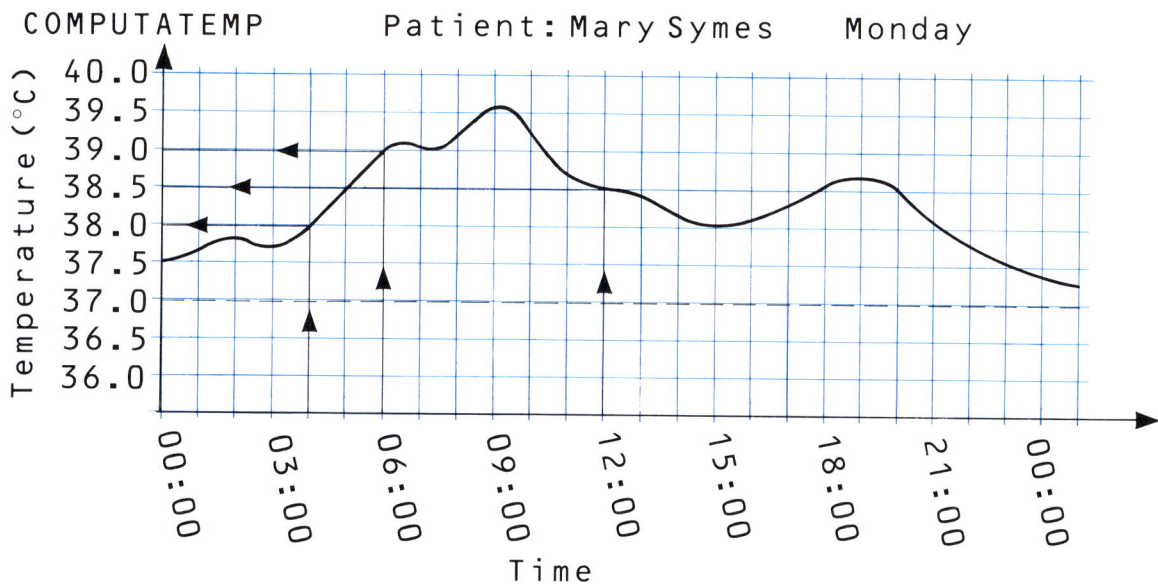

a) Follow the arrow from 06:00.
   What was Mary's temperature at 06:00?

b) What was it at 12:00?

c) At 04:00 Mary's temperature was 38.0°C.
   At what other times was it 38.0°C?

d) At which times did Mary's temperature start to fall?   (five different times)

e) At which times did Mary's temperature start to rise?   (three times)

f) Roughly, what was Mary's highest temperature on Monday?

g) Roughly, what was Mary's lowest temperature on Monday?

h) At about what time was Mary's temperature highest?

Don't count midnight.

● 247

6   While the new machine is being tried,
    nurses take temperatures in the
    usual way.
    They draw a graph for each patient.
    This is the nurses' graph for Mary Symes.

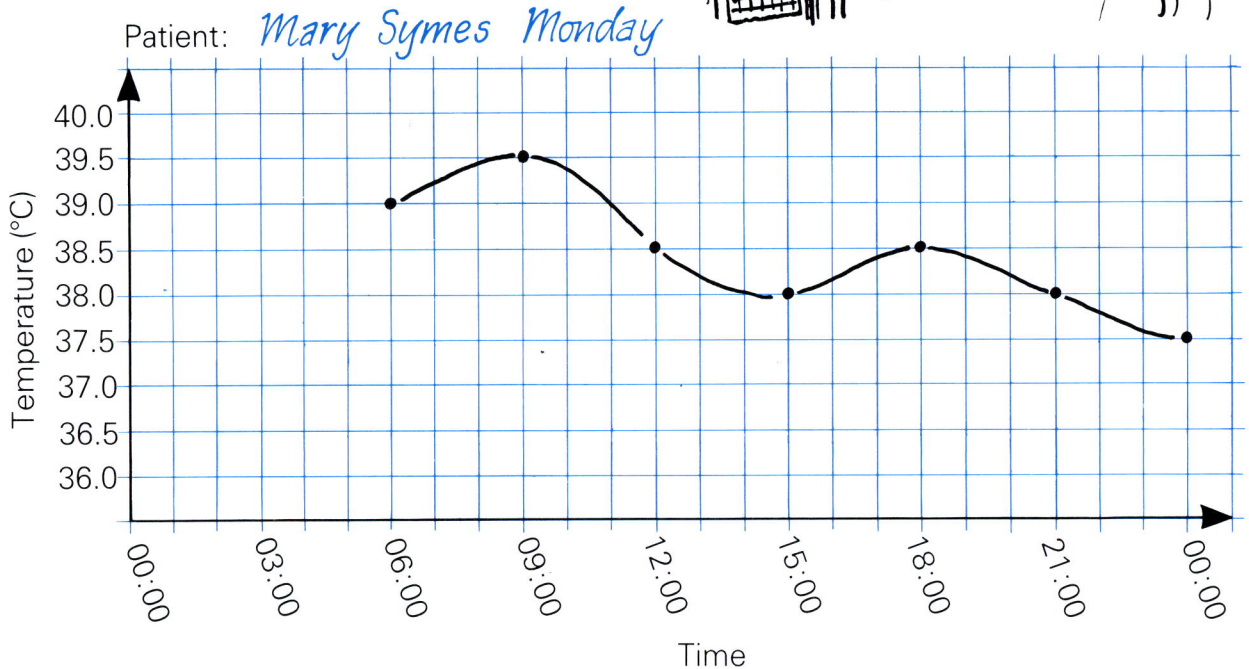

Patient: *Mary Symes Monday*

a) Each dot shows a temperature taken by a nurse.
   How many times did a nurse take Mary's temperature?

b) How many times did a nurse take Mary's temperature between
   midnight and 6 am?

c) During the day, nurses took Mary's temperature every . . .
   how many hours?

d) Compare the seven readings with the machine readings on page 247.
   Which readings
    (i) agree?
    (ii) disagree?

e) Write down why you think the nurses have
    (i) drawn curved lines between readings.
    (ii) **not** joined them to the dots.

Think it through

7   The hospital decides to buy the new machine.
    Write down **two** reasons why you think it is better
    than the old method.

27

8 This kettle is heating up.

This is a graph of the
water temperature.

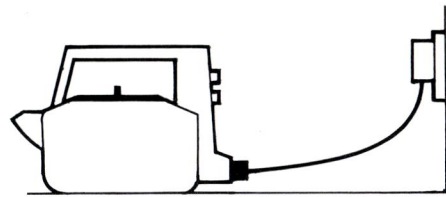

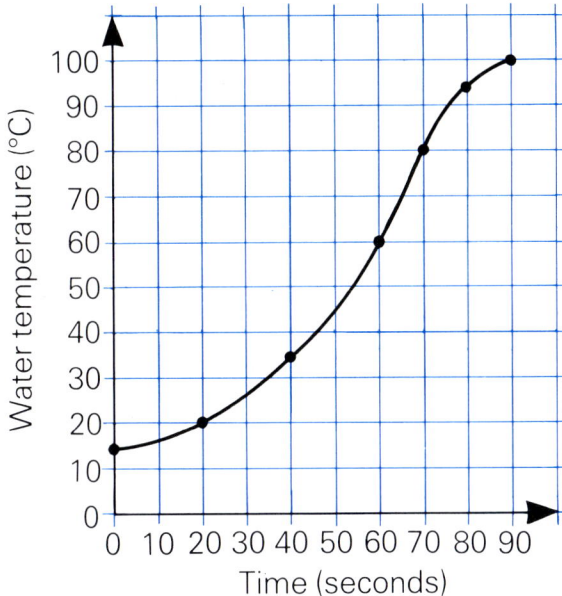

a) What was the water temperature
   (i) when the kettle was switched on?
   (ii) after 70 seconds?

b) After how many seconds was the
   temperature
   (i) 20°C?
   (ii) 35°C?

c) How many °C did the temperature rise
   (i) between the 20th and 40th seconds?
   (ii) between the 60th and 80th seconds?

d) When was the kettle heating up
   more quickly,
   A between the 10th and 20th seconds,
or **B** between the 70th and 90th seconds?

e) Copy and complete these tables.

(i)

Time (seconds)	0			60	70		90
Temperature (°C)		20	35			95	

(ii)

Time interval	0th–20th seconds	20th–40th seconds	40th–60th seconds	60th–80th seconds
Temperature change (to nearest 5°C)	5°C			

With a friend

9 We can record measurements in tables

or on graphs.

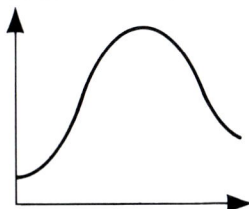

Think of **three** good reasons for using graphs.
Each of you write them down.

B ## *Activity:* *Topple a ruler*

1 Work with a friend.
   You need a 30 cm ruler and six coins of the same value.
   You also need 2 mm squared paper.

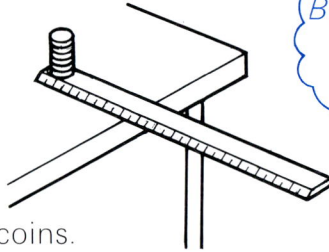

*six 1p,
or six 5p,
or six 10p,
...*

*Borrow from your
teacher if you
have to!*

a) Set up your ruler and coins like this.
   Use all six coins to balance it.
   Gently push the ruler.
   Stop just before it topples
   over the edge.
   Record the length of ruler
   touching the table.

b) Repeat (a) for different numbers of coins.
   Record your results like this:

Number of coins	0	1	2	3	4	5	6
Length of ruler touching table (cm)							

c) Draw a graph of your results on
   2 mm squared paper.

d) Predict the dots for seven,
   eight and nine coins.

e) Get three more coins.
   Check your predictions.

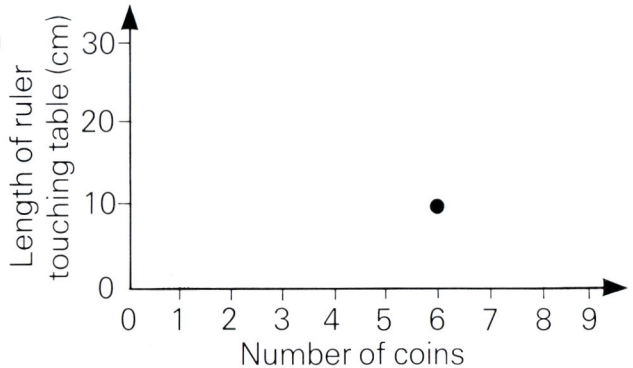

Challenge

2 You need 1 mm squared paper.
   A superball is dropped onto a hard floor.
   This table tells you how high it reached after each bounce.

Bounce number	1	2	3	4	5	6
Height reached (cm)	160	128	102	82	66	52

a) Draw a graph for the bouncing ball.

b) From what height do you think the ball was dropped?

c) How high do you think the ball reached after
   (i) the 7th bounce   (ii) the 10th bounce?

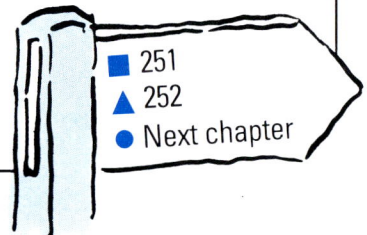

■ 251
▲ 252
● Next chapter

27

Graphs and tables

A
B

1 These sketches show a seedling . . . full size.

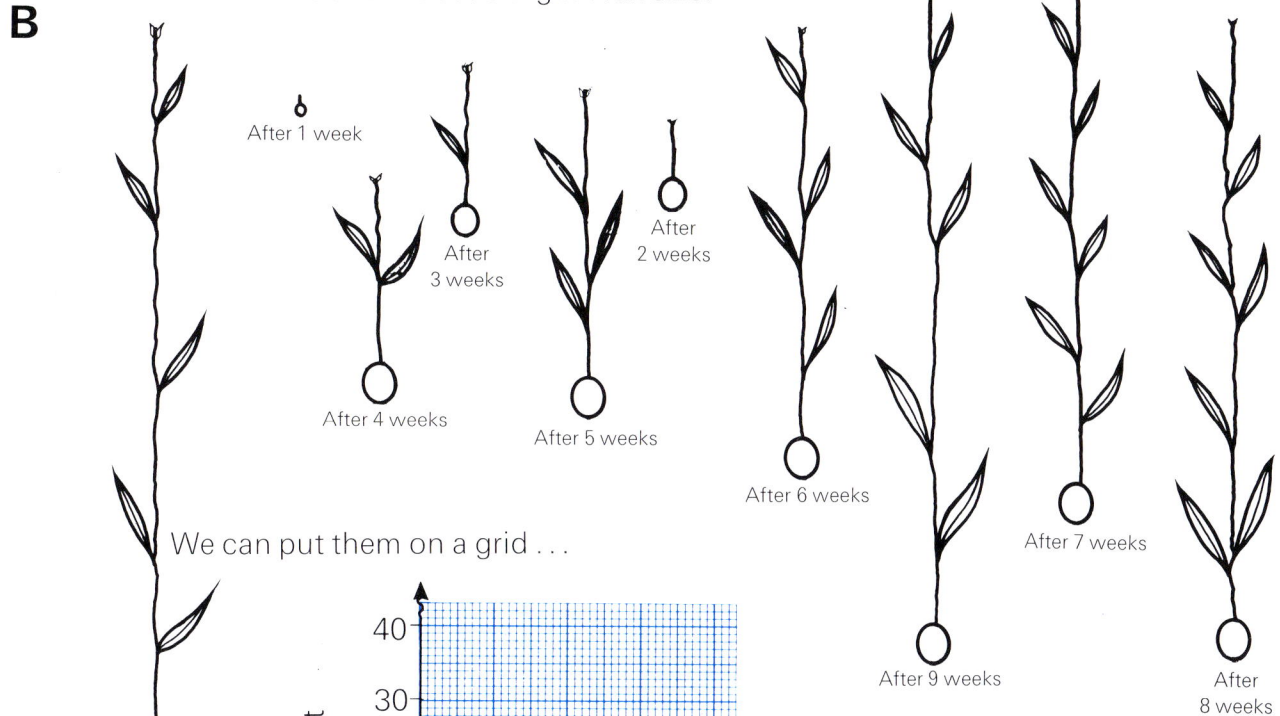

After 1 week

After 2 weeks

After 3 weeks

After 4 weeks

After 5 weeks

After 6 weeks

After 7 weeks

After 8 weeks

After 9 weeks

After 10 weeks

We can put them on a grid . . .

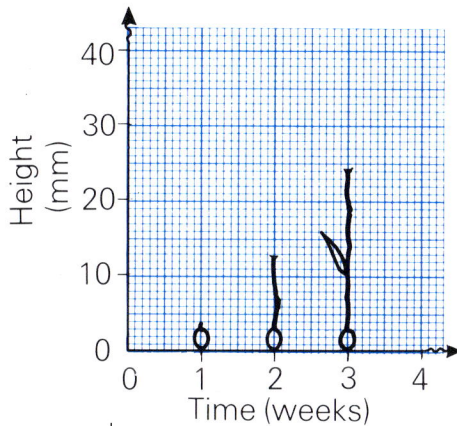

. . . and mark dots to show the seedling's height.

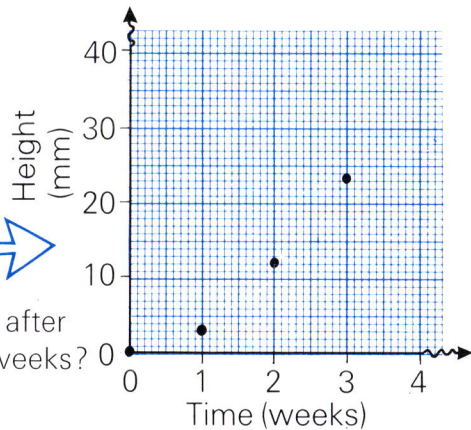

a) Use 2 mm squared paper.
Copy and complete this record of the seedling's height.

b) About how tall do you think the seedling was after
(i) $1\frac{1}{2}$ weeks (ii) $5\frac{1}{2}$ weeks (iii) $8\frac{1}{2}$ weeks?

Think it through

2 This table shows Baby Buster's height.

At birth	After 1 month	After 2 months	After 3 months	After 4 months	After 5 months	After 6 months
50 cm	52 cm	54 cm	58 cm	64 cm	71 cm	78 cm

Use 2 mm squared paper.
Mark axes for time (months) and height (cm).
Mark dots to record Baby Buster's height
from birth to 6 months.

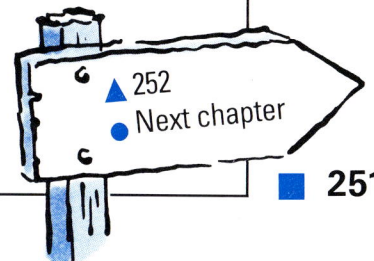

▲ 252
● Next chapter

Which graph is which?

C ___With a friend___

Work with a friend on questions 1 and 2.
Each of you write down what you decide in each question.

1 This table is for a free-fall jump from an aircraft.

Time (seconds)	0	5	10	15	20
Height (metres)	3000	2875	2500	1875	1000

a) How high above the ground was the aircraft when the free-fall began?

b) How many metres did the jumper fall during the first 5 seconds?

c) One of these graphs records the jump. Which one?

d) Explain why you decided against the other two.

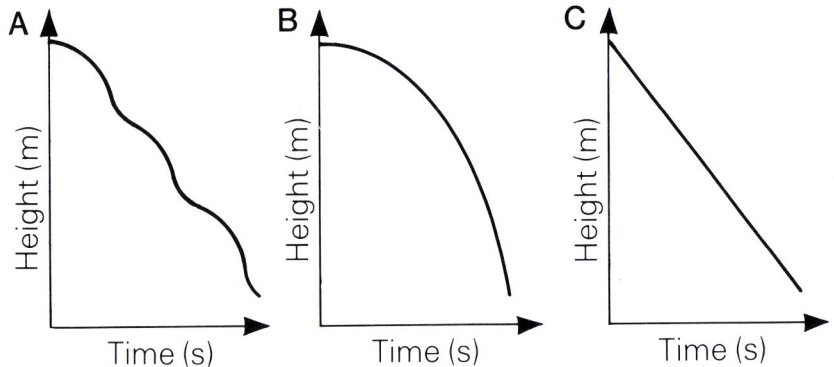

A   B   C

Height (m) — Time (s)   Height (m) — Time (s)   Height (m) — Time (s)

2 There are six graphs on page 253.
Their titles and axis labels have been left off.
Decide which of these goes with which graph.

a) Title: Graph for a squirrel
        climbing a tree.
   Axis 1: Time (in seconds)
   Axis 2: Height above ground
        (in metres)

b) Title: Graph of the growth
        of a seedling.
   Axis 1: Time (in days)
   Axis 2: Height of seedling
        (in cm)

c) Title: Graph of measurements for a
        1 cm wide rectangle.
   Axis 1: Length of rectangle (in cm)
   Axis 2: Area of rectangle (in cm²)

d) Title: Graph of an ant walking
        across a counter top.
   Axis 1: Distance from left edge (in cm)
   Axis 2: Distance from front edge (in cm)

e) Title: Graph for buying lawn
        edging strip (sold only
        in 10 m rolls)
   Axis 1: Length needed (in m)
   Axis 2: Length to be bought (in m)

f) Title: Graph of a panful of snow
        being heated over a camp
        stove in a wind.
   Axis 1: Time (in seconds)
   Axis 2: Temperature (in °C)

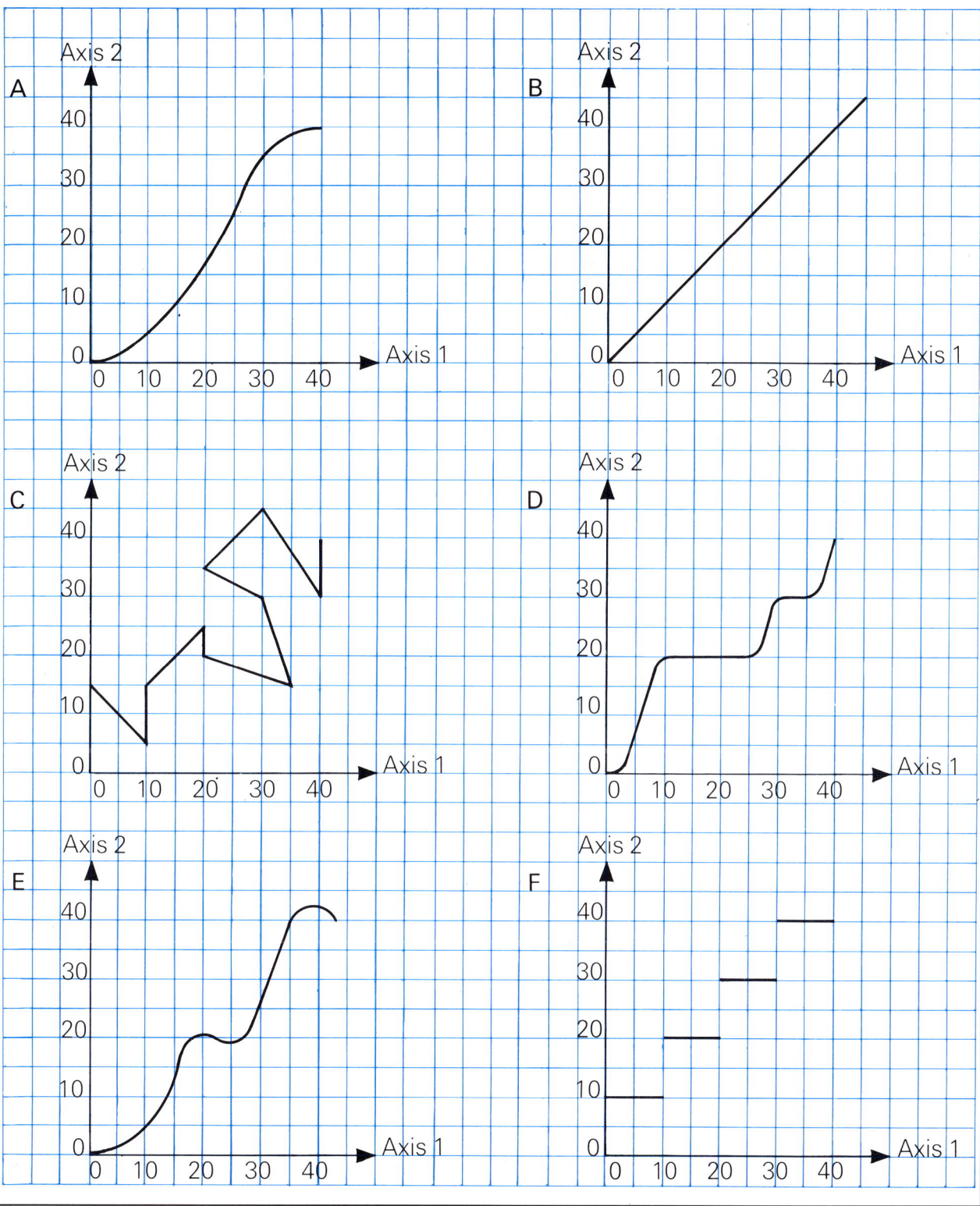

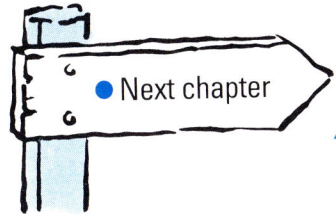

Next chapter

27

28 What do you expect?

A *With a friend*

Discuss each of these with your friend.
Each of you record your answers to the questions.

1   Lenny spins a 10p coin
    600 times.
    The number of times *heads*
    turns up is one of these:

    A 319   B 308

    C 284   D 217

    a) Which do you think it is?

    b) Explain why you chose
       this answer.

2   Fiona spins a 5p coin
    many times.
    She gets *heads* 147 times.
    The number of times she spins
    the coin is one of these:

    A 600   B 500

    C 400   D 300

    a) Which do you think it is?

    b) Explain why you chose
       this answer.

3   Horace is going to throw
    a dice 1000 times.

    a) About how many times
       should he expect
       to turn up?

    b) Why?

4   Midge throws a dice many times.

    She scores   77 times.

    The number of times she throws
    the dice is one of these:

    A 800   B 700

    C 500   D 400

    a) Which do you think it is?

    b) Explain why you chose
       this answer.

5   Horace spins this spinner
    1000 times.
    The number
    of times
    red turns
    up is one
    of these:

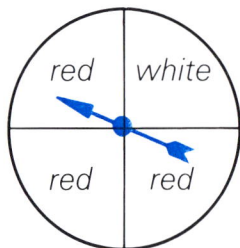

    A 900

    B 700   C 600   D 500

    a) Which do you think it is?

    b) Explain why you chose
       this answer.

Stay with your friend for the next page too!

Activity: _Choosing letters_

6 Work with a friend.
You need some card and a pair of scissors.

about 2 cm square;
all the same size

a) Make ten letter cards like these:

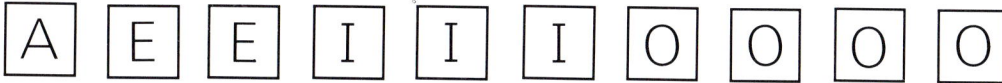

| A | E | E | I | I | I | O | O | O | O |

They must all be the same size, shape and colour.

b) Put the cards face down on the desk.
Each of you is going to choose a card 50 times.
Each time you choose a card, you will
- put it back, face down
- shuffle the cards again.

Before you begin,
guess how many times

| A | will turn up
| E | will turn up
| I | will turn up
| O | will turn up

Each of you write down
what you decide.

c) Now start choosing the letter cards.
One of you choose;
the other record the results.
Change over after 50 choices.

d) Draw a bar chart of your
results for 100 choices.

e) Which letter is most likely to be chosen in any choice?

f) Which letter is least likely to be chosen in any choice?

g) Each of you write one or two sentences about your guesses
in (b) and your results.
Would you change your guesses next time?
Why?

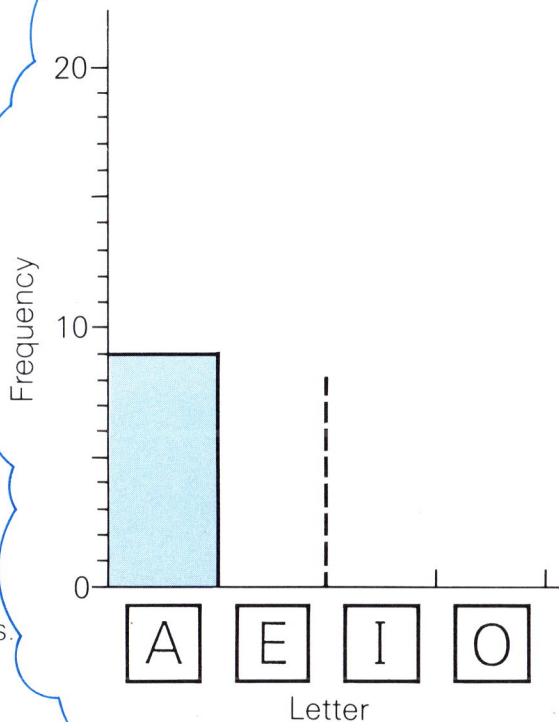

28

Think it through

7 Meg and Winston are doing an experiment.
   They have lots of coloured beans in a bag.
   They take out a bean,
       record what colour it is,
    and put it back in the bag.
   They repeat this over and over again.

> The most likely colour is *red*.
> The least likely colour is *white*.
> The other two colours, *green* and *blue*, are equally likely.

a) This is the bar chart after 200 choices.
   Copy it.
   On your copy, complete the axes to show what you think happened.

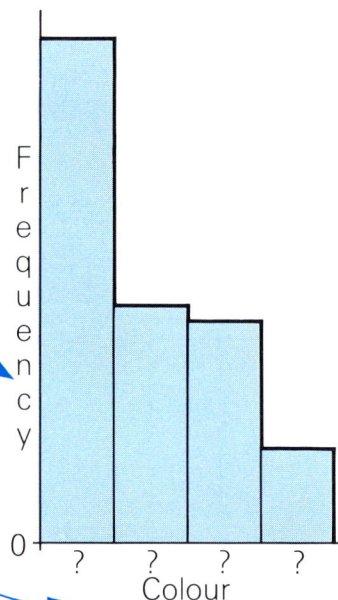

Show what you think the scale is.

Decide which bar is for which colour.

b) One of these is correct:
   A *red* is three times more likely than *white*.
   B *red* is four times more likely than *white*.
   Which do you think it is?
   Explain why you made this choice.

c) There are 20 red beans in the bag.
   About how many beans do you think there are of each of the other colours?

8 a) Horace spins an ordinary 50p coin.
      Which of these is true?

   A *Heads* is more likely to turn up than *tails*.

   B *Tails* is more likely to turn up than *heads*.

   C *Heads* and *tails* are equally likely to turn up.

   b) Horace spins the coin 100 times.
      One of these is a bar chart of the results:

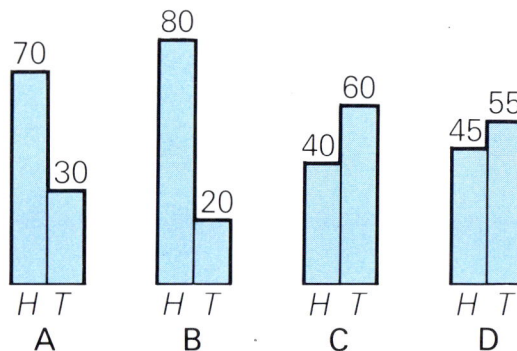

70 30   80 20   40 60   45 55
H T   H T   H T   H T
 A   B   C   D

Which do you think it is?
Explain why you made this choice.

28

Heads and tails are equally likely to turn up.

In 100 spins we would expect about 50 *heads* about 50 *tails*.

For this spinner, *red* is twice as likely as *yellow*.
In 300 spins we would expect
red about 200 times
yellow about 100 times.

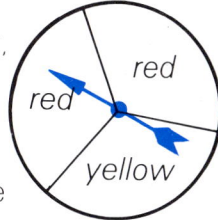

9   Midge made this spinner.

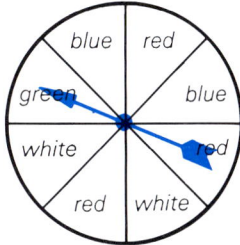

a) On which colour is the arrow most likely to stop?

b) On which colour is the arrow least likely to stop?

c) On which two colours is the arrow equally likely to stop?

d) Midge spins her spinner 800 times.
  (i)   About how many times would you expect the arrow to stop on *green*?
        Explain your answer.
  (ii) About how many times would you expect the arrow to stop on *blue*?
        Explain your answer.
  (iii) Midge draws a bar chart for her results.
        It is one of these:

A
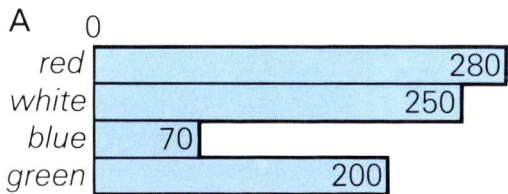

B
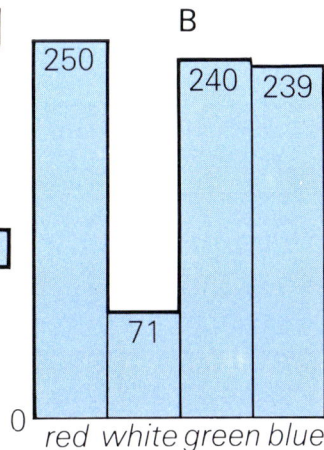

C
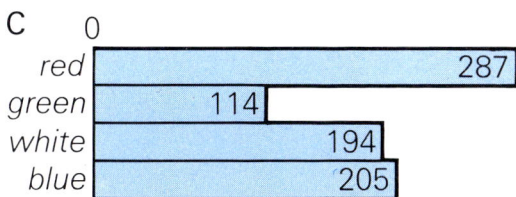

Which do you think it is?
Explain why you made this choice.

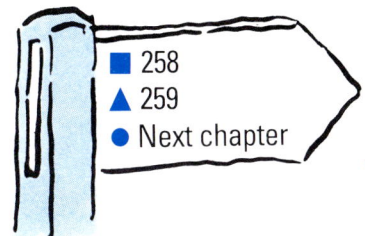

■ 258
▲ 259
● Next chapter

28

The chance bag

A *With a friend*

1 Ask your teacher for the *chance bag*.
Tip out the contents.
They might be beans, beads, marbles,
counters, ...
We will call them **colours**.
Each of you will make 35 choices of a colour.
After each choice, you will put the colour
back in the bag.

a) Before you begin,
guess how many times
   red will be chosen
   white will be chosen
   black will be chosen.
Each of you write down
what you decide.

b) Now put the contents back
in the bag.
One of you make choices;
the other record the results.
Change over after 35 choices.

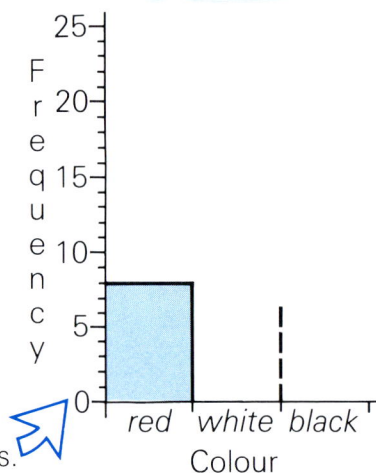

c) Draw a bar chart of your results for 70 choices.

d) One of these is correct:
   A *Black* is four times more likely to be chosen than *white*.
   B *Black* is twice as likely to be chosen as *white*.
   Which do you think it is?
   Explain why you made this choice.

e) Each of you write one or two sentences about
your guesses in (a) and your results.
Would you change your guesses next time?
Why?

Take note

A bag contains 4 black beans and 2 white beans.
We are more likely to choose a black bean than a white bean.
In 120 choices we would expect to choose
   a black bean about 80 times
   a white bean about 40 times.

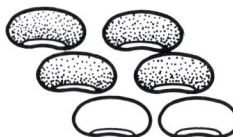

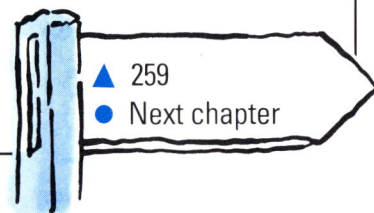

▲ 259
● Next chapter

What are the chances?

B 1 Horace spins an ordinary 50p piece.
The **chance** that it will turn up *heads* is **1 out of 2.**

   a) What is the chance it
   will turn up *tails*?
   Explain your answer.

   b) In 1000 spins, about how
   many times would you expect
   Horace to get *heads*?

*Heads is **one** out of **two** equally likely possibilities.*

heads and tails

2 Midge throws a dice.
The **chance** that it will turn
up ⚅ is **1 out of 6.**

   a) What is the chance
   it will turn up

      (i) ⚀ ?

      (ii) ⚂ ?

      (iii) ⚅ ?

   b) In 600 throws, about how many times
   would you expect Midge to get ⚀ ?

⚅ *is one out of **six** equally likely possibilities*

⚀ , ⚁ , ⚂ , ⚃ ⚄ *and* ⚅

3 Midge throws a dice.
The **chance** that she will get an even number is **3 out of 6.**

   a) What is the chance she will
   get an odd number?
   Explain your answer.

   b) In 1000 throws, about how
   many times would you
   expect Midge to get
   an even number?

Three possibilities (⚁ , ⚃ , ⚅)

*out of **six** equally likely*

possibilities (⚀ , ⚁ , ⚂ , ⚃ , ⚄ , ⚅).

28

Think it through

4 Horace spins this spinner.

   a) What is the chance the arrow will land on
      (i) *red*?
      (ii) *blue*?
      (iii) *white*?

   b) In 800 spins, about how many times would you
   expect Horace to get each colour?

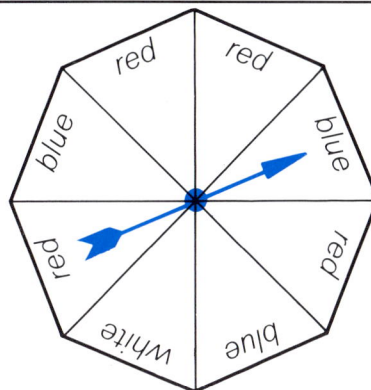

Chancy situations

EXPLORATIONS

Work with a friend.
Investigate one or more of these three situations.
You may decide to do an experiment,
or simply to think about the situation.
Write down what you find out, and how you reached your results.

1 Two dice are thrown at once.

   a) What is the most likely total score?

   b) What is the chance of getting double six?

Total = 3 + 6 = 9

2 You drop a 1p coin onto this grid.
How likely is it to land completely
inside a square?

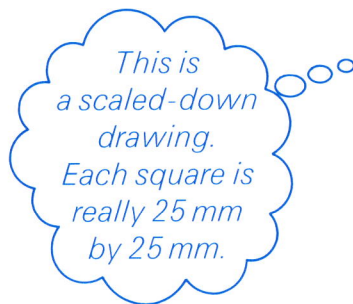

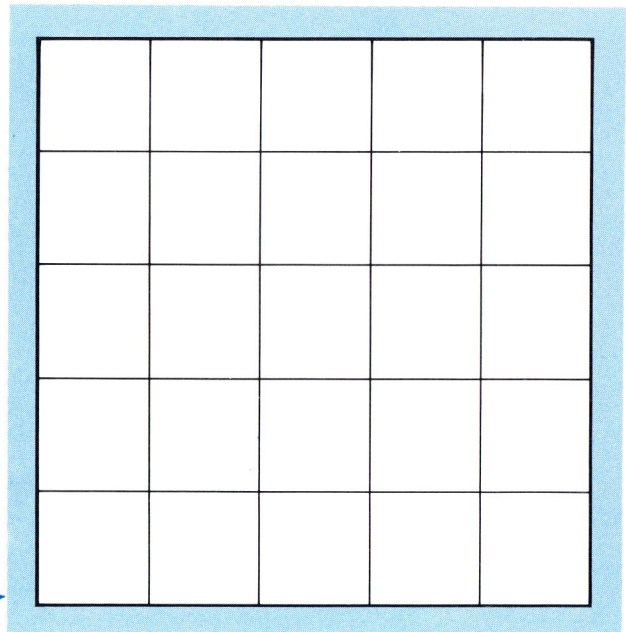

This is a scaled-down drawing. Each square is really 25 mm by 25 mm.

raised border

3 Ask your teacher for the *sampling bottle*.
Pick one of the colours it contains.
Tip up the sampling bottle.
How likely is it that the colours
showing are all different
from the one you chose?

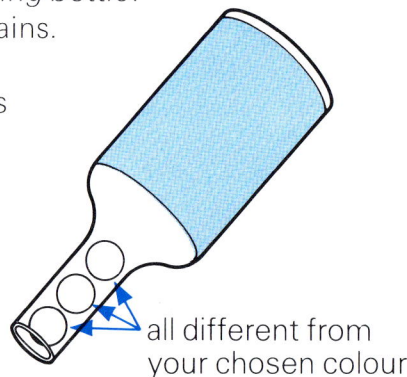

all different from
your chosen colour

● Next chapter

29 Working with numbers

A Do not use a calculator on this page.

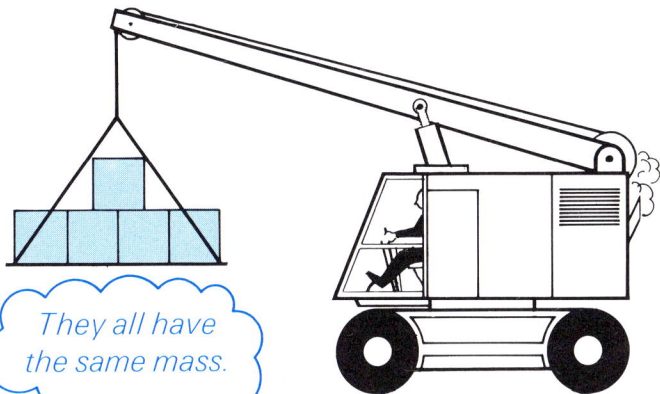

1 The five marble blocks weigh 4 tonnes altogether.

a) How many kilograms is this?

b) How many kilograms does each block weigh?

They all have the same mass.

Take note

To divide amounts . . .

3 tonne ÷ 4

. . . we can change to smaller units.

3000 kg ÷ 4 = 750 kg

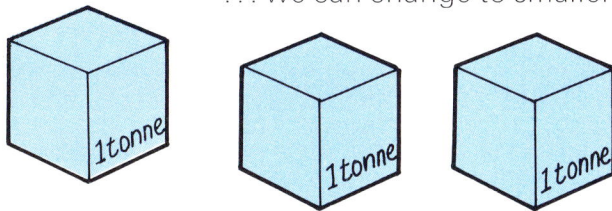

2 The ski lift travels at a steady speed. Each car travels 1 km in 5 minutes.

a) How many metres is this?

b) How many metres does a car travel in 1 minute?

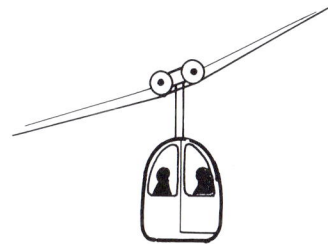

3 Lenny wants to make five rolls from this flour. How many grams should he weigh out for each?

The same amount for each roll.

FLOUR ½ kg

Challenge

4 Horace is making rock cakes. He uses this flour to make four cakes. How many grams does he use for each?

The same amount for each one.

DO YOU REMEMBER . . . ?
wholes . tenths
☐ . ☐

FLOUR 0·8 kg

Think it through

Do not use a calculator.

5   Maeve has to saw this plank into seven equal parts.

3 m

a) Which of these is the best estimate for the length of each part?

A   about 40 cm

B   about 50 cm

C   about 30 cm

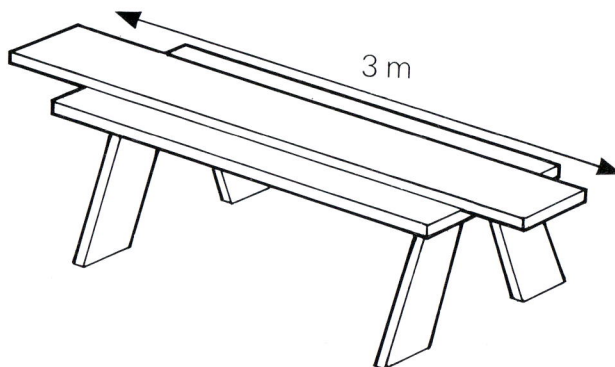

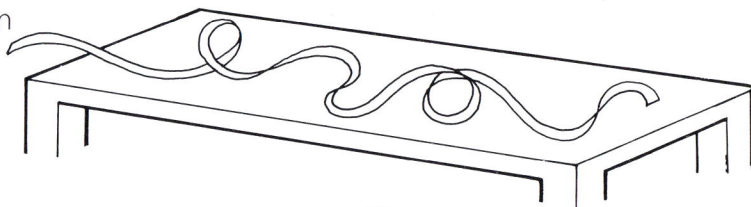

b) Copy and complete:

$$3\text{ m} \div 7 \approx \square \text{ cm}$$
$$3\text{ m} \div 7 \approx 0.\square \text{ m}$$

REMEMBER...?
$\approx$ means 'is about' or 'is approximately'.

6   Meg has to cut this plastic edging strip into eleven equal parts.

12 m

a) Which of these is the best estimate for the length of each part?

A   about 1 m

B   about 1 m 10 cm

C   about 1 m 20 cm

b) Copy and complete:

$$12\text{ m} \div 11 \approx \square \text{ m} \square \text{ cm}$$
$$12\text{ m} \div 11 \approx \square.\square \text{ m}$$

Challenge

7   Lenny has to cut the melon into nine equal parts.

a) About how many grams will each part weigh?

7 kg

b) Copy and complete: $7\text{ kg} \div 9 \approx \square.\square \text{ kg}$

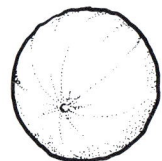

29

Mental division

B 1   Meg uses the flour to make five loaves.
She works out how many kilograms
for each loaf like this:

1 kg IS THE SAME AS 10 TENTHS kg...

SO EACH LOAF IS 2 TENTHS kg...

FLOUR 1 kg

2 tenths kg

2 tenths kg

2 tenths kg

2 tenths kg

2 tenths kg

wholes . tenths
0 . 2

WHICH CAN BE WRITTEN LIKE THIS : 0.2 kg.

Do these in your head.
Think in tenths.
Write the results as decimals.

a) **1 kg ÷ 10**

b) **1 kg ÷ 2**

c) **1 m ÷ 5**

d) **2 kg ÷ 10**

e) **2 l ÷ 5**

f) **2 km ÷ 4**

g) **3 kg ÷ 5**

h) **3 tonne ÷ 6**

i) **3 m ÷ 2**

Think it through

2   Meg writes down her division to check her result.
She writes:

$$\begin{array}{r} 0.\,2\text{ kg} \\ \hline 5\,\overline{)\,\cancel{1}.\,{}^{1}0\text{ kg}} \end{array}$$

a) Why do you think she crosses out **1** and writes **¹0**?

b) Check your results in question **1** (h) and (i) by
writing out divisions.

c) Check all your results in question **1**, using your calculator.

3 Do these in your head.
Write your results as decimals.

a)
$$0.6 \text{ kg} \div 2$$

Think:
*0.6 kg is
6 tenths kg.*

b)
$$0.8 \text{ m} \div 4$$

c)
$$1.2 \text{ kg} \div 2$$

d)
$$1.6 \text{ cm} \div 4$$

e)
$$2.5 \text{ l} \div 5$$

f)
$$3.6 \text{ km} \div 9$$

g)
$$8.4 \text{ cm} \div 2$$

h)
$$12 \text{ m} \div 10$$

4 Meg does this in her head.

$$1.6 \text{ kg} \div 4$$

THAT'S 0.4 kg.

She checks her result:

$$\begin{array}{r} 0{\cdot}4\,\text{kg} \\ 4\overline{)1{\cdot}{}^{1}6\,\text{kg}} \end{array}$$

a) Why does she cross out the **1** and write **¹6**?

b) Check your results in question **3** by writing out the divisions.

c) Check the results again using your calculator.

Think it through

5 First try these in your head.
Write down your answers.
Then
- check your results by writing out the divisions
- check again using your calculator.

a)
$$8.4 \text{ kg} \div 6$$

b)
$$4.8 \text{ km} \div 3$$

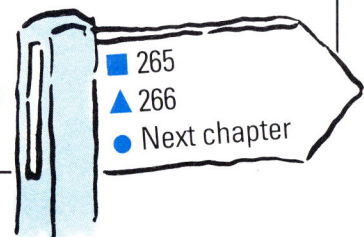

■ 265
▲ 266
● Next chapter

29

Dividing decimals

A
B

1 Do these in your head.
  Write your results as decimals.

a) 2 km ÷ 5

b) 4 cm ÷ 10

c) 4 km ÷ 5

d) 2 m ÷ 10

e) 6 l ÷ 10

f) 3 kg ÷ 5

g) 1.2 m ÷ 2

h) 5.0 cm ÷ 2

i) 8.6 g ÷ 2

j) 3.2 m ÷ 8

k) 6.4 kg ÷ 2

l) 8.5 m ÷ 5

2 Use your calculator.
  Check each of your results in question **1**.

3 Copy and complete each division:

a) 4 | 0.8 kg

b) 5 | 2.5 m

c) 3 | 8.4 l

d) 9 | 1.8 tonne

e) 7 | 9.8 m

f) 3 | 13.8 km

29

4 Use your calculator.
  Check each of your results in question **3**.

5 For each division, choose the closest estimate from the list.
  Do not use your calculator.

a) 11 m ÷ 8   1.2 m   2.1 m

b) 7 m ÷ 6   1.5 m   2.4 m

c) 7 m ÷ 3   1.3 m   2.3 m   2.2 m

d) 17 m ÷ 8   1.4 m

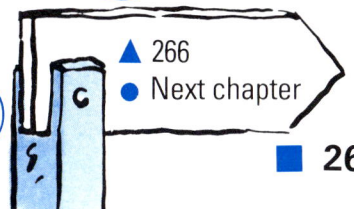

▲ 266
● Next chapter

Written and mental division

C 1 Meg is sawing the
copper tube into
eight equal lengths.
She works out the lengths
of the parts in her head:

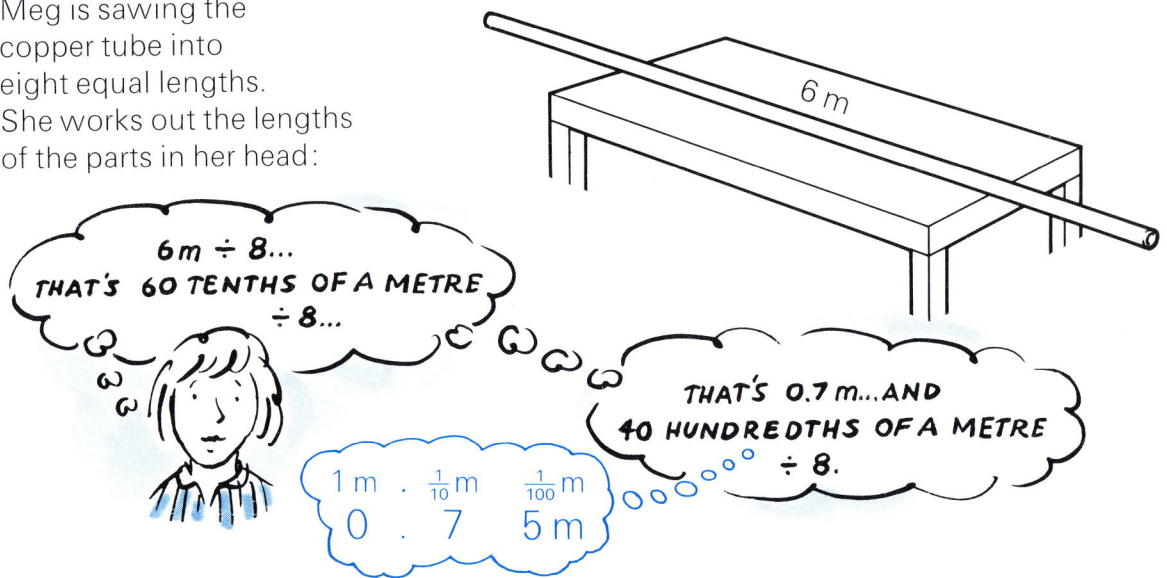

6 m

$6m \div 8...$ THAT'S 60 TENTHS OF A METRE $\div 8...$

THAT'S 0.7 m... AND 40 HUNDREDTHS OF A METRE $\div 8$.

1 m	.	$\frac{1}{10}$ m	$\frac{1}{100}$ m
0	.	7	5 m

She starts to check her result:

a) What does ^{6}0 mean?

b) This is how she continues:

   Why does she cross out the
   ^{6}0 and write ^{4}0?

$$\begin{array}{r} 0.7 \quad m \\ 8 \overline{\smash)6.^600m} \end{array}$$

$$\begin{array}{r} 0.75\,m \\ 8 \overline{\smash)6.^6\cancel{0}^40m} \end{array}$$

2 a) Try these in your head.
   Write your results as decimals.

   (i) $7\,kg \div 4$

   (ii) $7\,cm \div 5$

   (iii) $15\,m \div 8$

   b) Check your results in (a) by writing out the divisions.

   c) Check your results again using your calculator.

Think it through

3 a) **Estimate** these in your head.
   Try to give results to the nearest tenth of a kilogram.

   (i) $4\,kg \div 6$

   (ii) $5\,kg \div 7$

   b) Check your estimates with your calculator.
   Mark them *Excellent, Good, OK*, or *Abysmal*.

29

266 ▲

Equivalent divisions

D   EXPLORATION

1 a) These divisions have something in common.
      Use your calculator to find out what it is.
      Write down what you find out.

$$19 \div 5 \qquad 57 \div 15 \qquad 95 \div 25 \qquad 190 \div 50$$

  b) Find two more divisions which belong to the same
      'family' as those in (a).

  c) **Do not use your calculator.**
      Find **five** divisions which will give the same result as $42 \div 3$.

  d) Check your results in (c) with your calculator.

Take note

Every decimal goes with a whole family of divisions . . .

$$0.1 \quad \text{goes with}$$
$$\{1 \div 10,\ 2 \div 20,\ 0.1 \div 1,\ 0.2 \div 2,\ 11 \div 110,\ \ldots\}$$

$$0.4 \quad \text{goes with}$$
$$\{4 \div 10,\ 2 \div 5,\ 0.4 \div 1,\ 200 \div 500,\ 16 \div 40,\ \ldots\}$$

2 Do not use your calculator.

  a) Write down which of these divisions belong
      to the same family as $11 \div 5$.

    A $\quad 12 \div 6 \qquad\qquad$ B $\quad 10 \div 4$

                              C

    D $\quad 9 \div 4 \qquad\qquad\qquad 22 \div 10$

  b) Use your calculator to check your results in (a).

  c) Write down three divisions of your own which belong to the
      same family as $11 \div 5$.

29

▲ 267

Think it through

3 This is a cannonball rolling down a ramp.
   How far above the ground is the cannonball
   as it passes each flag?

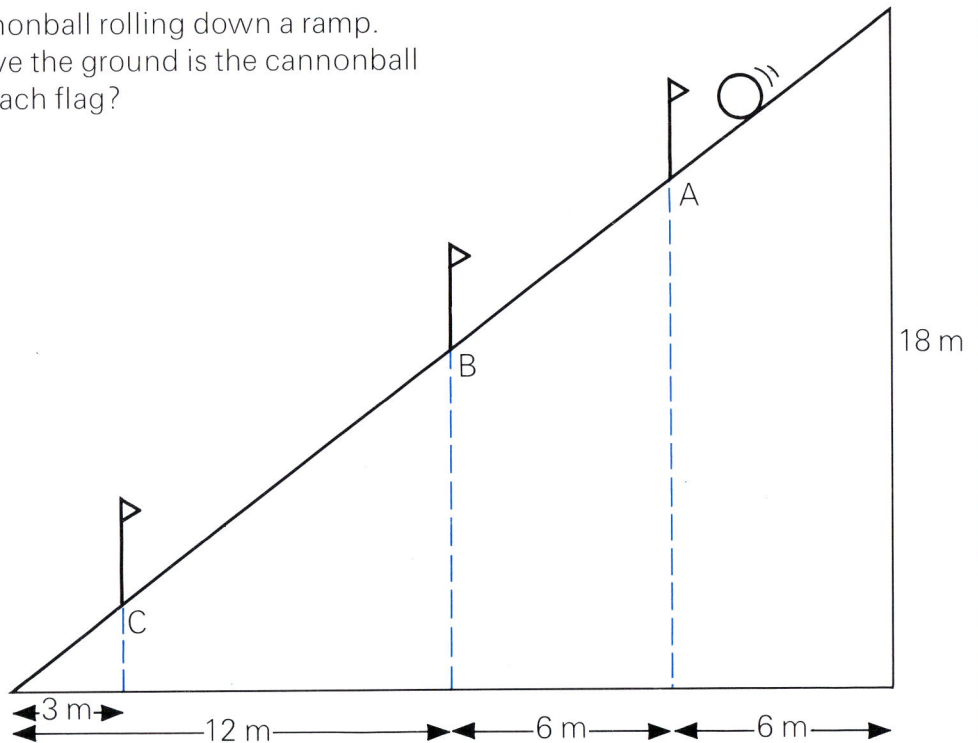

18 m

◄3 m►

├──────12 m──────►◄───6 m───►◄───6 m───►

Challenge

A   B   C

E   D

4 a) You may use a ruler to measure.
      Do not use a protractor.
      Three of the ladders are leaning
      at the same angle.
      Which are they?

   b) Explain how you decided.

● Next chapter

29

30 Journeys

A EXPLORATION

1 One way of getting from START to FINISH is:

along	2
up	3
along	1

You can only move **along** → and **up** ↑.

a) List **five** more ways of getting from START to FINISH.

b) How many ways are there altogether?

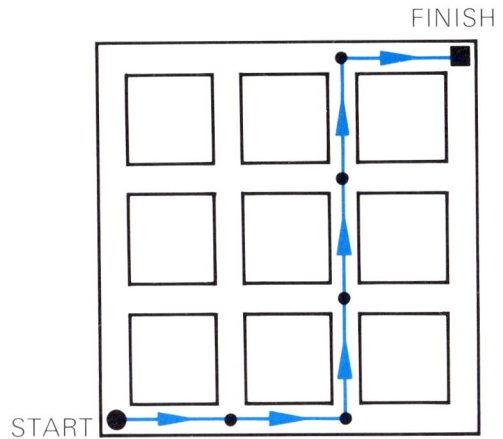

Activity: Protector

2 Play Protector with a friend.
Make the pieces you need from card.
Draw the board on squared paper.

Start with the pieces like this:

PIECES

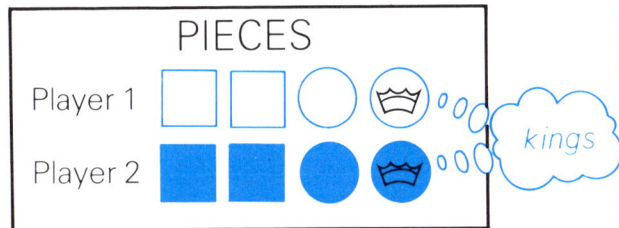

kings

MOVES

Round pieces and kings

Square pieces

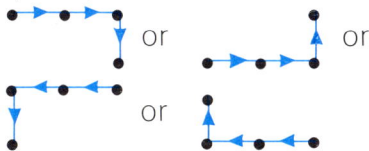

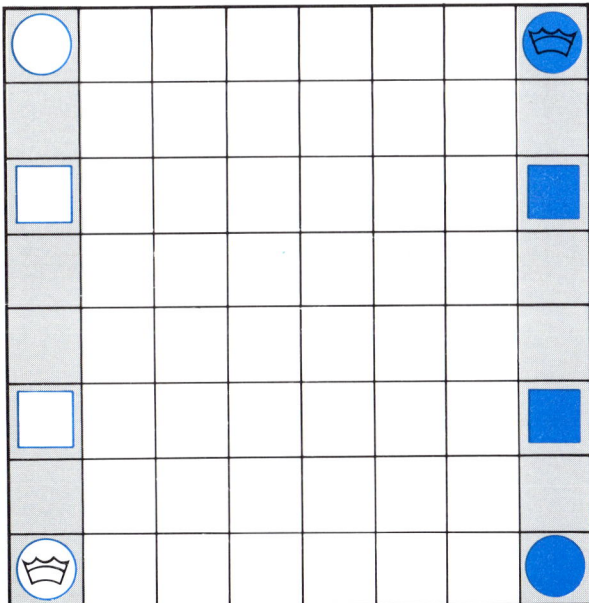

RULES

- Take turns to move one piece.

- Capture a piece by landing on its square.

- You must not move one piece **over** another.

THE WINNER

is the first player to get the king safely across the board into the shaded area.

Play the game several times.

B 1 **DO YOU REMEMBER...?**

We can write the position of oil rig Alpha from Aberdeen as (120 km, 060°).

For bearings we always use three digits.

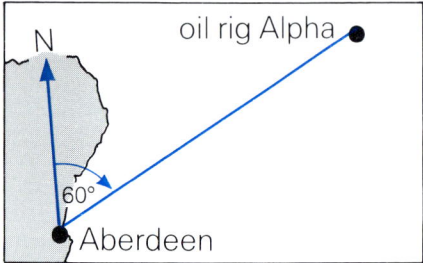

a) Look at the map.
   Use (☐ km, ☐°) to describe the positions of these from Aberdeen:
   (i) oil rig Derek   (ii) Elgin   (iii) Dundee.

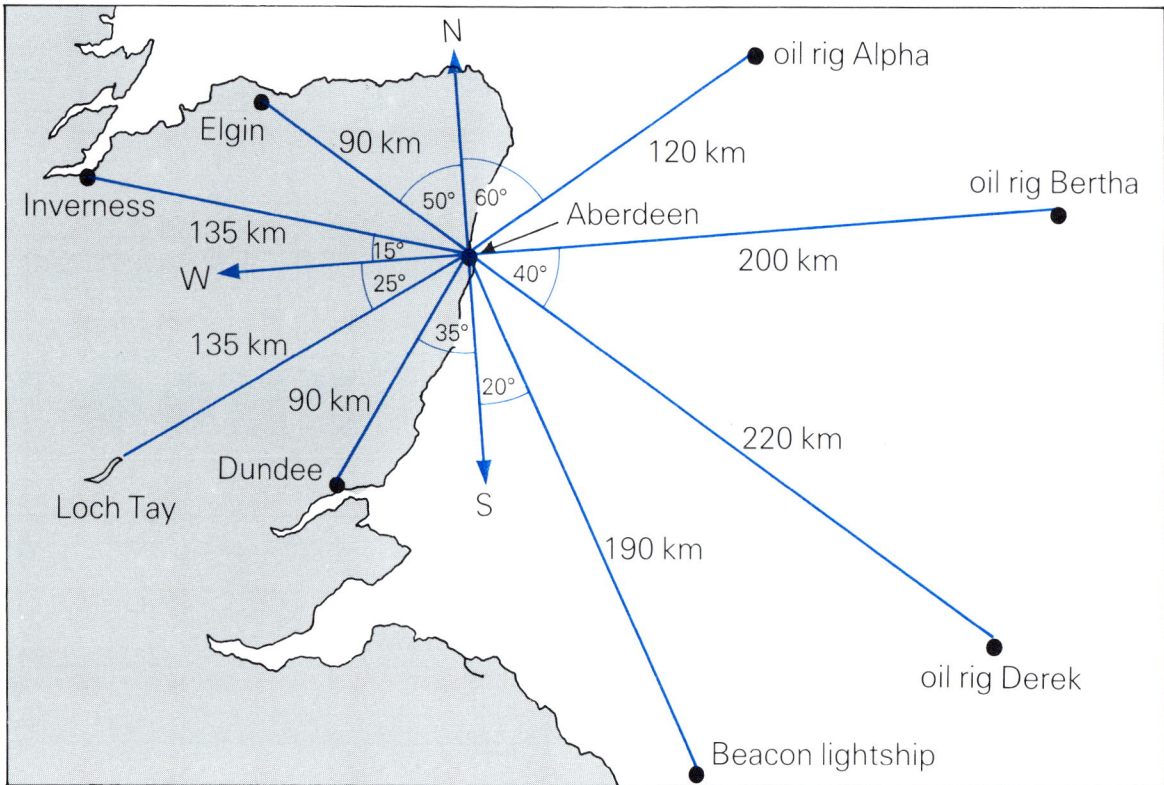

b) Think about a helicopter **journey** from Aberdeen to oil rig Alpha.

We can describe the journey like this: $\begin{pmatrix} 120\,\text{km} \\ 060° \end{pmatrix}$

Use $\begin{pmatrix} \square\,\text{km} \\ \square° \end{pmatrix}$ to describe these journeys from Aberdeen:

(i) to oil rig Derek   (ii) to Elgin   (iii) to Dundee.

Take note

We use (90 km, 215°) to represent the **position** of Dundee from Aberdeen.

We use $\begin{pmatrix} 90\,\text{km} \\ 215° \end{pmatrix}$ to represent the **journey** from Aberdeen to Dundee.

Think it through

2   We can describe the journey from Aberdeen to Elgin by $\begin{pmatrix} 90\,km \\ 310° \end{pmatrix}$.

Use $\begin{pmatrix} \square\,km \\ \square° \end{pmatrix}$ to describe the journey from Elgin to Aberdeen.

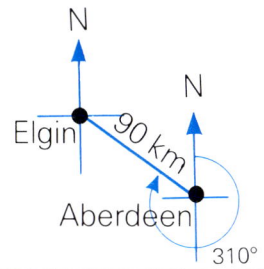

DO YOU REMEMBER ...?

We can use two distances to write the position of one thing relative to another. Relative to Aberdeen, oil rig Gamma is (130 km E, 90 km N).

3   a)   Look at the map.
Use ($\square$ km $\square$, $\square$ km $\square$) to describe the positions of these relative to Aberdeen:
(i) oil rig Beta   (ii) HMS *Honour*   (iii) Aberfoyle

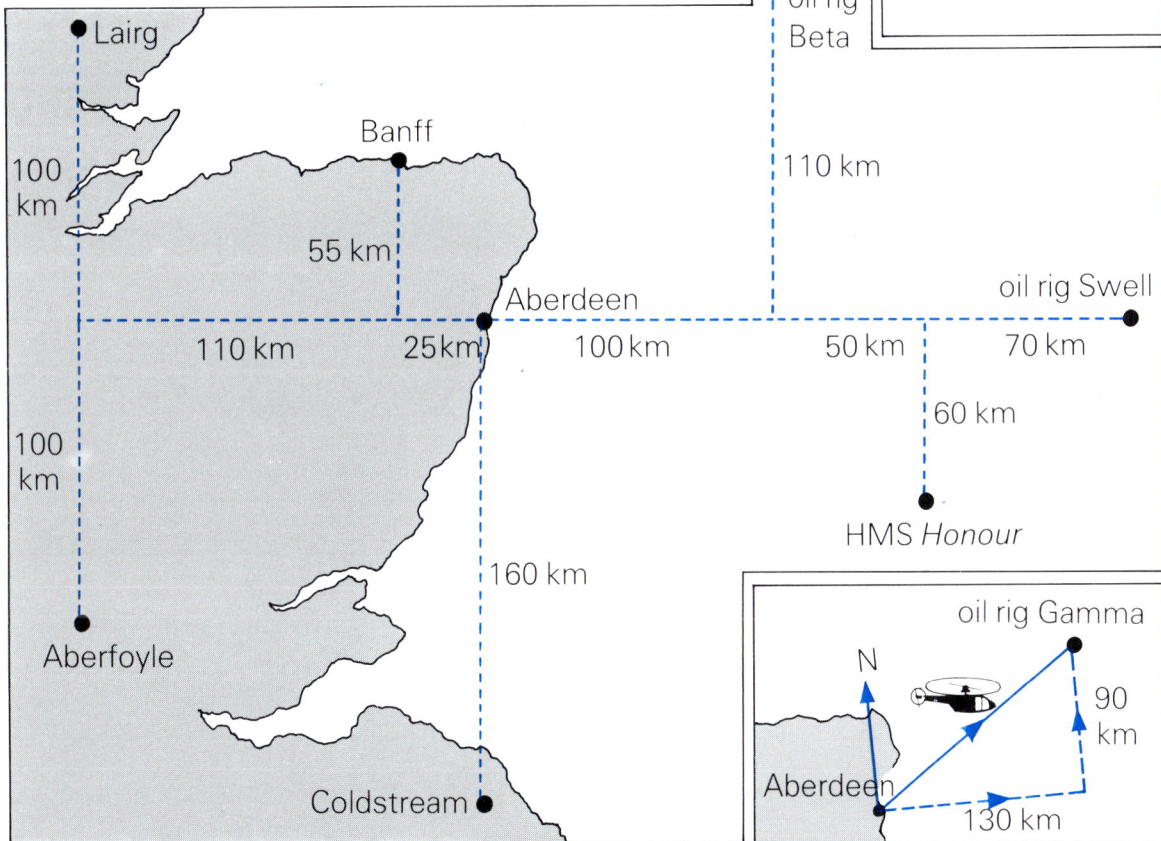

b)   Think about a helicopter flying from Aberdeen to oil rig Gamma.
We can describe its journey by $\begin{pmatrix} 130\,km\ E \\ 90\,km\ N \end{pmatrix}$.

always E or W

always N or S

Use $\begin{pmatrix} \square\,km\ \square \\ \square\,km\ \square \end{pmatrix}$ to describe these journeys from Aberdeen:
(i) to oil rig Beta   (ii) to HMS *Honour*   (iii) to Aberfoyle.

30

Take note

We use (130 km E, 90 km N) to represent the **position** of oil rig Gamma from Aberdeen.

We use $\begin{pmatrix} 130\,\text{km E} \\ 90\,\text{km N} \end{pmatrix}$ to represent the **journey** from Aberdeen to oil rig Gamma.

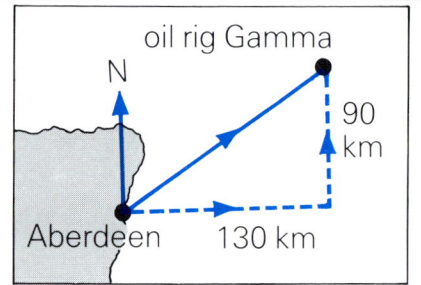

Think it through

4   We can describe the journey from Aberdeen to Lairg by $\begin{pmatrix} 135\,\text{km W} \\ 100\,\text{km N} \end{pmatrix}$.

Use $\begin{pmatrix} \square\,\text{km}\,\square \\ \square\,\text{km}\,\square \end{pmatrix}$ to represent the journey from Lairg to Aberdeen.

5   You need a piece of string and a protractor.

a) About how far is it from Hull to York
   (i) if you travel through Beverley?
   (ii) if you travel through Selby?

b) Describe the approximate position of York from Hull using

   (i) ($\square$ km W, $\square$ km N)

   (ii) ($\square$ km, $\square$°).

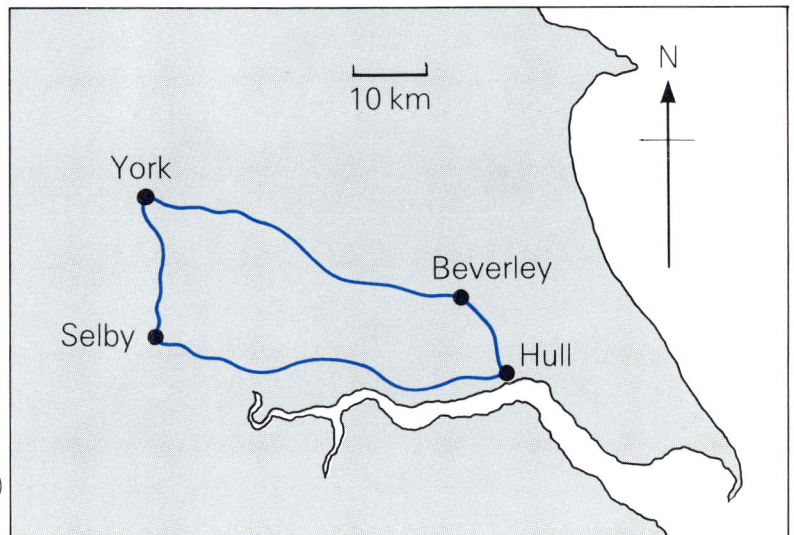

c) Describe the journey from Hull to York using   (i) $\begin{pmatrix} \square\,\text{km W} \\ \square\,\text{km N} \end{pmatrix}$ (ii) $\begin{pmatrix} \square\,\text{km} \\ \square° \end{pmatrix}$.

d) Describe the position of Hull from York using

   (i) ($\square$ km $\square$, $\square$ km $\square$)   (ii) ($\square$ km, $\square$°).

e) Describe the journey from York to Hull using   (i) $\begin{pmatrix} \square\,\text{km}\,\square \\ \square\,\text{km}\,\square \end{pmatrix}$ (ii) $\begin{pmatrix} \square\,\text{km} \\ \square° \end{pmatrix}$.

Take note

When we use $\begin{pmatrix} \blacksquare \\ \blacksquare \end{pmatrix}$ to describe a journey, we ignore the route we take.

The description is the same for all routes.

30

Journeys on grids

C 1 Here are three different journeys from A to B.
We can describe all of them like this:

$$\begin{pmatrix} {}^+5 \\ {}^+4 \end{pmatrix}$$ ← *5 right* / *4 up*

a) Describe all the possible journeys from A to D in the same way.

b) Describe all the possible journeys from D to C using

$$\begin{pmatrix} {}^+\square \\ {}^+\square \end{pmatrix}.$$

2 We can describe the journeys from B to A like this:

$$\begin{pmatrix} {}^-5 \\ {}^-4 \end{pmatrix}$$ ← *5 left* / *?*

a) What does the $^-4$ tell us?

b) Describe these journeys in the same way:
   (i) B to D (ii) D to A (iii) C to D (iv) C to A

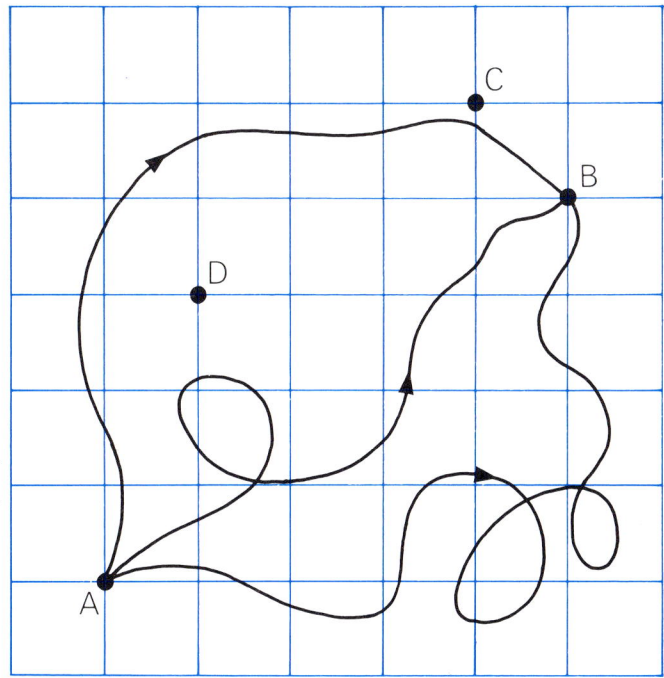

Notice

Take note

We use $^+$ and $^-$ to tell us the directions on a grid.

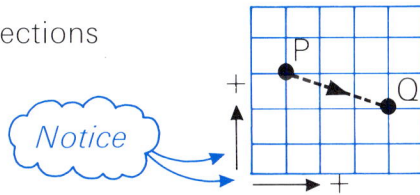

Notice

On this grid we can describe the journey P → Q

as $\begin{pmatrix} {}^+3 \\ {}^-1 \end{pmatrix}.$

Think it through

3 a) Five of the journeys on the grid have something in common. What is it?

b) Describe each journey using $\begin{pmatrix} \square \\ \square \end{pmatrix}.$

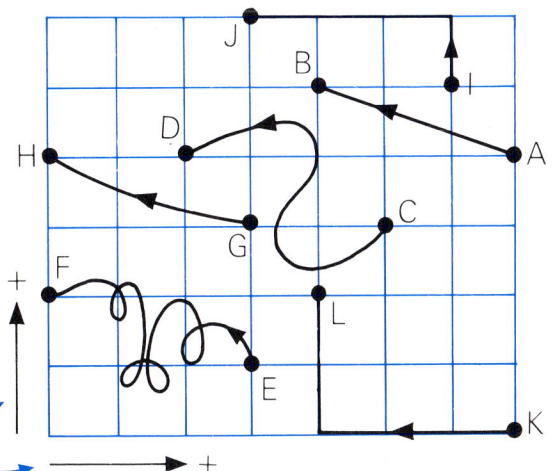

Notice

4 a) Look at the grid.
Millie is a programmed
electronic mouse.
These instructions will send
Millie all the way around the
rectangle ABCD:

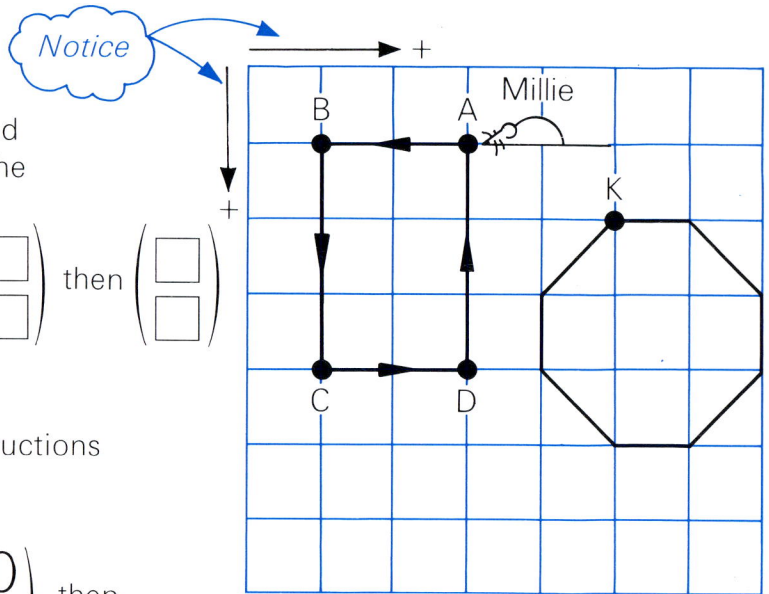

$$\begin{pmatrix} ^-2 \\ 0 \end{pmatrix} \text{ then } \begin{pmatrix} 0 \\ ^+3 \end{pmatrix} \text{ then } \begin{pmatrix} \square \\ \square \end{pmatrix} \text{ then } \begin{pmatrix} \square \\ \square \end{pmatrix}$$

Copy and complete them.

b) Starting from K, these instructions
will send Millie all the
way around the octagon:

$$\begin{pmatrix} ^+1 \\ 0 \end{pmatrix} \text{ then } \begin{pmatrix} ^+1 \\ ^+1 \end{pmatrix} \text{ then } \begin{pmatrix} 0 \\ ^+1 \end{pmatrix} \text{ then } \quad \ldots$$

 (i) Copy and complete them.
 (ii) In which direction would Millie travel, clockwise or anticlockwise?
 (iii) Write down instructions which will send Millie around the octagon
 ... in the opposite direction.

Start at K.

5 Copy the grid.
Travel only in straight lines.
What kind of shape do these
instructions trace out?

$$\begin{pmatrix} ^+3 \\ 0 \end{pmatrix} \text{ then } \begin{pmatrix} 0 \\ ^-3 \end{pmatrix} \text{ then } \begin{pmatrix} ^-3 \\ 0 \end{pmatrix} \text{ then } \begin{pmatrix} 0 \\ ^+3 \end{pmatrix} .$$

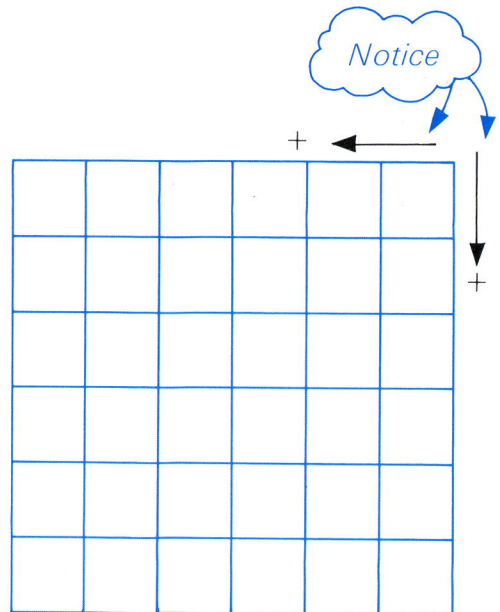

Notice

30

Challenge

6 Write out a set of instructions to trace out

 a) a letter  b) a trapezium.

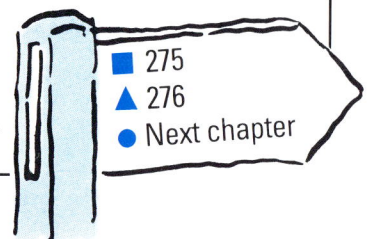

■ 275
▲ 276
● Next chapter

Describing journeys

1 Manchester is about 35 km south and 45 km west of Leeds.

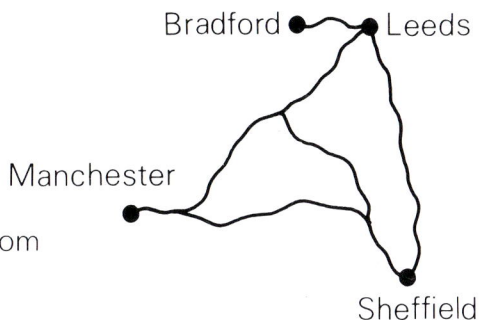

a) Use (☐ km ☐, ☐ km ☐) to describe
   (i) the position of Manchester relative to Leeds.

   (ii) the position of Leeds relative to Manchester.

b) Use $\left(\begin{array}{c}\square\ \text{km}\ \square\\ \square\ \text{km}\ \square\end{array}\right)$ to describe journeys from

   (i) Manchester to Leeds.
   (ii) Leeds to Manchester.

2 The grid shows six journeys on a grid. The journey from A to B

can be described by $\left(\begin{array}{c}+5\\-1\end{array}\right)$

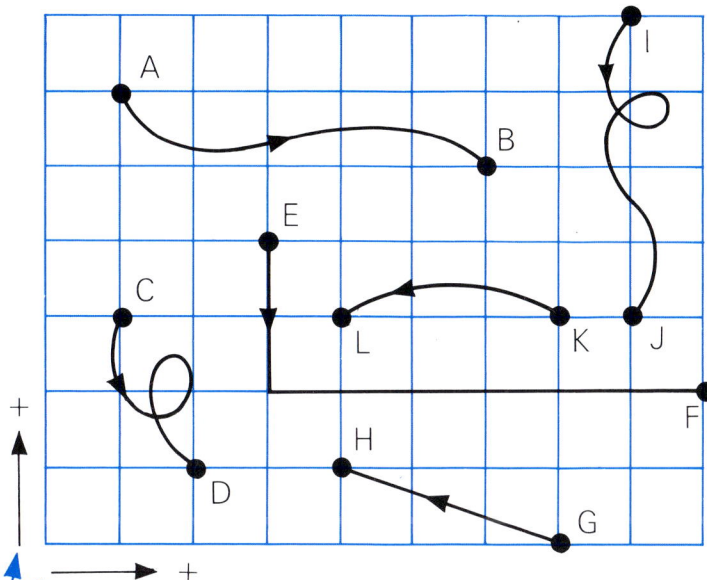

a) Describe the other journeys using $\left(\begin{array}{c}\square\\\square\end{array}\right)$.

b) Describe these journeys using $\left(\begin{array}{c}\square\\\square\end{array}\right)$:

   (i) A → I
   (ii) I → F
   (iii) F → G
   (iv) C → I

Notice

Notice

3 The grid shows a mouse's journey around a kitchen floor. The first part of his journey

can be described by $\left(\begin{array}{c}+2\\+1\end{array}\right)$

a) Describe the other four parts using $\left(\begin{array}{c}\square\\\square\end{array}\right)$.

b) Describe the five parts of the journey in the opposite direction using $\left(\begin{array}{c}\square\\\square\end{array}\right)$.

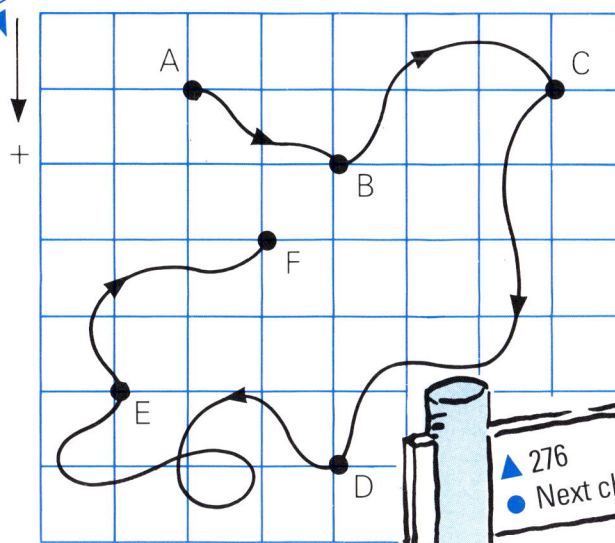

30

▲ 276
● Next chapter

Journeys on different grids

D 1   On this grid, any journey from

A to D can be written as $\begin{pmatrix} ^+5 \\ ^-2 \end{pmatrix}$.

a) Write these journeys
   in the same way:

     (i)   A → G
     (ii)   B → G
     (iii)   C → E
     (iv)   E → G

b) Andrew Ant starts
   at E.
   He goes on this
   two-stage journey:

$\begin{pmatrix} ^+4 \\ ^+3 \end{pmatrix}$ then $\begin{pmatrix} 0 \\ ^-4 \end{pmatrix}$.

     (i)   Where does he finish?
     (ii)   Describe the journey from E to Andrew's finishing point.

         Use $\begin{pmatrix} \square \\ \square \end{pmatrix}$.

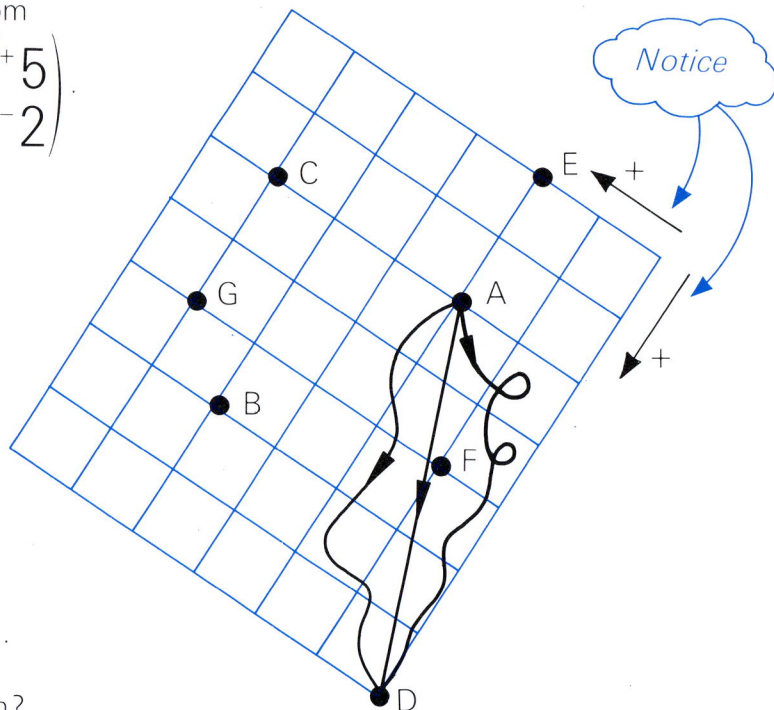

Notice

Think it through

2   a)   Check that any journey from
       A to B can be described

       by $\begin{pmatrix} ^-2 \\ ^-3 \end{pmatrix}$

b) Write these journeys
   in the same
   way:
     (i)   A → D
     (ii)   H → G
     (iii)   B → D
     (iv)   E → C

Notice

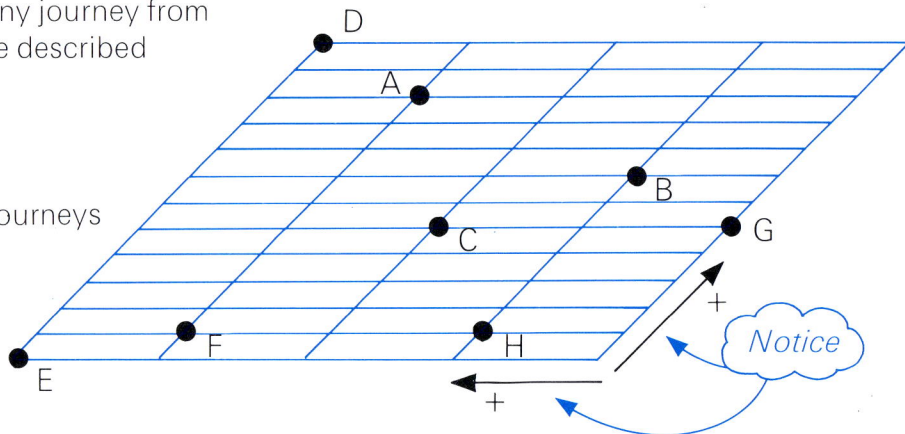

c) Here is an eight-stage journey:
     D → F → C → H → B → G → C → E → D

     (i)   Describe the journey, using $\begin{pmatrix} \square \\ \square \end{pmatrix}$ for each part.

$\begin{pmatrix} ^-1 \\ ^-11 \end{pmatrix}$ then $\begin{pmatrix} ^-1 \\ ^+4 \end{pmatrix}$ then ...

     (ii)   Describe the whole journey using just one $\begin{pmatrix} \square \\ \square \end{pmatrix}$.

30

Making tracks

Challenge

You can only travel along the grid lines.
You can move **up**, **down**, **left** or **right**.
You cannot retrace your steps.

1 How many different journeys are there
which can be described

by $\begin{pmatrix} +2 \\ +2 \end{pmatrix}$?

Sketch them all like this:

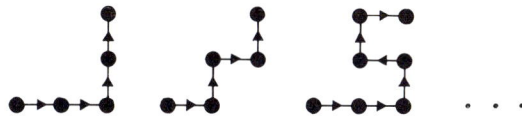

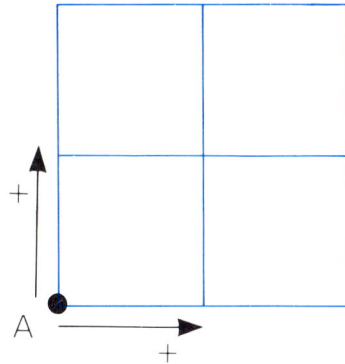

2 a) How many different journeys are there starting from A which

can be described by $\begin{pmatrix} +1 \\ +1 \end{pmatrix}$?

b) How many different $\begin{pmatrix} +1 \\ +1 \end{pmatrix}$ journeys are there if you can start
from anywhere?

3 How many different journeys are there which can be described by $\begin{pmatrix} +1 \\ +2 \end{pmatrix}$

a) if you can only start at A?

b) if you can start from anywhere?

4 How many different journeys are there for each of these?

a) $\begin{pmatrix} +2 \\ -2 \end{pmatrix}$

b) $\begin{pmatrix} -1 \\ -1 \end{pmatrix}$, if you start from A

c) $\begin{pmatrix} -1 \\ -1 \end{pmatrix}$, if you can start from anywhere

d) $\begin{pmatrix} +2 \\ +1 \end{pmatrix}$, if you start from A

e) $\begin{pmatrix} +2 \\ +1 \end{pmatrix}$, if you can start from anywhere

30

● Next chapter

▲ 277

31 Special numbers

A 1 a) This **number star** shows the numbers which divide exactly into 28.
   Copy and complete it.

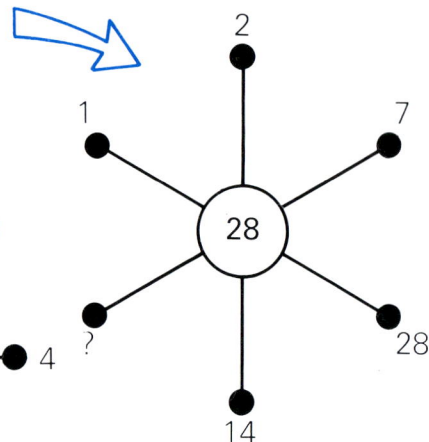

   b) Copy and complete the star for 24.

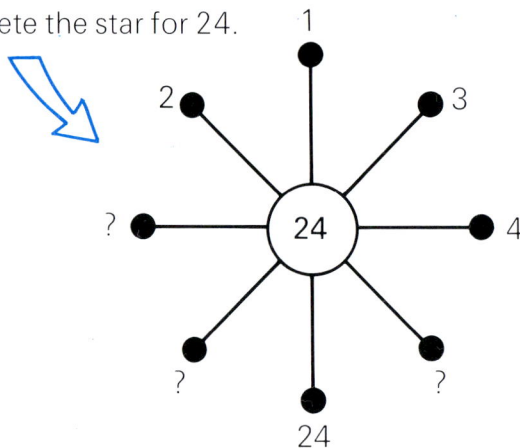

Take note

1, 2, 3, 4, 6 and 12 are called the **factors** of 12.
Each one divides exactly into 12.

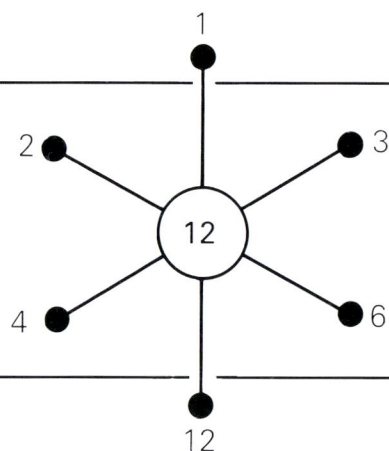

2 The factors in the **18** star and **30** star are arranged in a special way.

   a) Write down what is special.

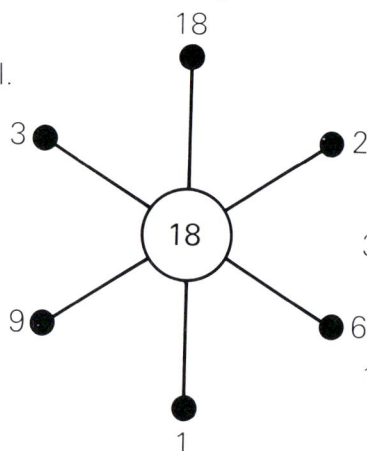

   b) Redraw the **20** star.
      Arrange the factors in the same special way.

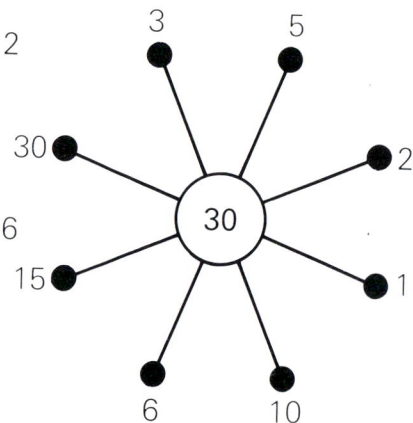

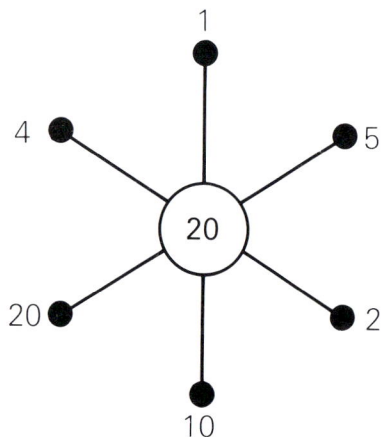

3 **a)** These stars are special:

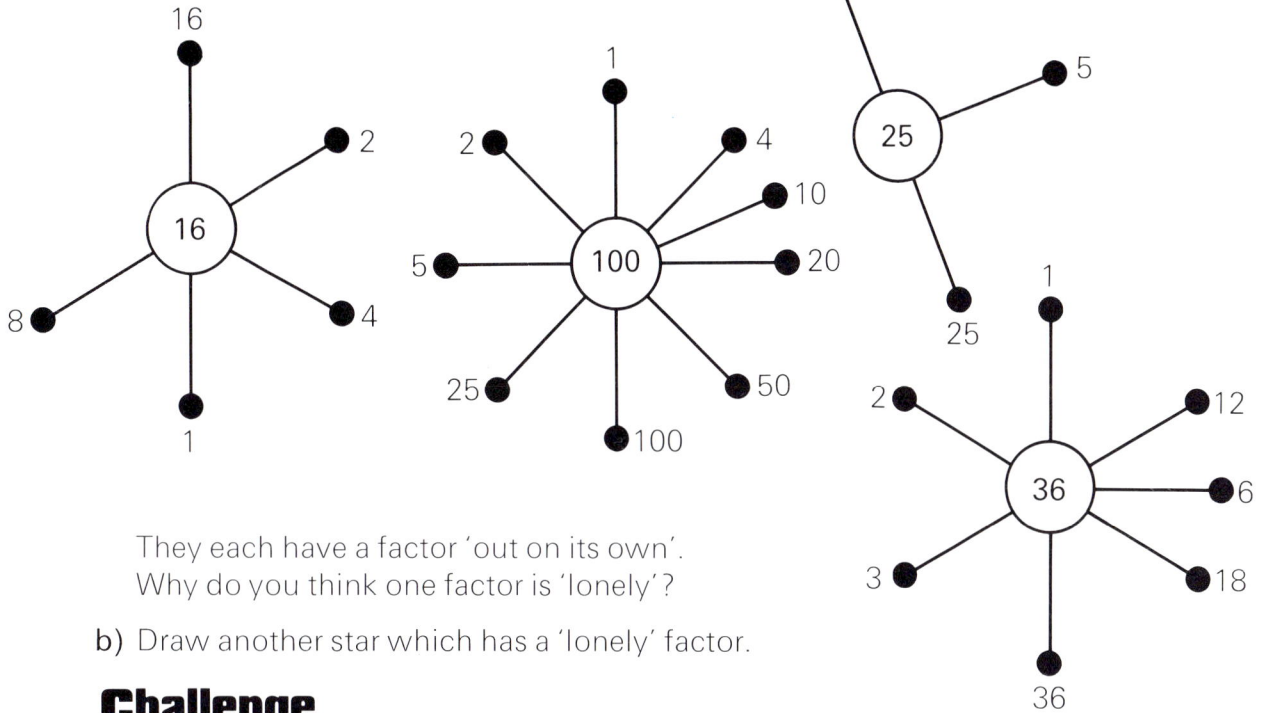

They each have a factor 'out on its own'.
Why do you think one factor is 'lonely'?

b) Draw another star which has a 'lonely' factor.

Challenge

4 Numbers with 'lonely-factor stars' have
an odd number of factors.

a) Write down the first **ten** numbers which
have an odd number of factors.

b) This is the lonely factor for a number.
What is the number?

 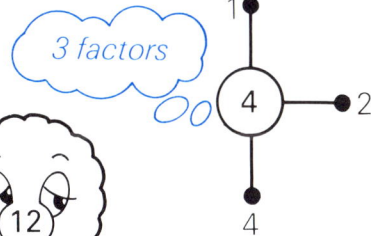

5 20 has an even number of factors.
Write down the first **ten** numbers which have an even number of factors.

6 How many numbers have only **one** factor?
List them all.

31

Challenge

7 Find the first number which has

a) 2 factors.

b) 3 factors.

c) 4 factors.

d) 12 factors.

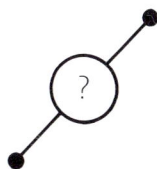 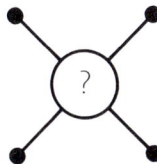 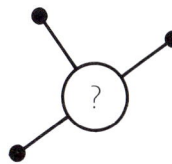 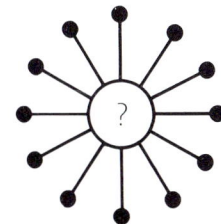

8 Work with a friend.

Start with any number: ⑨

Write down all its factors: $1, 3, 9$

Add all but the largest: $1 + 3 = ④$

Write down the factors of the new number: $1, 2, 4$

Add all but the largest: $1 + 2 = ③$

Write down the factors of the new number: $1, 3$

Add all but the largest: ①

This is called the **factor chain** for 9:

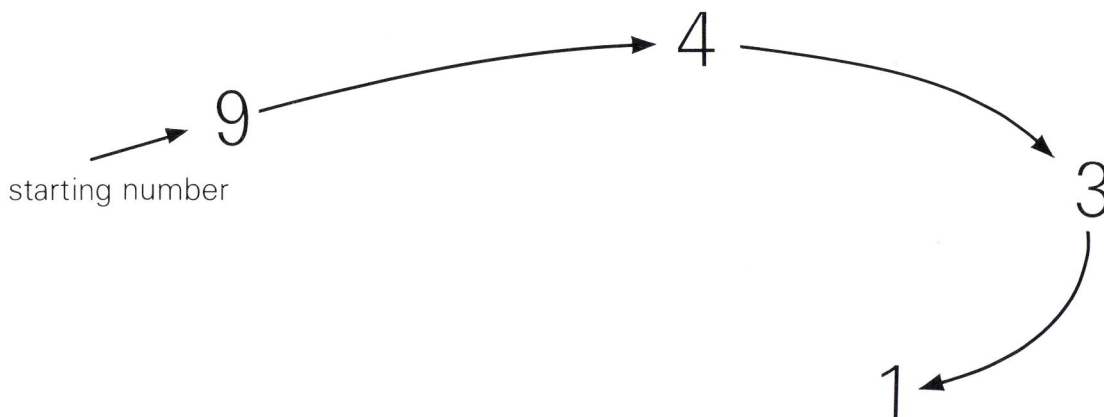

9 → 4 → 3 → 1

starting number

a) Find out what happens with

(i) 16   (ii) 29   (iii) 12

b) Find out what happens with 6

c) Look for another number which ends like 6

d) Find out what happens with 102 ₒₒₒ *Its factor chain is long!*

31

Multiples

B 1 Read the Auto Bank instructions.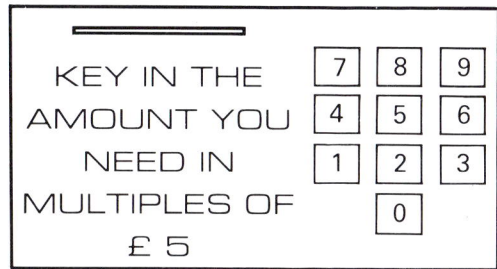

  a) What is the smallest amount
     you can withdraw?

  b) What is the next smallest amount?

  c) ... and the next?

Take note

£10, £20, £30, ... are all **multiples** of £10.

2 The first multiple of 8 is 8.
   The second is 16.

  a) What is the third?

  b) ... and the tenth?

3 There is a multiple of 18 between 85 and 97.
   What number is it?

4 12 is a multiple of the numbers
   which surround it.
   Surround 20 by the numbers
   whose multiple it is.

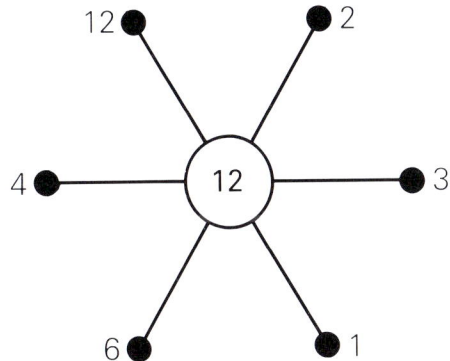

5 a) Write down all the numbers of which 36 is a multiple.

  b) Write down all the factors of 36.

Take note

12 is a **multiple** of 1, 2, 3, 4, 6 and 12.

1, 2, 3, 4, 6 and 12 are **divisors** of 12.

or factors

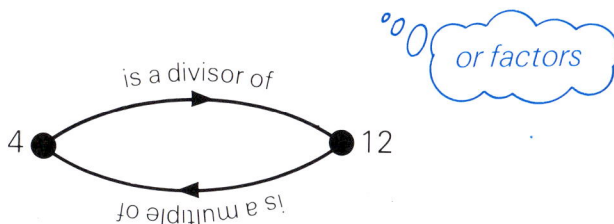

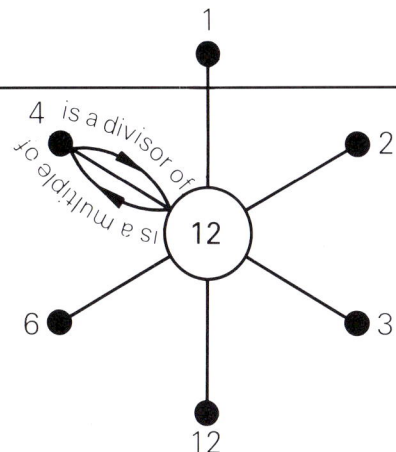

31

Primes and squares

C 1 The first number with exactly **two** factors is 2.
The next is 3.

List the next **nine**.

Challenge

2 Which is the first number larger than 100 which has only two factors?

Take note

Numbers which have exactly **two** factors are called **prime** numbers.

3 a) Which number is the first prime number?

b) Is 84 a prime number?

c) Is 466 a prime number?

d) How many prime numbers are even?
Explain your answer.

Take note

Read this as 'seven squared'.

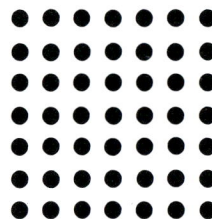

We write 7×7 in shorthand as 7^2.

We call 49 a **square number** because . . .
. . . it can be written as 7×7
and because . . .

. . . we can show it as a square of dots.

4 The first square number is 1 . . . 1×1 . . . 1^2 . . .

The second square number is 4 . . . 2×2 . . . 2^2 . . .

List the next **nine** square numbers.
Write each one in three different ways.

Example: 81, 9×9, 9^2

5 a) Draw factor stars for the
first five square numbers.

b) There is something the same
about the number of factors in
all five stars.
What is it?

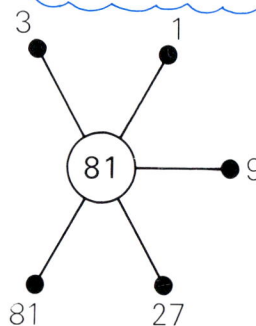

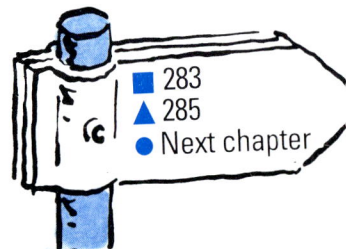

31

283
285
Next chapter

282

Factors and multiples

A
B

1 Copy and complete each factor star:

a)

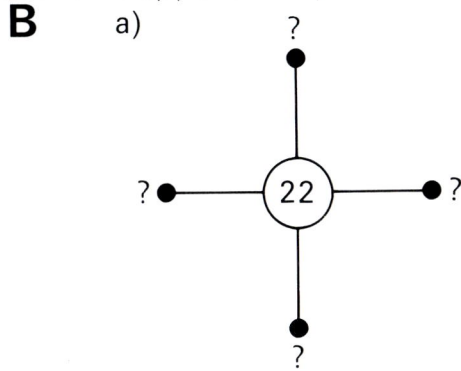

b)

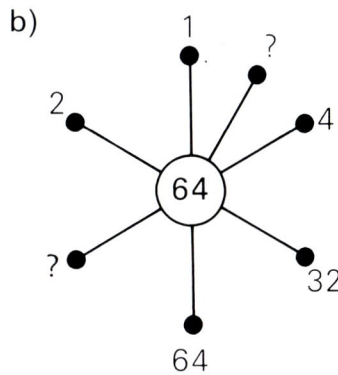

c)

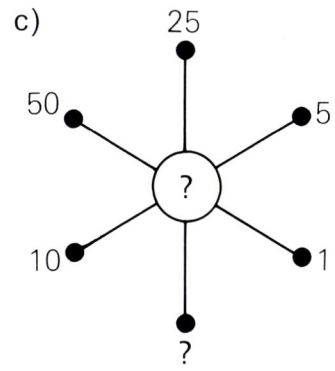

2 1 is the first number with an odd number of factors.

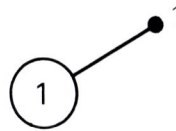

a) What is the next?

b) What is the fifth?

3

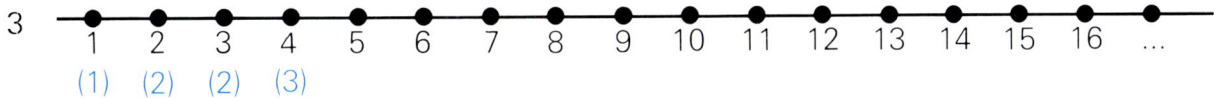

a) Draw a line.
   Mark the numbers from 1 to 20.
   Underneath each number write in brackets how many factors it has.

b) What is the first number with more than five factors?

c) Which number from 1 to 20 has the most factors?

4 These are the first three multiples of 12:
              12, 24, 36, . . .
These are the first three multiples of 15:
              15, 30, 45, . . .

a) What is the first number which is a multiple of both 12 and 15?

b) What is the next?

c) . . . and the next?

d) What is the tenth?

e) The numbers you are finding follow a pattern.
   What is the pattern?

Challenge

5 Which numbers have exactly two multiples between 90 and 100?

Do not include 90 and 100 as multiples.

31

■ 283

Dot patterns

C 1 These dot patterns are for multiples of 5:

a) Draw dot patterns for the next three multiples of 5.

b) One multiple is a square number.
   Which one?

2 These are dot patterns for multiples of 17:

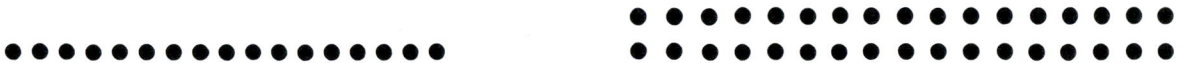

a) Draw the next dot pattern.

b) Which multiple of 17 is a square number?

You don't have to draw the dot pattern.

3 This dot pattern is for 30:

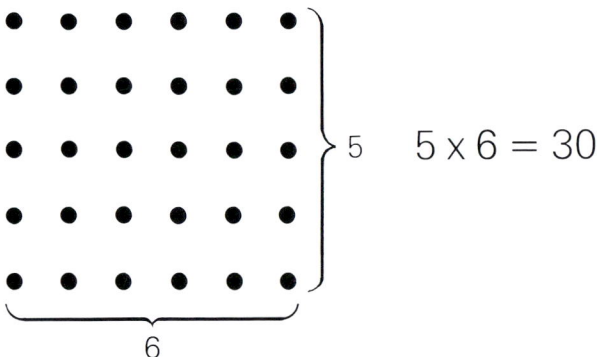

$5 \times 6 = 30$

6

Draw **three** more dot patterns for 30. Write a sentence for each one:

$\square \times \square = 30$

with more than one row

4 Think of all the prime numbers: 2, 3, 5, 7, ...
  It is not possible to draw a rectangular dot pattern for any of them.
  Explain why not.

5 8^2 and 2×8 are not the same.
  In what way are they different?

6 $\square^2 = 81$.
  What is the missing number?

Challenge

7 Subtract 1 from each square number less than 200.

  How many of the resulting numbers are primes?

▲ 285
● Next chapter

Pot pourri

Challenge

Do these questions with a friend.

1 A number with exactly three factors is 25.
   What is the first number larger than 1000 with exactly three factors?

2 These members of the Jones family all have their birthdays today.
   How old is each one, do you think?

YESTERDAY I WAS A SQUARE, BUT NOW I AM IN MY PRIME!

Great-grandfather
Jones

Henrietta
Jones ...
and her children

George

Lilly

Baby Peanut

... who can talk, and who is a gifted mathematician!

3 Which of these numbers are multiples of 217?

   1085   4614   2710
       1953   2094

4 a) These are multiples of 26, written in order:
       26, 52, 78, 104, ...
       These are multiples of 14:
       14, 28, 42, 56, ...
       (i) Are any numbers multiples of both 26 and 14?
       (ii) If yes, what is the first one, and how many are there?

   b) Can you think of two numbers which have **no** multiples in common?
       If so, write them down.

5 Lighthouse Hurricane flashes its light
   towards Crab Rock once every 20 seconds.
   Lighthouse Sundial flashes its light
   towards Crab Rock once every 15 seconds.
   You are shipwrecked on Crab Rock.
   Sometimes you can see both lights at once.
   How often do you think this happens?

31

6 Some numbers can be shown as dot triangles, like this:

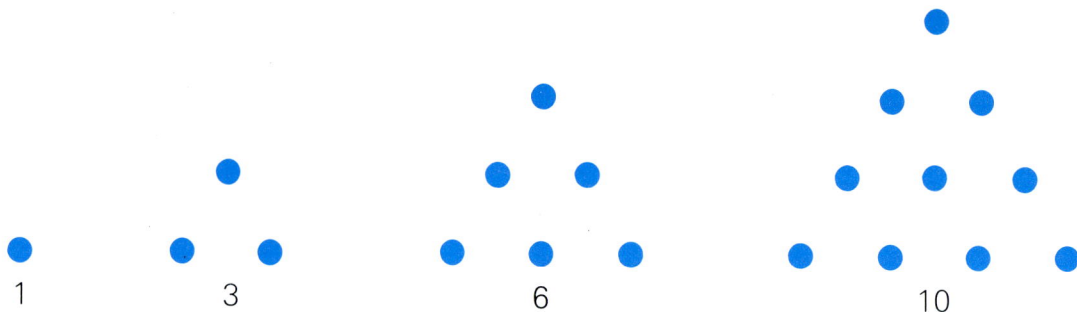

1   3   6   10

a) What is the next **triangle number** after 10?

b) ... and the next?

c) The third triangle number is 6.
How many dots are added to make the fourth?

d) How many dots are added to the fifth triangle number to make the sixth?

e) How many are added to the sixth to make the seventh?

f) The 99th triangle number is 4950.
What is the 100th?

Think it through

7 This is a chemist's pill counter.

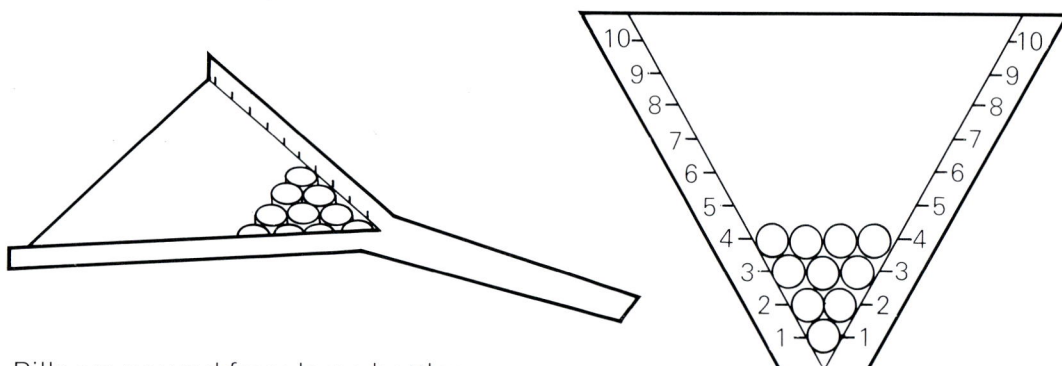

Pills are poured from large bottles into the tray.

a) How does the chemist use it to count out 50 pills quickly?
Write one or two sentences to explain.

b) Make up a chart which would help the chemist to use the tray.

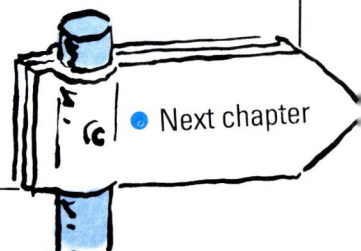

● Next chapter

31

32 Rate and speed

A 1 Dave is an oilman.
He earns £10 for every hour he works.
We say his **rate of pay** is £10 per hour.

Write down the rate of pay for these:

I EARN £20 FOR 5 HOURS' WORK — Fred

I EARN £56 FOR 8 HOURS' WORK — Tina

2 Mary works a five-day week.
She earns £400 each week.
What is her rate of pay per day?

4 weeks

3 Every month, Sara is paid £900.
What is her rate of pay per week?

4 Sylvie earns £16 800 annually. *per year*
She gets the same amount each month.
What is her rate of pay per month?

Think it through

5 Alan is paid £5 per hour on weekdays and £8 per hour on
Saturdays and Sundays.
Last week he worked 8 hours per day, Monday to Friday.
On Saturday he worked 6 hours.
How much did he earn?

6 Rula worked $1\frac{1}{2}$ hours.
She earned £6.75.
What is her rate of pay per hour?

7   We measure rates of pay in £ per hour, £ per week, and so on.
    Here are some units for other kinds of rates.
    Write down which one you think has to do with

    a) typing.

    b) the cost of wood.

    c) the price of lemonade.

    d) pop records.

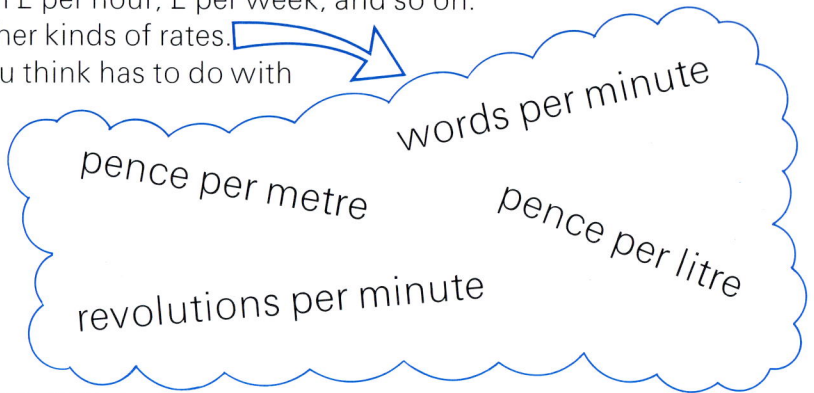

pence per metre

words per minute

pence per litre

revolutions per minute

With a friend

8   Here are some units for rates.
    For each unit, discuss
    what it might measure.

Example:
pence per litre –
price of lemonade

Write down
two examples
for each unit.

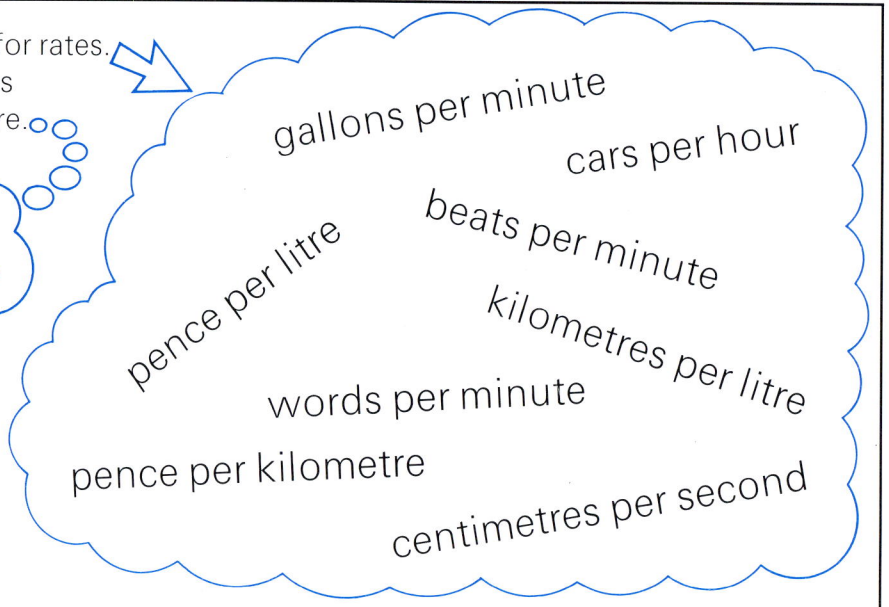

gallons per minute

cars per hour

beats per minute

pence per litre

kilometres per litre

words per minute

pence per kilometre

centimetres per second

9   a) Estimate your normal breathing rate in breaths per minute.

    b) About how many breaths do you take per hour?

when you are resting

10 Write down **two** possible units for each of these:

    a) the price of curtain material

    b) the rate at which air is pumped into a balloon

11 The chart shows the hire rate for taxis.
    How much would

    a) a 15 km journey cost?

    b) a 25 km journey cost?

RANK BAD TAXIS

Charges

	Hire rate
First 20 km	20p per km
Over 20 km	15p per km

32

Speed

B

Take note

The steamroller is travelling 2 metres every second.
We say its **speed** is 2 m/s.

metres per second

Think it through

1 Jim and Sara are at the airport.
The check-in gate is 540 m away.
Jim stands on the moving pavement.
Its speed is 1 m/s.
Sara walks steadily.
She arrives in 6 minutes.

a) How far does Jim travel every second?

b) Who is the first to arrive
at the check-in gate?

c) How far does Sara walk every minute?

d) How far does she walk every second?

e) What is Sara's steady walking speed?

2 Here are two things which normally travel at a steady speed.

moving pavement
at an airport

sound along a
telephone wire

Here are two things which normally do not.

cars along a road

birds flying

List two more things which normally

a) move at a steady speed.

b) do not move at a steady speed.

3 Concorde flies for $2\frac{1}{2}$ hours at a steady speed of 2300 km/h.
How far does it travel?

4 Estimate the speed of top-class athletes in these events:

Assume they run at a steady speed.

a) the women's 100 m

b) the men's 1500 m

just over 42 km

c) the women's marathon

32

● **289**

5 Each of these travelled at a steady speed.
What was the speed in each case?

a) cheetah

540 metres in 20 seconds

b) marathon runner

120 metres in 30 seconds

c) ski chair-lift

480 metres in 4 minutes

d) high-speed train

75 kilometres in $\frac{1}{2}$ hour

6 A lift carries Horace 36 m.
Its speed is 3 m/s.
How long does the journey take?

7 An ocean liner travels 960 km at a steady speed of 60 km/h.
How many hours long is the journey?

8 Water flows into a tub at a rate of 36 litres/min.
The tub holds 180 litres of water.
How long does it take to fill the tub?

9 How many days would it take you to walk 500 km at 20 km per day?

10 a) Estimate your normal steady walking speed.

b) About how long would it take you to walk 10 km?

11 A cable car travels 750 metres at a steady speed.
The journey takes $\frac{1}{4}$ hour.
What is its speed

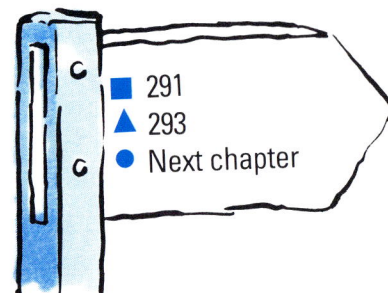

a) in m/min?

b) in km/h?

■ 291
▲ 293
● Next chapter

Working with rates

A 1 You need a watch which shows seconds.
Find the pulse in your wrist.

   a) Count the number of beats in 30 seconds.

   b) What is your pulse rate in beats per minute?

   c) About how many times does your pulse beat in an hour?

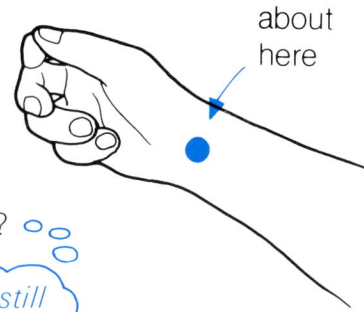

   when you are sitting still doing nothing

about here

2 Horace can write his full name *neatly* at the rate of one full name every 8 seconds.

   Horace Egbert Hornblower

   How quickly can **you** write his full name?

3 Copy and complete this table:

Name	Rate of pay (per hour)	Number of hours worked	Total pay
Penny	£2	4	
Tariq	£3.50	6	
Winston	£5		£35
Fiona	£2.50		£10
Lenny		3	£12
Meg		$\frac{1}{2}$	£3.25

Think it through

4 You want to earn £30 for your holiday.
You can work 2 hours every Saturday.
There are six Saturdays left before your holiday.
What rate of pay must you get to earn the money?

32

5 A bottle-washing machine washes 600 bottles per hour.

   a) What rate is this per minute?

   b) How long does it take to wash each bottle?

Thinking about speed

B 1 Here are some speeds.

2 m/s
6 km/h
90 km/h
110 km/h
330 m/s
820 m/s
300 000 km/s

Match each speed with one of these :

car travelling along motorway

girl hiking

bullet from a gun

sound of thunder

cheetah running

light travelling from the sun

man in a lift

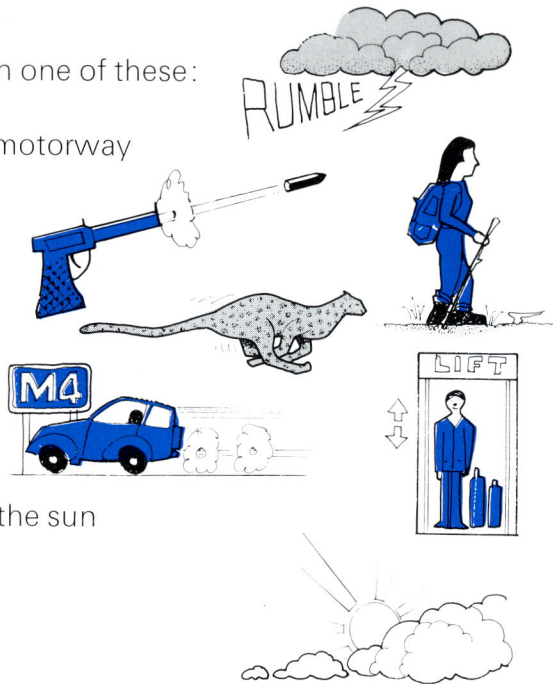

With a friend

2 Discuss how you would measure these.
   a) Write down what you decide.

   b) List everything you would need to make the measurements.

A **the speed at which a stream is flowing**

B **the rate at which water flows out of a sink**

3 This table is about people travelling on escalators.
   Write down what is missing for each person.

Person	Speed (m/s)	Distance (m)	Travelling time (s)
a)	1	20	
b)	$1\frac{1}{2}$	24	
c)	2		12
d)	$2\frac{1}{2}$		8
e)		30	15
f)		18	12

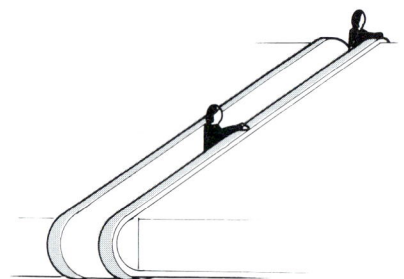

▲ 293
● Next chapter

32

Using rates

C 1 Two of these metal rods are made of the same metal.
Which do you think they are?
Explain your answer.

They all have the same cross-section.

    A   2.3 m, mass 22 kg

    B   5.4 m, mass 43 kg

    D   2.76 m, mass 26.4 kg

    C   3.1 m, mass 32 kg

With a friend

2 Frank and Freda estimate the area of their feet:

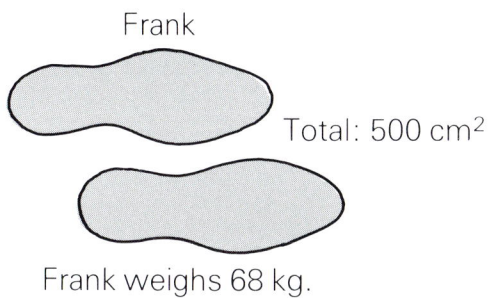

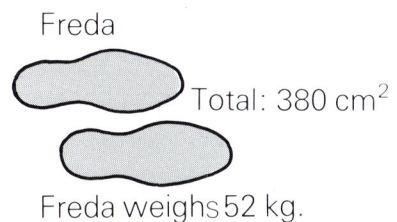

Frank

Total: 500 cm^2

Frank weighs 68 kg.

Freda

Total: 380 cm^2

Freda weighs 52 kg.

a) What is the pressure on Frank's feet in g/cm^2?

grams per square centimetre

b) What is the pressure on Freda's feet in g/cm^2?

c) Help each other to estimate the pressure on your feet in g/cm^2.
Whose feet have more pressure on them?

3 Here are some population figures about two towns.

Valley Heath	Halesover
Population (end of last year): 29 000	Population (end of last year): 78 000
Number of births this year: 420	Number of births this year: 1025
Number of deaths this year: 200	Number of deaths this year: 270

a) Which town had more births this year?

b) Which town's population is growing at a faster rate?

c) Is it possible to say which town's families have more children, on average?
Explain your answer.

32

Speed and graphs

D 1 Here are some speedometer readings for two steamrollers.

Steamroller 1

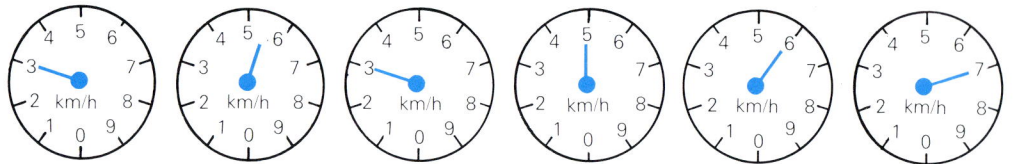

After 1 min After 2 min After 3 min After 4 min After 5 min After 6 min

Steamroller 2

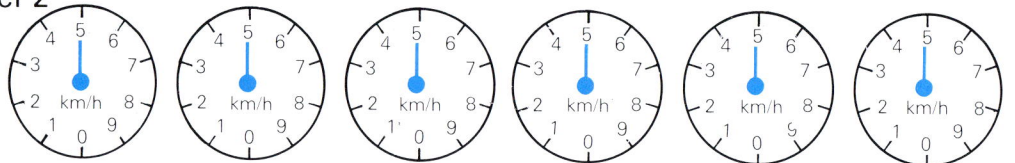

After 1 min After 2 min After 3 min After 4 min After 5 min After 6 min

a) One of them is definitely not travelling at a steady speed.
Which one?

b) We cannot tell whether the other is travelling at a steady speed.
Explain why not.

c) This graph is for steamroller 1.
Copy the graph on squared paper and complete it.

d) The dots of the graph are joined with a dashed line.
Why do you think this is?

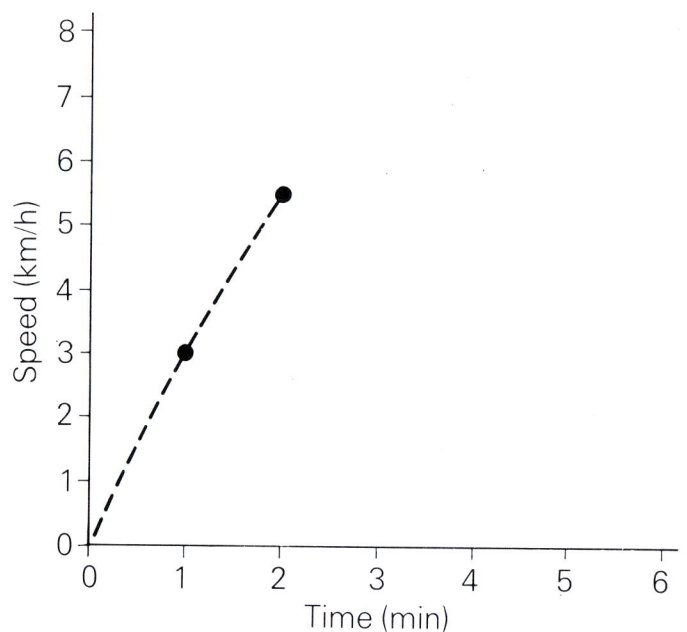

32

Think it through

e) Draw **two** possible graphs for steamroller 2.
Make one of them suggest the steamroller travelled at a steady speed between the 1st and 6th minutes.
Make the other suggest that the steamroller's speed was
6 km/h after $2\frac{1}{2}$ minutes and 4 km/h after $4\frac{1}{2}$ minutes.

Think it through

2 a) Two of these graphs are for cars moving at a steady speed.
   Which ones must they be? ○○○○○ _Look carefully at the axes._

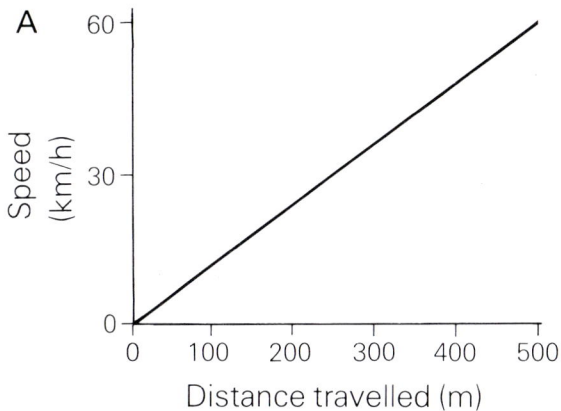

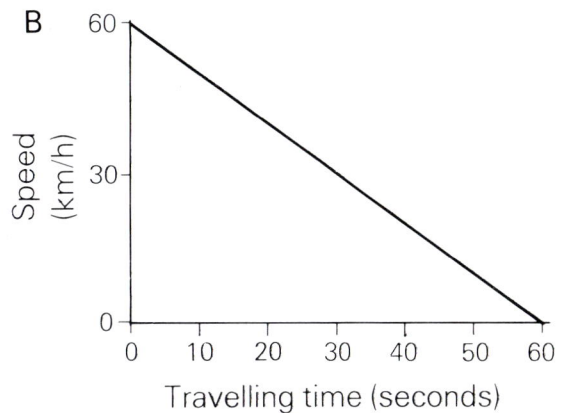

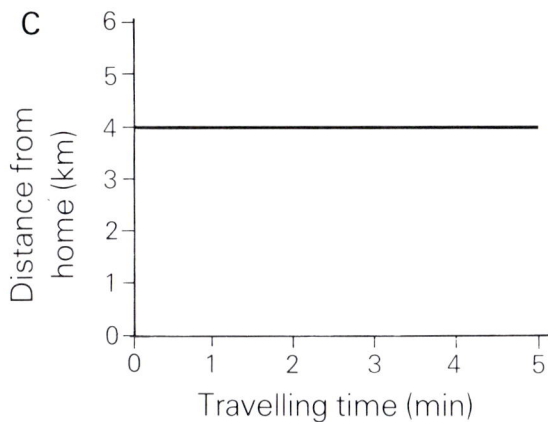

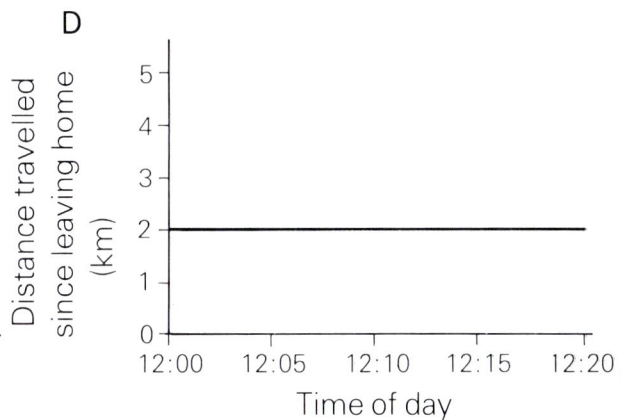

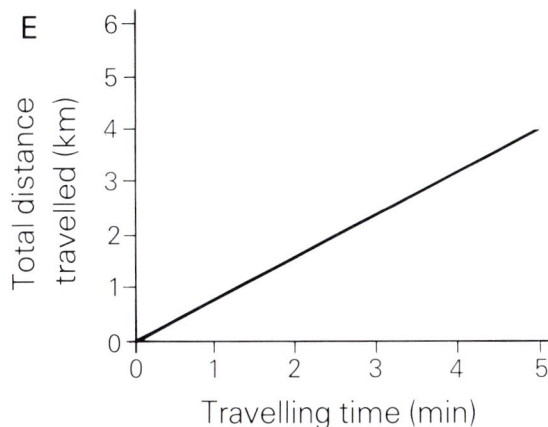

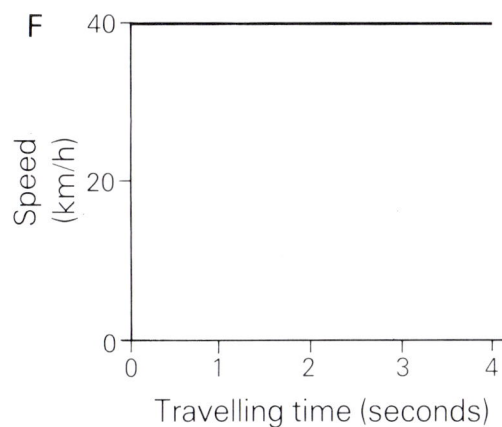

A

B

C

D

E

F

b) Which graphs could be for
   (i) a car sitting in a traffic jam? (2 graphs)
   (ii) a car travelling in a circle? (at least 2 graphs)
   (iii) a car accelerating away?

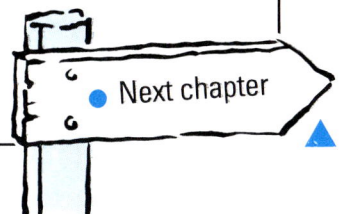

● Next chapter

33 Volume and capacity

A 1 Rupert works in a supermarket.
He stacks these boxes on a shelf.

a) How many boxes are there?

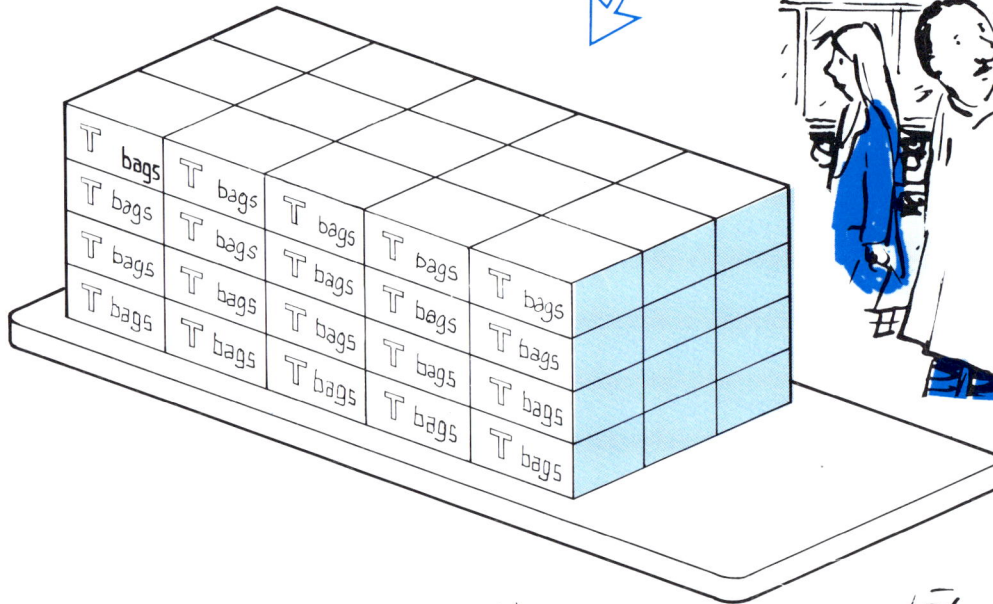

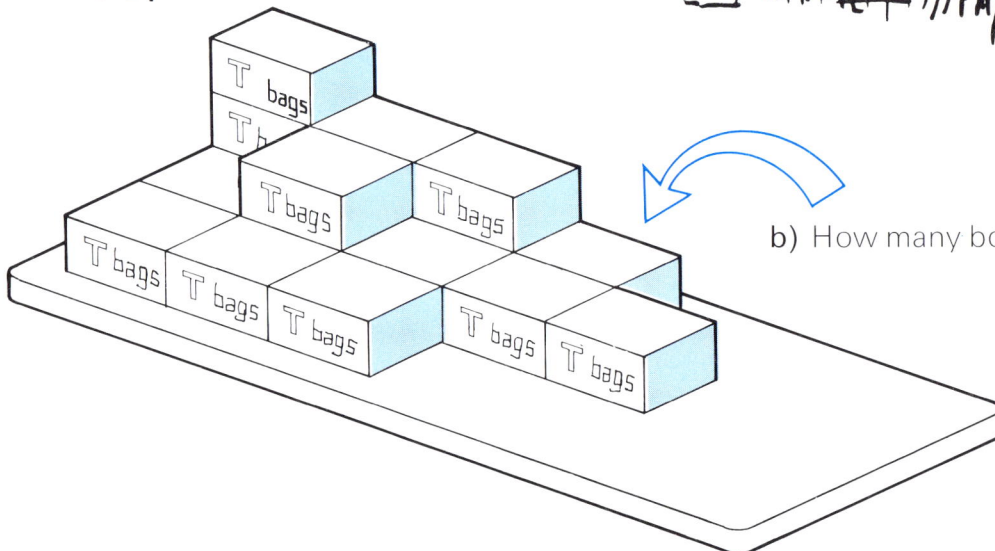

b) How many boxes are there now?

2 Now Rupert is stacking tissue boxes.

a) How many are there here?

 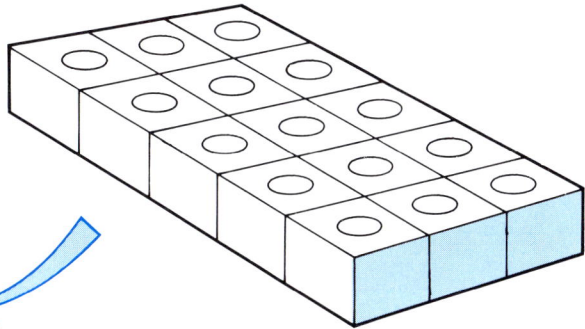

b) ... and here?

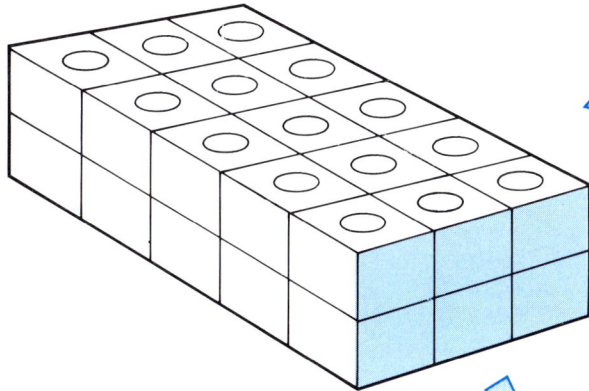 

c) ... and here?

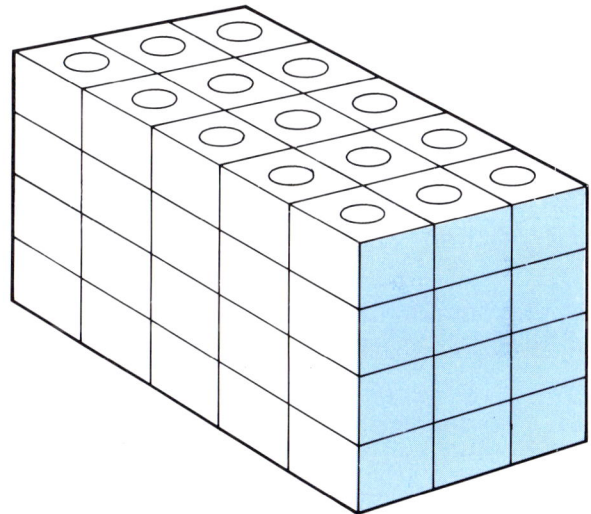

3 Copy and complete the **Take note**.

Take note

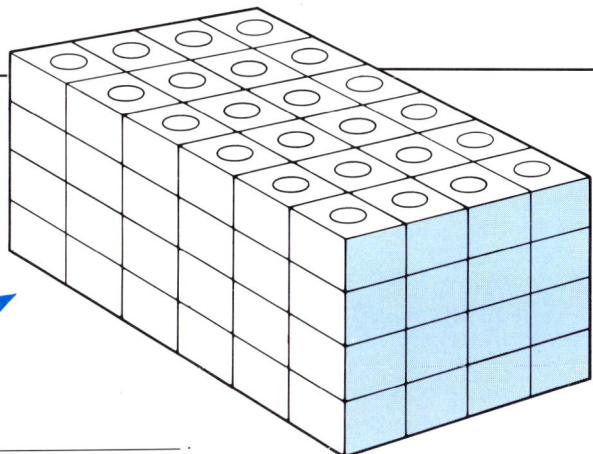

The number of boxes in this stack is

number in bottom layer × _____ .

33

Prisms

B 1 a) Ken has some 1 cm cubes.

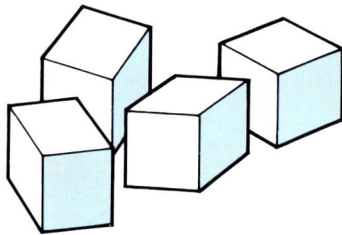

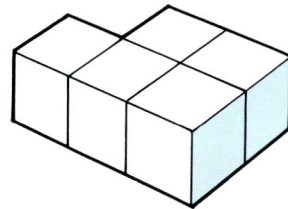

He sticks them together to make L-shapes like these.
How many cubes does he need for each L-shape?

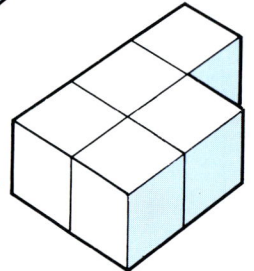

b) Ken makes this prism.
   (i) How many L-shapes
       does he need?
   (ii) How many 1 cm cubes
        does he need?

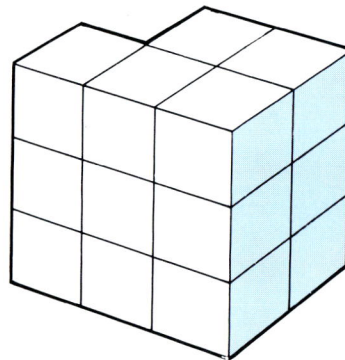

DO YOU REMEMBER ...?
*A **prism** can be cut into layers which are all the same shape and size.*

Think it through

2 a) Ken uses some more
    of his L-shapes to
    make this prism.

    How many 1 cm cubes
    does he use?

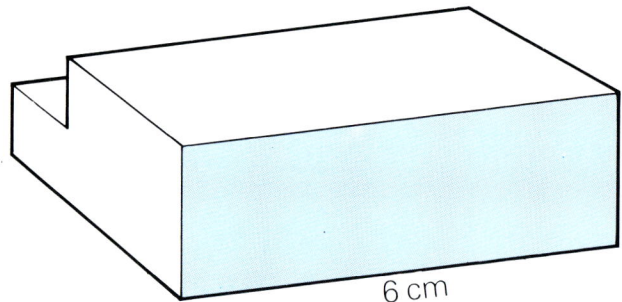

6 cm

b) This is another of Ken's prisms.
   He has dipped one end in some paint.
   Copy and complete:

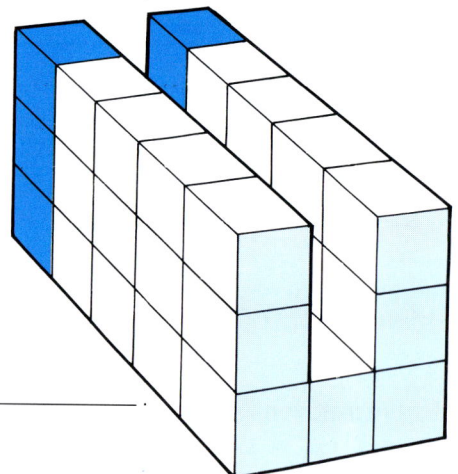

number of cubes Ken uses

=

number in the painted layer × the number _____ .

33

298

3 Ken made these solids from 1 cm cubes.

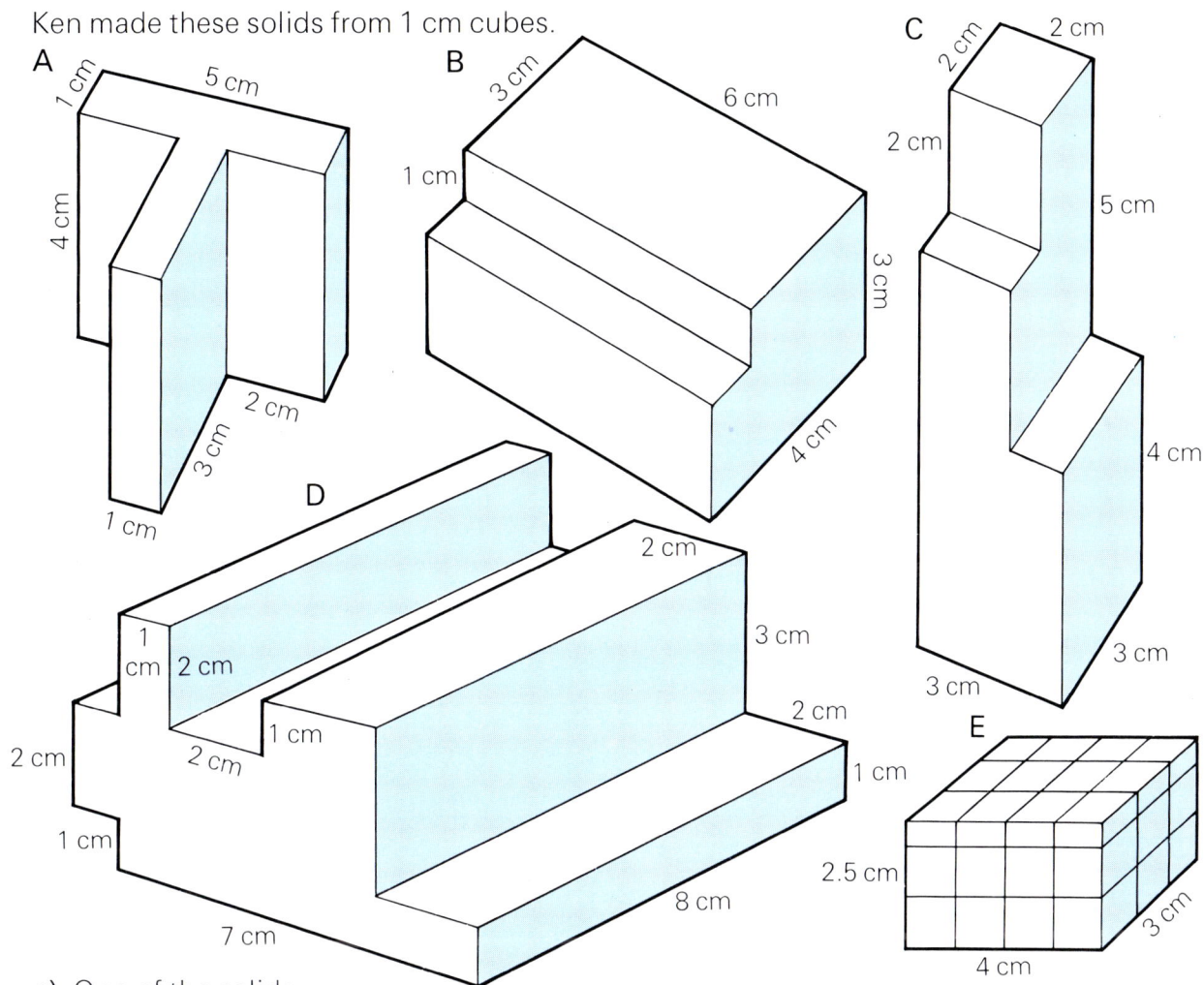

a) One of the solids
is **not** a prism.
Which is it?

b) Ken made the T-shaped prism from four of these.
How many cubes did he use to make
  (i) each **T**?
  (ii) the whole prism?

c) Ken used 8 cm³ of wood to make each **T**.
What volume of wood did he use to make
  (i) the T-shaped prism?
  (ii) each of the solids **B**, **C**, **D** and **E**?

d) Copy and complete the **Take note**.

REMEMBER ...?
Volume:
1 cm³

Take note

Volume of prism = volume of one layer of 1 cm cubes × _____ .

33

Capacity

C 1 a) Rupert removes the packets of soap powder from this box. How many packets fit into the box?

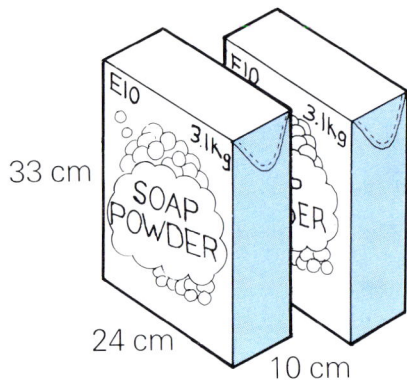

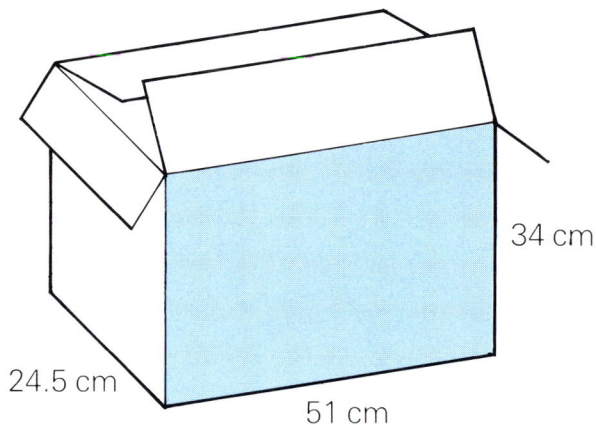

33 cm

SOAP POWDER

24 cm

10 cm

34 cm

24.5 cm

51 cm

b) How many of these tissue boxes fit into this box?

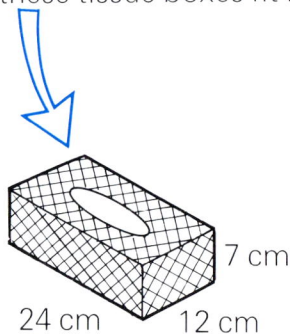

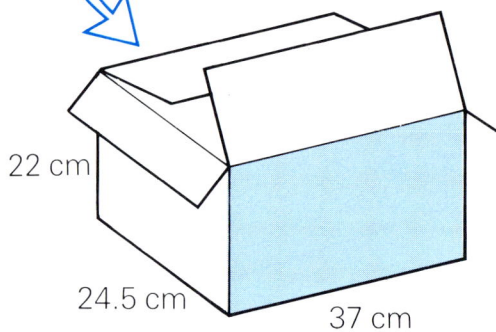

7 cm

24 cm   12 cm

22 cm

24.5 cm   37 cm

c) These bottles of washing-up liquid are 7 cm in diameter. How many fit into this box?

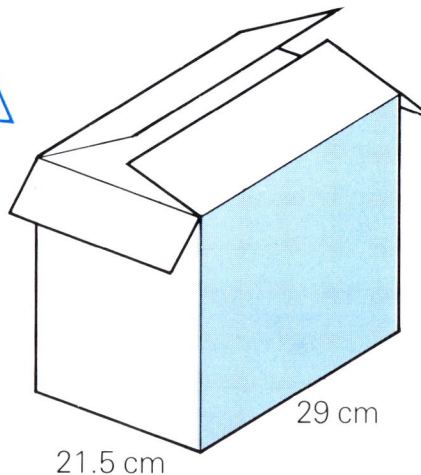

LIQUID

29 cm

21.5 cm

2 Melanie made this perspex container.
Ken filled it with 1 cm cubes.

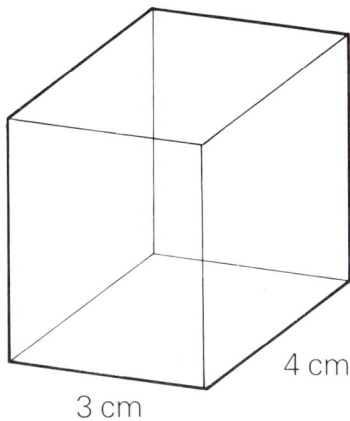

a) How many cubes did he use?

b) What is the capacity of the container
  (i) in cm³?
  (ii) in millilitres?
  (iii) in litres?

5 cm

4 cm

3 cm

REMEMBER ...?

This little glass container holds 1 ml of water.

Its capacity is 1 ml (or 1 cm³).

1 cm
1 cm
1 cm

3 Melanie made these containers.

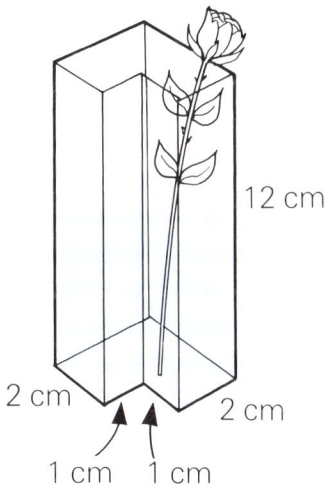

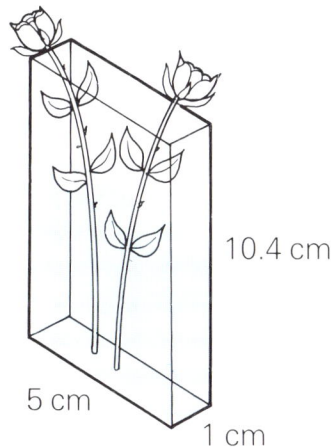

12 cm

2 cm   2 cm

1 cm   1 cm

10.4 cm

5 cm

1 cm

What is the capacity of each one
(i) in millilitres?   (ii) in litres?

With a friend

4 Work together on this problem.
About how many litres of water would you need to turn your classroom into a gigantic fish tank?

■ 302
▲ 304
● Next chapter

33

Volumes of prisms

A
B

1 a) Prism A has been dipped in paint. How many cubes have paint on them?

 b) What is the volume of the whole prism in cubic centimetres?

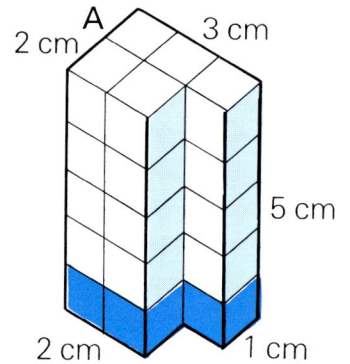

2 Prisms **B**, **C** and **D** are made from 1 cm cubes. They have been dipped in paint 1 cm deep.

 a) In each prism, how many cubes have paint on them?

 b) What is the volume of each prism in cubic centimetres?

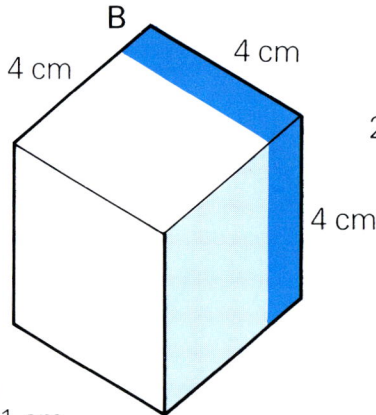

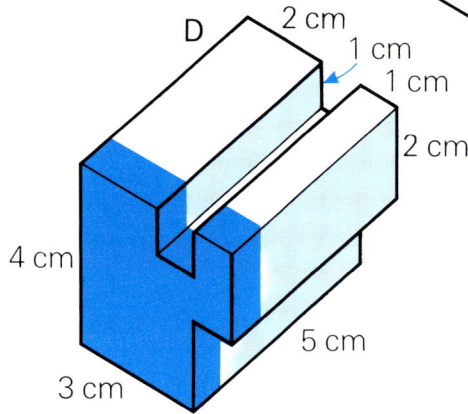

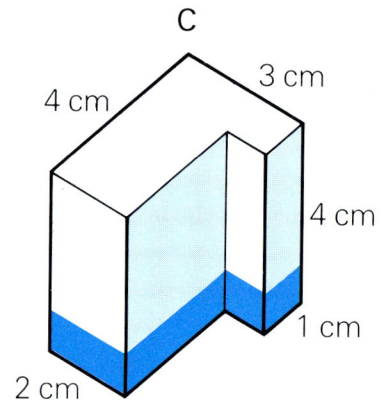

3 What is the volume of each of these prisms in cubic centimetres?

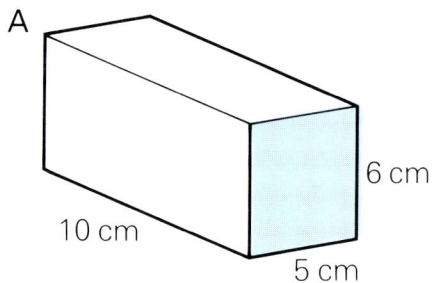

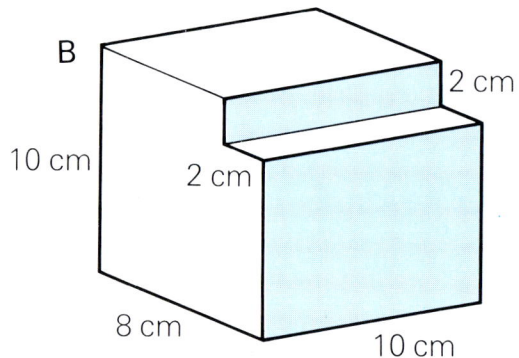

Think it through

4 The volume of a cube is 125 cm³. How long are its sides?

How much will it hold?

C 1 Lenny pours the cola into ice trays, like this:

Each section measures 1 cm × 1 cm × 1 cm.

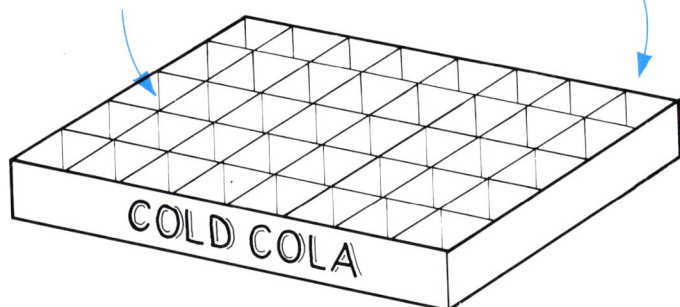

How many 'cola-cubes' can he make?

2 Lenny is filling his fish tank.
The water is 1 cm deep.

a) How many cubic centimetres of water are there?

b) How many millilitres of water is this?

c) Now the water is 5 cm deep. How many millilitres of water is this?

d) When the tank is full, the water is 20 cm deep.
What is the capacity of Lenny's tank
   (i) in millilitres?
   (ii) in litres?

24 cm   40 cm

Challenge

3 Design a flower vase which will hold exactly 480 ml of water.

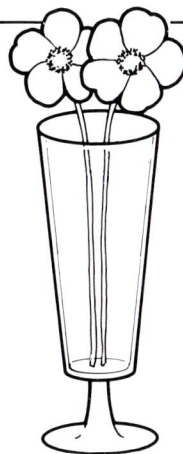

▲ 304
● Next chapter

33

Interesting prisms

D 1   Horace is making a prism from 1 cm cubes.
What is its volume?

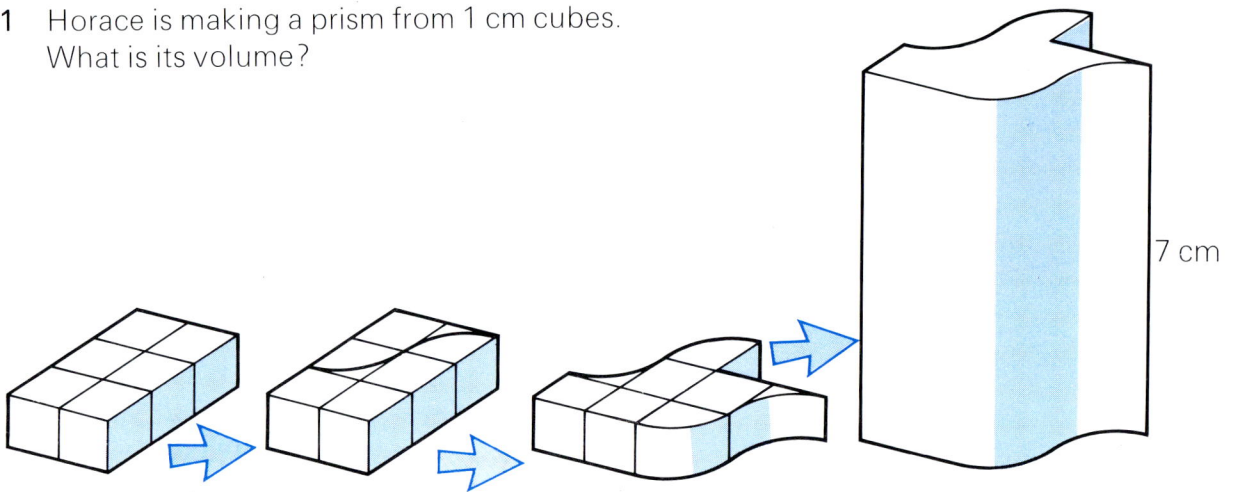

7 cm

2   Meg is also making a prism from 1 cm cubes.
What, roughly, is its volume?

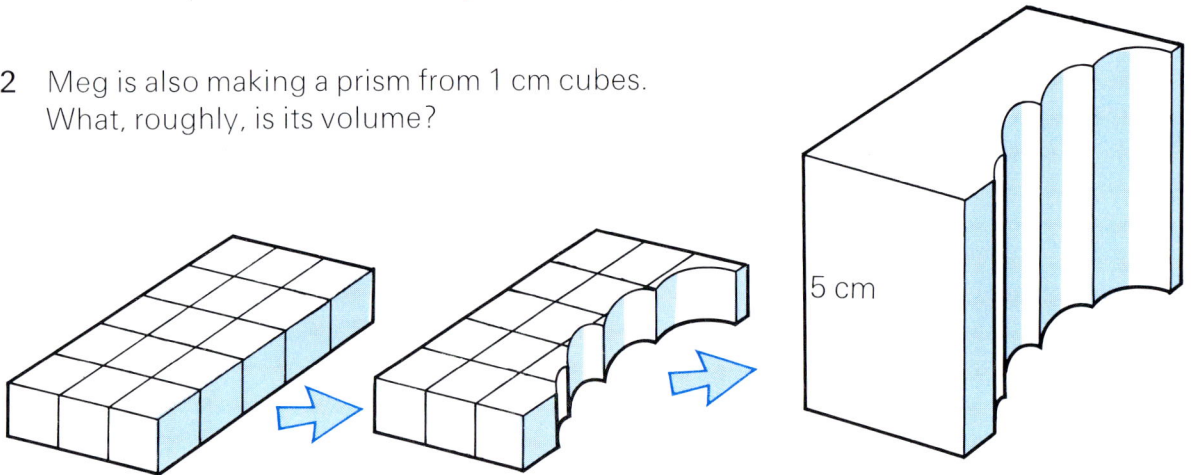

5 cm

Think it through

3   The area of this face is 9 cm².
What is the volume of the prism?

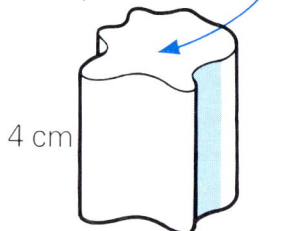

4 cm

4   Here is a pile of 10p pieces.
The area of the circular face is about 6 cm².
What is the volume of the pile?

2 cm

5   The volume of this prism is 24 cm³.
What is the area of this face?

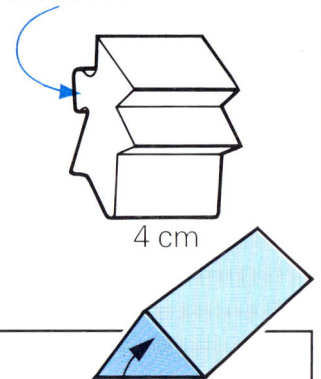

4 cm

6   Copy and complete the **Take note**.

Take note

Volume of prism = area of constant cross section × _____

cross section

A capacity problem

EXPLORATION

1   Work with a friend.
You need a calculator.

Glenda has a 21 cm square sheet of card.

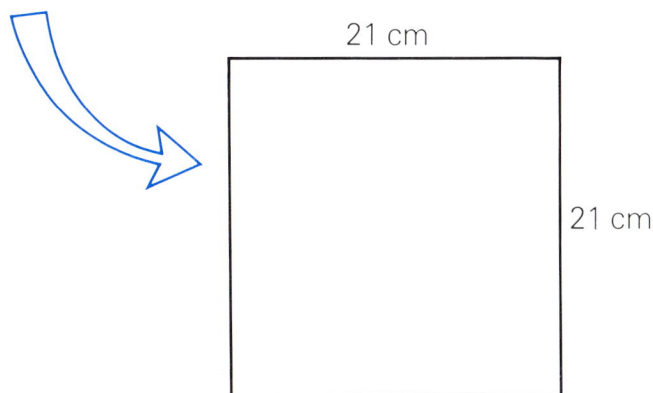

21 cm

21 cm

She cuts a 6 cm square from each corner.

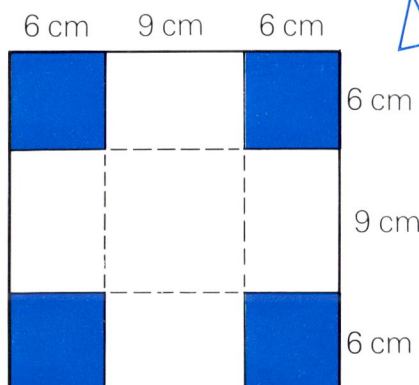

| 6 cm | 9 cm | 6 cm |

6 cm

9 cm

6 cm

She folds the remaining card
to make an open box.

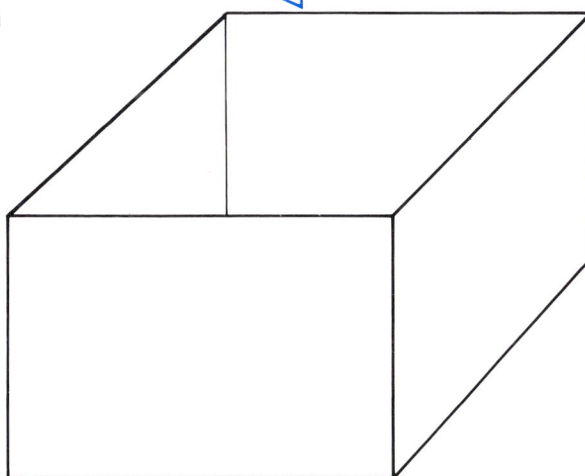

a) What is the capacity of Glenda's box?

b) What is the largest-capacity open box Glenda can make from
a 21 cm × 21 cm piece of card?

c) What is the largest-capacity open box
Glenda can make from a 25 cm square sheet of card?

33

● Next chapter

34 Generalising

A 1 These tile patterns are made from blue and white tiles.

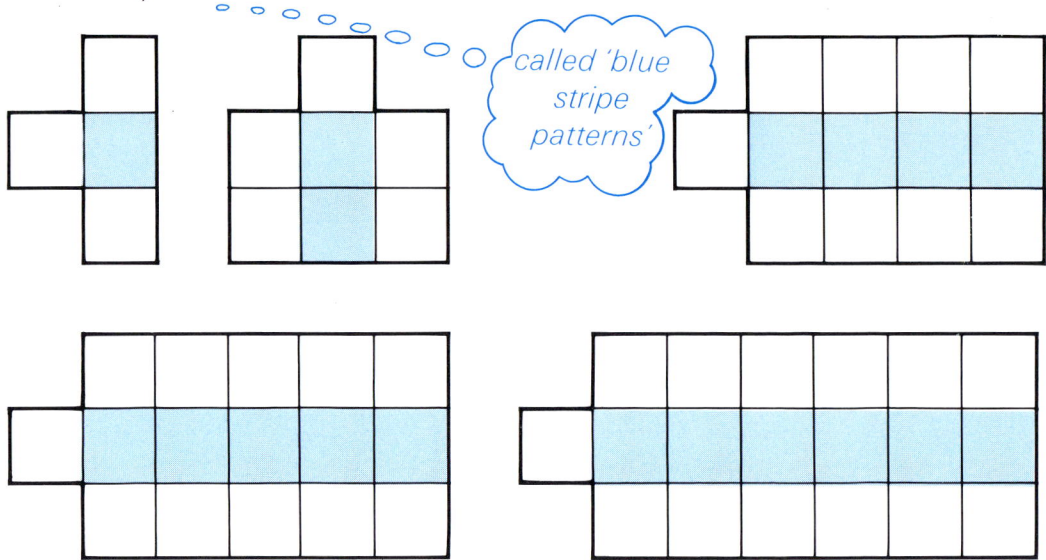

called 'blue stripe patterns'

a) This is part of a blue stripe pattern.
It has all of its blue tiles.
How many **white** tiles will it have altogether?

b) A blue stripe pattern has 3 blue tiles.
How many white tiles does it have?

c) A blue stripe pattern has 19 white tiles.
How many blue tiles does it have?

d) One of these rules tells you how many white tiles are needed
for any blue stripe pattern.
Which rule is correct?

A
Count the number of blue tiles. Add 3.

B
Count the number of blue tiles. Multiply by 3.

C
Count the number of blue tiles. Multiply by 2 and add 1.

D
Count the number of blue tiles. Multiply by 2 and add 2.

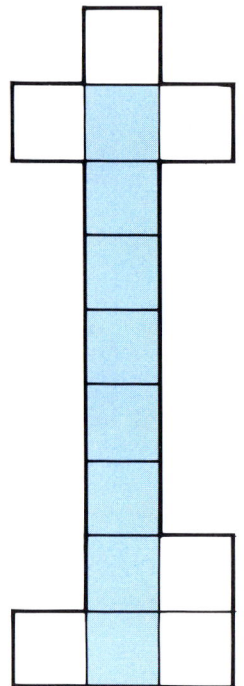

Think it through

e) Complete this to tell you how many blue tiles are needed:

Count the number of white tiles. Subtract ☐, then divide by ☐.

f) When there are 83 white tiles, there are 41 blue tiles.
Check that your rule in (e) gives the correct result.

2 These are doodles.
Lenny does them during Maths lessons.

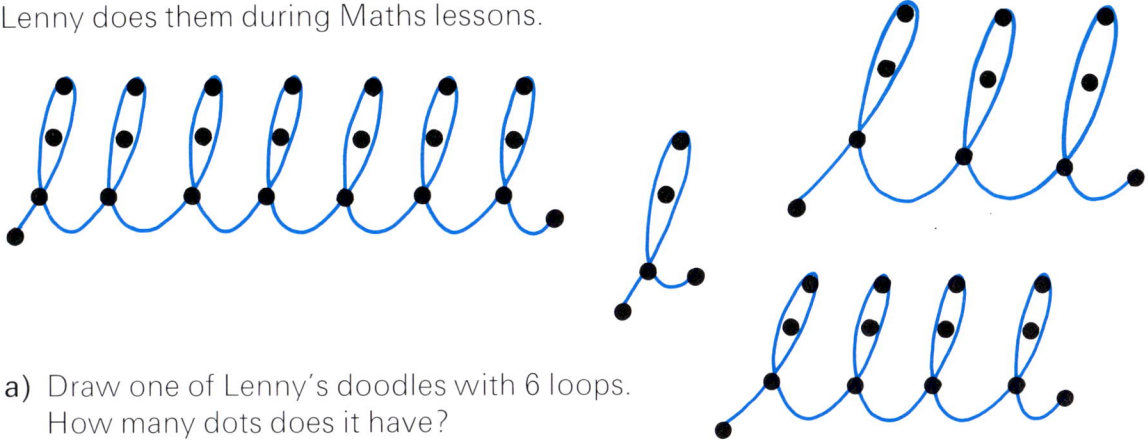

a) Draw one of Lenny's doodles with 6 loops.
   How many dots does it have?

b) Draw one of Lenny's doodles with 17 dots.
   How many loops are there?

c) One of these rules tells you how many dots there will be in
   any of Lenny's doodles.
   Which rule is it?

A Count the number of loops.
  Add 5.

B Count the number
  of loops.
  Multiply by 3
  and add 2.

C Count the number of loops.
  Multiply by 6.

D Count the number
  of loops.
  Multiply by 2, then
  add the number
  of loops.

Think it through

d) A Lenny-doodle has 40 loops.
   How many dots are there?

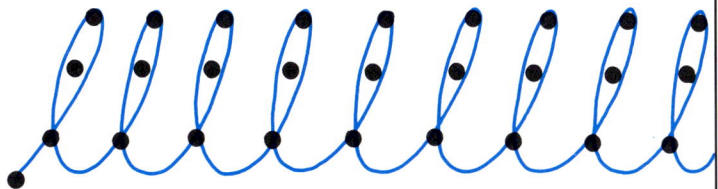

e) Another Lenny-doodle has 62 dots.
   How many loops are there?

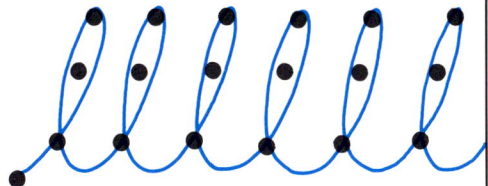

34

Writing rules

B 1   This rule tells you how many horses' legs there are in a horse race.

> Count the number of horses.
> Multiply by 4.

a) Write down a rule which tells you
 (i) how many horses' ears there are.
 (ii) how many jockeys' toes there are.

b) This rule will tell you how many horses there are in a horse race.
 Copy and complete it.

> Count the number of horses' legs.
> _____ by ☐.

2 a) This rule tells you how to find the perimeter of rectangle shapes.
 Copy and complete it.

> Add together the
> lengths of a shorter side
> and a longer side.
> _____ by ☐.

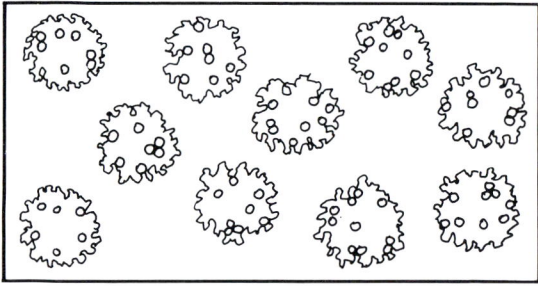

b) Write down your own rule for the **area** of rectangle shapes.

With a friend

3   Work together.
 Write down rules which tell you about

a) the number of wings in a flock of birds.

b) the number of wheels (including the spare) on some cars.

c) the number of spaces between any two fence posts.

34

308

4 a) These strip patterns are made from buttons.
One is not finished.
Copy and complete it.

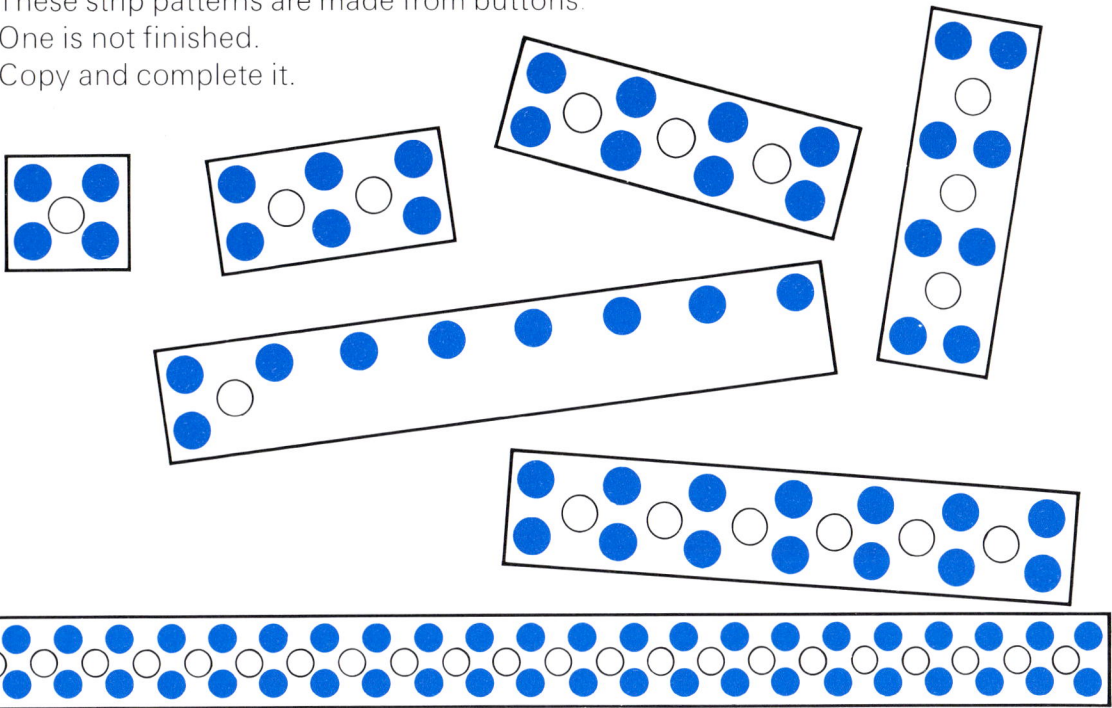

b) A button-strip has 10 white buttons.
How many blue buttons does it have?

c) Another button-strip has 26 blue buttons.
How many white buttons does it have?

d) One of these rules it correct.
Which one?

A	B	C
To find the number of white buttons:	To find the number of white buttons:	To find the number of white buttons:
• Multiply the number of blue buttons by 2.	• Subtract 2 from the number of blue buttons, then divide by 2.	• Divide the number of blue buttons by 2, then subtract 2.

Think it through

e) Write down your own rule for the number of blue buttons.
Start like this:

To find the number of blue buttons:

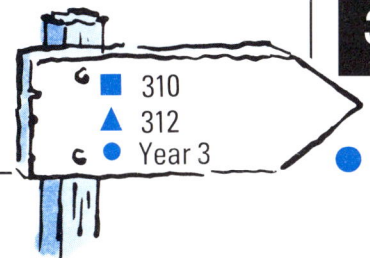

34

■ 310
▲ 312
● Year 3

● **309**

Patterns and rules

A 1

a) Draw a pattern like these which has 5 circles.
How many crosses are there?

b) How many crosses are needed for 10 circles?

c) This pattern is partly hidden.
It has 25 circles.
How many crosses does it have?

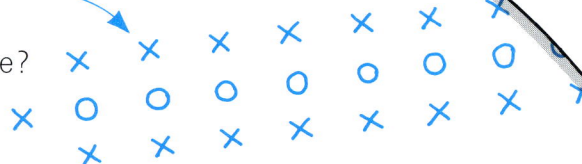

d) Lenny uses 36 crosses
to draw a pattern.
How many circles does he draw?

e) One of these rules is correct for the number of crosses.
Which one is it?

A

> Count the number
> of circles.
>
> Multiply by 2.

B

> Count the number
> of circles.
>
> Multiply by 2 and
> add 1.

C

> Count the number
> of circles.
>
> Multiply by 2 and
> add 2.

D

> Count the number
> of circles.
>
> Multiply by 2, then
> subtract 2.

34

B 1   Copy and complete these rules.

a)

To find the number of days:

Multiply the number of weeks by ☐.

b)

To find the number of weeks:

Divide the number of days by ☐.

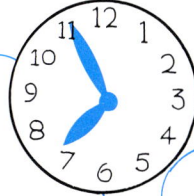

c)

To find the number of minutes:

Divide the number of _____ by 60.

d)

To find the number of minutes:

Multiply the number of _____ by 60.

e)

To find the number of days:

_____ the number of hours by ☐.

f)

To find the number of petals:

_____ the number of ___ by ☐.

g)

To find the total number of eggs:

_____ the number of _____ by ☐.

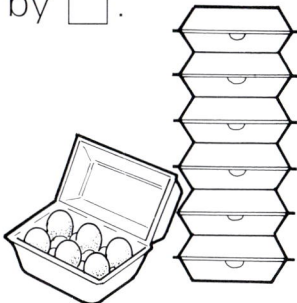

h)

posts
spacers

To find the number of spacers needed:

_____ ☐ from the number of

_____.

34

▲ 312
● Year 3

311

Finding rules

C 1 a) All these polygons are divided into triangles.
      Draw your own 5-sided polygon.
      Divide it up in the same way.

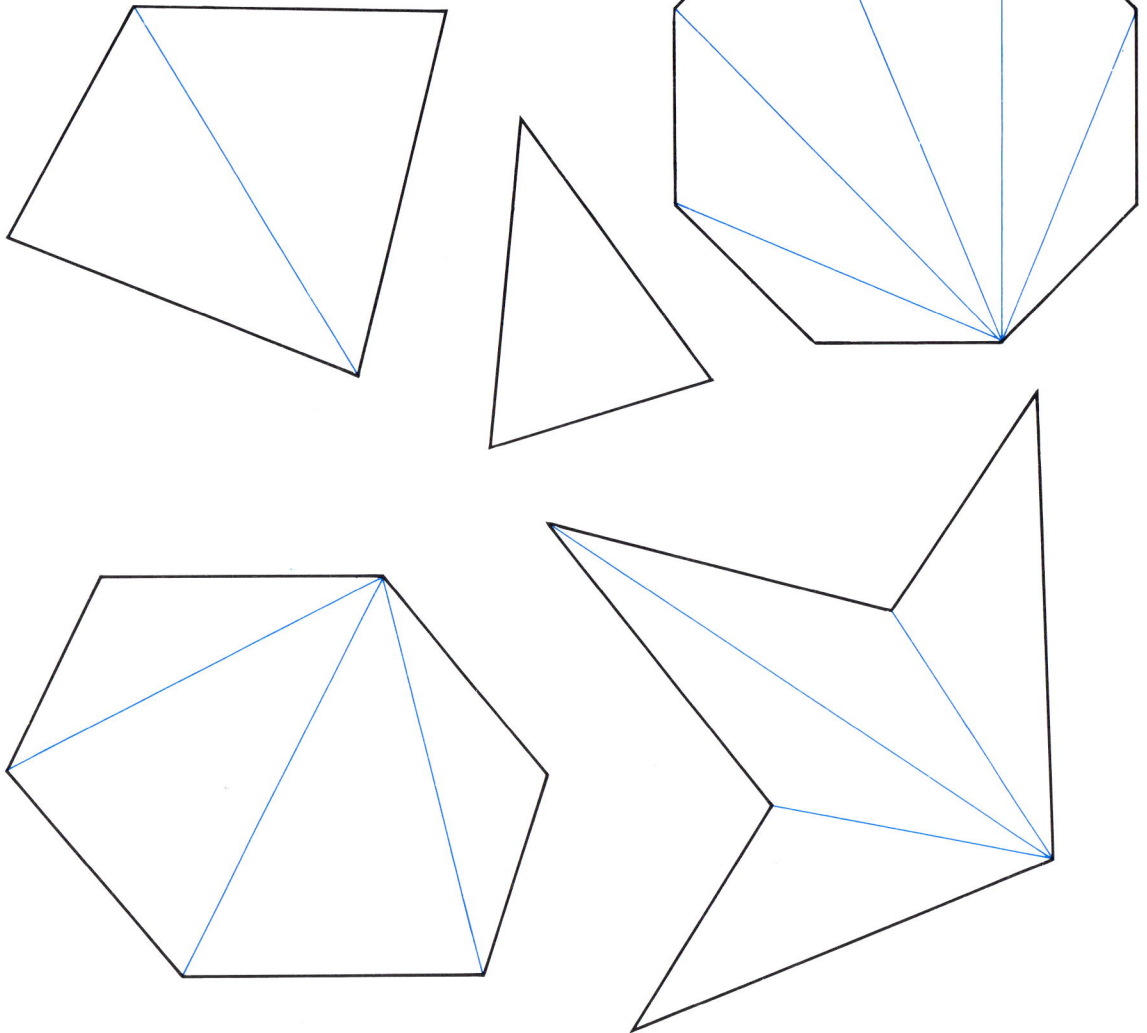

   b) Predict how many triangles are made in a 9-sided polygon.
      Check your prediction by drawing.

   c) You are told how many sides a polygon has.
      How can you calculate the number of triangles?
      Write down a rule.

   d) A 25-sided polygon has 23 triangles.
      Use this to check your rule in (c).

   e) How many sides has a polygon which divides into 98 triangles?

   f) Someone tells you the number of triangles.
      Write down your own rule for calculating the number of sides.

34

2 a) Each dot is joined to every other dot.

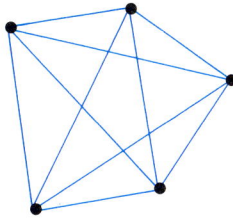

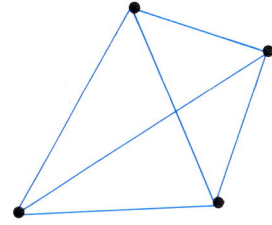

Ask yourself this question:

Is there a rule which tells me how many lines for how many dots?

Think about it, then read on.
If you think you know a rule, write it down.

b) Horace draws 10 dots to find the rule.
Complete his second try for him.

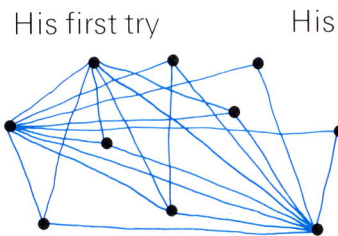

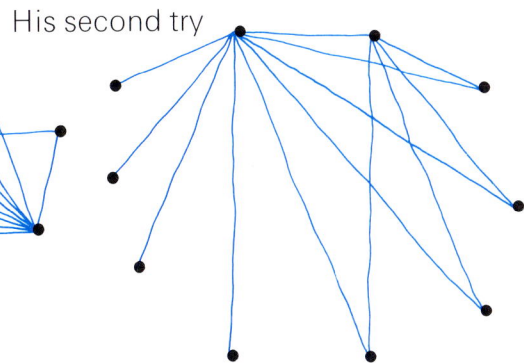

His first try   His second try

c) Horace says there are 9 lines from every dot.
So there are 10 × 9 lines.
Is he correct?

d) Meg says, 'Every line joins 2 dots . . .'.
She writes this rule:
Number of lines = number of dots × number of dots ÷ 2.
Does Meg's method work for 3 dots?

e) Is Meg's method correct?
If you say **No**, write down your own rule.

Take note

When you think there is a pattern . . .

draw good diagrams ─────▶ predict the pattern

check your prediction with examples . . .

. . . until you are sure you are correct.

34

3

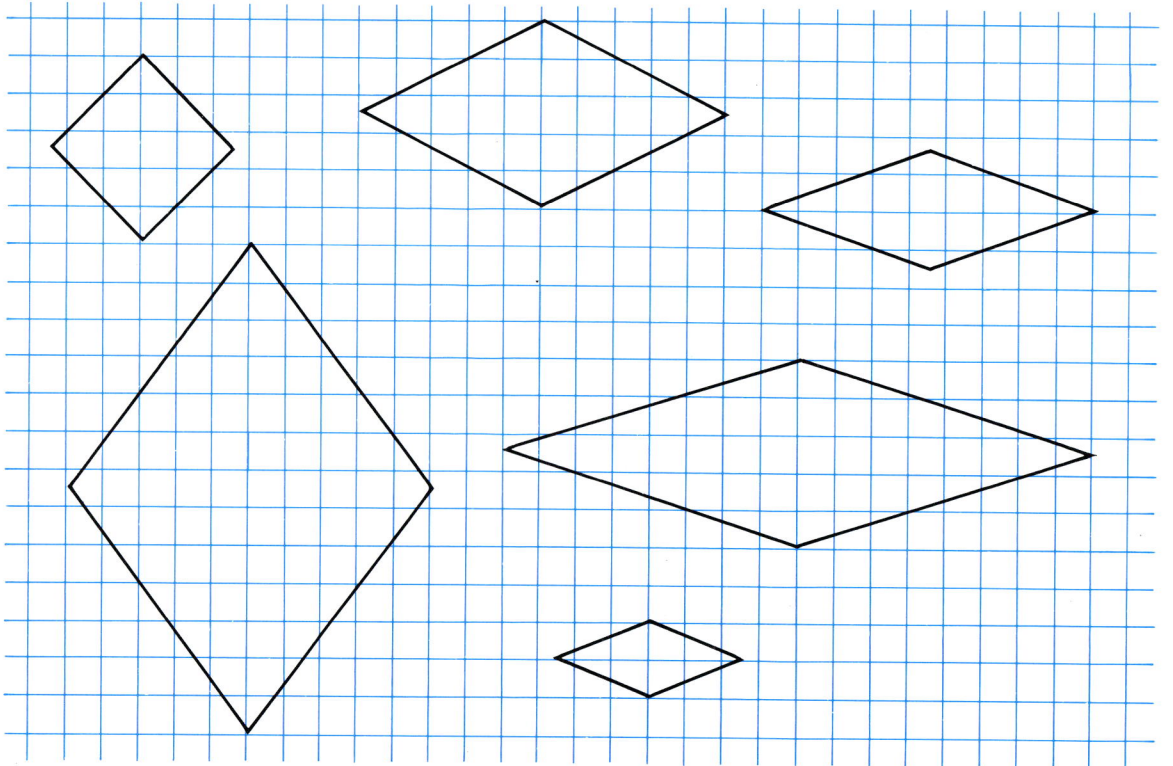

a) There is a connection between the lengths of the diagonals of rhombuses and their area.
What is it?

b) Write down a rule for finding the area of a rhombus.

Challenge 2

4 The decorations are made from beads and straws.

a) What is the connection between the number of diamonds and the number of straws?

b) Someone tells you how many diamonds to make.
Write down a rule which tells you how many beads you need.

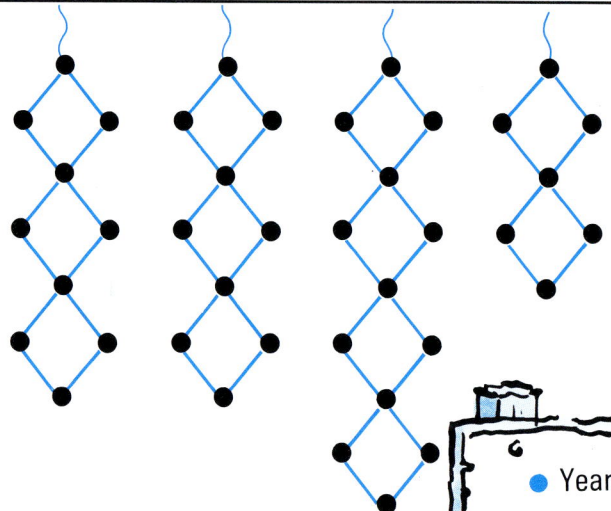

34

● Year 3